Corrosion and Protection of Steels in Marine Environments: State-of-the-Art and Emerging Research Trends

Corrosion and Protection of Steels in Marine Environments: State-of-the-Art and Emerging Research Trends

Editors

Philippe Refait
Igor Chaves

MDPI • Basel • Beijing • Wuhan • Barcelona • Belgrade • Manchester • Tokyo • Cluj • Tianjin

Editors
Philippe Refait
Université de La Rochelle
France

Igor Chaves
University of Newcastle
Australia

Editorial Office
MDPI
St. Alban-Anlage 66
4052 Basel, Switzerland

This is a reprint of articles from the Special Issue published online in the open access journal *Corrosion and Materials Degradation* (ISSN 2624-5558) (available at: https://www.mdpi.com/journal/cmd/special_issues/Corrosion_Steels_Marine).

For citation purposes, cite each article independently as indicated on the article page online and as indicated below:

LastName, A.A.; LastName, B.B.; LastName, C.C. Article Title. *Journal Name* **Year**, *Volume Number*, Page Range.

ISBN 978-3-0365-5861-5 (Hbk)
ISBN 978-3-0365-5862-2 (PDF)

Cover image courtesy of Philippe Refait

© 2022 by the authors. Articles in this book are Open Access and distributed under the Creative Commons Attribution (CC BY) license, which allows users to download, copy and build upon published articles, as long as the author and publisher are properly credited, which ensures maximum dissemination and a wider impact of our publications.
The book as a whole is distributed by MDPI under the terms and conditions of the Creative Commons license CC BY-NC-ND.

Contents

About the Editors vii

Preface to "Corrosion and Protection of Steels in Marine Environments: State-of-the-Art and Emerging Research Trends" ix

Philippe Refait, Anne-Marie Grolleau, Marc Jeannin, Celine Rémazeilles and René Sabot
Corrosion of Carbon Steel in Marine Environments: Role of the Corrosion Product Layer
Reprinted from: *Corros. Mater. Degrad.* **2020**, *10*, 10, doi:10.3390/cmd1010010 1

Robert E. Melchers
Experience-Based Physico-Chemical Models for Long-Term Reinforcement Corrosion
Reprinted from: *Corros. Mater. Degrad.* **2021**, *2*, 6, doi:10.3390/cmd2010006 23

Hoang C. Phan, Linda L. Blackall and Scott A. Wade
Effect of Multispecies Microbial Consortia on Microbially Influenced Corrosion of Carbon Steel
Reprinted from: *Corros. Mater. Degrad.* **2021**, *2*, 8, doi:10.3390/cmd2020008 43

Julien Duboscq, Julia Vincent, Marc Jeannin, René Sabot, Isabelle Lanneluc, Sophie Sablé and Philippe Refait
Influence of Organic Matter/Bacteria on the Formation and Transformation of Sulfate Green Rust
Reprinted from: *Corros. Mater. Degrad.* **2022**, *3*, 1, doi:10.3390/cmd3010001 61

Xiaolong Zhang, Nanni Noël-Hermes, Gabriele Ferrari and Martijn Hoogeland
Localized Corrosion of Mooring Chain Steel in Seawater
Reprinted from: *Corros. Mater. Degrad.* **2022**, *3*, 4, doi:10.3390/cmd3010004 77

Philippe Refait, Julien Duboscq, Kahina Aggoun, René Sabot and Marc Jeannin
Influence of Mg^{2+} Ions on the Formation of Green Rust Compounds in Simulated Marine Environments
Reprinted from: *Corros. Mater. Degrad.* **2021**, *2*, 3, doi:10.3390/cmd2010003 99

Kranthi Kumar Maniam and Shiladitya Paul
Corrosion Performance of Electrodeposited Zinc and Zinc-Alloy Coatings in Marine Environment
Reprinted from: *Corros. Mater. Degrad.* **2021**, *2*, 10, doi:10.3390/cmd2020010 115

Michael P. Milz, Andreas Wirtz, Mohamed Abdulgader, Anke Kalenborn, Dirk Biermann, Wolfgang Tillmann and Frank Walther
Corrosion Fatigue Behavior of Twin Wire Arc Sprayed and Machine Hammer Peened ZnAl4 Coatings on S355 J2C + C Substrate
Reprinted from: *Corros. Mater. Degrad.* **2022**, *3*, 7, doi:10.3390/cmd3010007 143

Alexis Renaud, Victor Pommier, Jérémy Garnier, Simon Frappart, Laure Florimond, Marion Koch, Anne-Marie Grolleau, Céline Puente-Lelièvre and Touzain Sebastien
Aggressiveness of Different Ageing Conditions for Three Thick Marine Epoxy Systems
Reprinted from: *Corros. Mater. Degrad.* **2021**, *2*, 39, doi:10.3390/cmd2040039 159

Philippe Refait, Anne-Marie Grolleau, Marc Jeannin and René Sabot
Cathodic Protection of Complex Carbon Steel Structures in Seawater
Reprinted from: *Corros. Mater. Degrad.* **2022**, *3*, 26, doi:10.3390/cmd3030026 181

About the Editors

Philippe Refait

Prof. Philippe Refait specializes in electrochemistry, materials science, and surface science. His main research axes are corrosion mechanisms (mainly of carbon steel and low alloy steel), including biocorrosion, and cathodic protection (more specifically the processes occurring under cathodic polarization).

He has published more than 175 research articles and 9 book chapters. He is currently the chair of the "Marine Corrosion" working party of the European Federation of Corrosion (EFC) and the head of the Department of Physics at the Faculty of Science and Technology, La Rochelle University.

Igor Chaves

Igor Chaves is a civil, structural, chartered materials professional engineer who specializes in material corrosion analysis and prediction. His main research focus includes the qualitative prediction and quantification of remaining structural service life and its implications towards the management of infrastructure assets.

As the Deputy Director of the Centre for Infrastructure Performance and Reliability at the University of Newcastle and the former National President of the Australasian Corrosion Association, he continues to strive for a strong academic and industry-focused career. His recent awards include the 2017 Marshal Fordham best Paper Award (ACA), and the 2020 simultaneous Industry Research Engagement and Teaching and Learning Excellence Awards (UON).

Preface to "Corrosion and Protection of Steels in Marine Environments: State-of-the-Art and Emerging Research Trends"

How and why did the book come about?

During times when so many academic publications are available to the public through open access, it is important for those who are more attuned to the cutting edge of a particular scientific field to put forward a formal recommendation on the works of those that have made a noticeable contribution to the niche field of marine corrosion science. Therefore, this book came about.

Why were the selected papers and authors chosen over others?

Unfortunately, there is a lack of appreciation for how much risk undetected marine corrosion could create for many of our existing and new steel infrastructures and devices. Because of the relatively stealthy and slow nature of corrosion reactions, it is difficult to quantitatively appreciate their consequences. To this end, various researchers around the world have researched systems, methods, and technologies that allow for not only early prediction but also timely intervention and thus protection for steel structures. Some researchers, such as the various authors and their teams selected for this Special Issue have accumulated so much experience and knowledge that a compilation of their work for exposure to a wider audience is certainly prudent.

How and why was the book organized/written?

Aiming to target a wider scientific and engineering-focused audience, this book gathered research papers focused on marine corrosion (and protection) of iron-based alloys, i.e., steel. The articles first cover fundamental scientific advancements and understanding of the chemical and electrochemical kinetics underpinning marine corrosion processes, followed by an advanced understanding of the uniquely specialized cases of microbiologically influenced corrosion with the ability to rapidly exacerbate durability losses, with closing works of various advanced protection systems that can be used in the pursuit of cost-effective practical solutions.

Philippe Refait and Igor Chaves
Editors

Review

Corrosion of Carbon Steel in Marine Environments: Role of the Corrosion Product Layer

Philippe Refait [1,*], Anne-Marie Grolleau [2], Marc Jeannin [1], Celine Rémazeilles [1] and René Sabot [1]

1. Laboratoire des Sciences de l'Ingénieur pour l'Environnement (LaSIE), UMR 7356 CNRS-La Rochelle Université, Av. Michel Crépeau, CEDEX 01, F-17042 La Rochelle, France; mjeannin@univ-lr.fr (M.J.); celine.remazeilles@univ-lr.fr (C.R.); rsabot@univ-lr.fr (R.S.)
2. Naval Group Research, BP 440, CEDEX 50104 Cherbourg-Octeville, France; anne-marie.grolleau@naval-group.com
* Correspondence: prefait@univ-lr.fr; Tel.: +33-5-46-45-82-27

Received: 5 May 2020; Accepted: 30 May 2020; Published: 3 June 2020

Abstract: This article presents a synthesis of recent studies focused on the corrosion product layers forming on carbon steel in natural seawater and the link between the composition of these layers and the corrosion mechanisms. Additional new experimental results are also presented to enlighten some important points. First, the composition and stratification of the layers produced by uniform corrosion are described. A focus is made on the mechanism of formation of the sulfate green rust because this compound is the first solid phase to precipitate from the dissolved species produced by the corrosion of the steel surface. Secondly, localized corrosion processes are discussed. In any case, they involve galvanic couplings between anodic and cathodic zones of the metal surface and are often associated with heterogeneous corrosion product layers. The variations of the composition of these layers with the anodic/cathodic character of the underlying metal surface, and in particular the changes in magnetite content, are thoroughly described and analyzed to enlighten the self-sustaining ability of the process. Finally, corrosion product layers formed on permanently immersed steel surfaces were exposed to air. Their drying and oxidation induced the formation of akaganeite, a common product of marine atmospheric corrosion that was, however, not detected on the steel surface after the permanent immersion period.

Keywords: carbon steel; seawater; localized corrosion; magnetite; green rust; iron sulfide

---

1. Introduction

Seawater is one of the most complex and aggressive media. Marine corrosion depends on numerous interdependent parameters and combines chemical, biological and mechanical factors. The understanding of the influence of each of these parameters and factors is the key to the optimization of the design of metal structures and devices used in marine environments. It is also the key to the optimization of anticorrosion methods and materials performance. Localized corrosion is a particularly insidious degradation phenomenon and thus a hazard for metal structures integrity. Severe localized degradations may induce major industrial failures while consuming a very small amount of materials. In seawater, owing to various combined factors such as materials heterogeneity (grain boundaries, inclusions, welds ...) [1–5], differential aeration [6–9], and biological activity [10–18], the corrosion processes of carbon steels, though commonly acknowledged as being mainly uniform, are often localized. The well-known ALWC (Accelerated Low Water Corrosion) phenomenon, which combines differential aeration and bacteria consortium activity [15,17], perfectly illustrates this point.

Carbon steels are widely used materials for marine applications. They are massively produced (~1.8×10^9 tons worldwide in 2018 [19]) and are actually inexpensive while they offer good mechanical properties. They also mainly suffer uniform corrosion and were studied for decades so that their corrosion rate can be modeled and their lifetime predicted with rather good accuracy in various marine environments. The phenomenological model proposed by Melchers et al. [20–22] was designed via a thorough analysis of numerous data. It is based on the description and understanding of the mechanisms of the uniform corrosion process of iron-based materials. It aims to predict the short and long-term behavior of steel structures permanently immersed in seawater and forecast reliably the lifetime of these structures.

Microbiologically influenced corrosion (MIC) is often considered the main reason for the non-uniformity of carbon steel corrosion in seawater [10–18]. The biofilm that forms on any material immersed in seawater is intrinsically heterogeneous and leads to the coexistence of aerated and deaerated zones, creating aeration cells and favoring the growth and activity of harmful microorganisms. Electroactive bacteria may even uptake electrons from the metal and thus influence directly the corrosion rate [23–25]. However, if sulfate-reducing bacteria (SRB) develop initially in deaerated zones because they are anaerobe microorganisms and favor temporarily localized corrosion processes, they rapidly colonize the whole metal surface [26], which ultimately leads again to more or less uniform corrosion [22,26].

The solid corrosion products forming on the steel surface are the primary consequences of the dissolution of the metal. They can strongly affect the ongoing corrosion process. First of all, they form a physical barrier between metal and environment and therefore contribute to the protection of the metal, because they hinder the transport of dissolved oxygen from seawater to the metal surface [20,21]. Secondly, they are porous and thus offer a unique habitat for microorganisms that can develop in a specific environment [26–31]. Thirdly, some phases are electronic conductors, e.g., magnetite [32–34] and iron sulfides [16,23], which can favor galvanic cells. The composition of the corrosion product layer also varies with the exposure zone [26,35–42] and may change in the long term [43].

This article reports recent advances in the understanding of localized corrosion processes of carbon steel permanently immersed in natural seawater, obtained through a detailed analysis of the corrosion product layers [26,30,36,41,42,44]. These layers, formed on steel surfaces in laboratory experiments or seaport exposure sites, reflect the complexity of iron chemistry in natural seawater. Their composition differs in anodic and cathodic zones [9,41,42] so that they participate actively in the persistence of corrosion cells and favor localized corrosion processes [41].

2. Materials and Methods

2.1. Materials and Exposure Sites

The article is focused on results obtained with natural seawater. Actually, the complex mechanisms involved during marine corrosion cannot be fully reproduced via laboratory experiments using artificial seawater.

The results discussed here were obtained with coupons immersed in three main exposure sites. Two of these sites were set in seaports of La Rochelle (Atlantic Ocean, France). In one site (Les Minimes marina), the coupons were immersed at constant immersion depth (~50 cm). In the other site (cargo port), the immersion depth, i.e., the distance between coupon and atmosphere, varied with the tide, between a few centimeters and ~5 m. The third site was the Naval Group Research laboratory in Cherbourg (English Channel, France). In this facility, natural seawater is pumped directly from the English Channel to flow continuously at a rate of 100 L/h through the electrochemical cells. In this last case, the steel coupons were 5–10 cm distant from the seawater/air interface so that differential aeration phenomena could be induced and favored localized corrosion. However, the results obtained in each site proved comparable so that general trends could be deduced. More detailed information on the exposure conditions can be found in the original articles [26,30,36,37,40–44].

It is also worthy to recall the average composition of seawater. The study of the water flowing through the Naval Group Research lab in Cherbourg led to the following composition (main ionic species only): $[Cl^-]$ = 19 g kg^{-1}, $[SO_4^{2-}]$ = 2.7 g kg^{-1}, $[HCO_3^-]$ = 0.14 g kg^{-1}, $[Na^+]$ = 10 g kg^{-1},

[Mg^{2+}] = 1.3 g kg^{-1}, [Ca^{2+}] = 0.41 g kg^{-1} and [K$^+$] = 0.4 g kg^{-1}. It clearly shows that seawater is not, in any case, a simple NaCl solution, and the role of various other species, e.g., SO$_4^{2-}$, Ca^{2+}, Mg^{2+}, and HCO$_3^-$, is clearly illustrated by the composition of the mineral layers covering carbon steel in seawater [26,30,36,41,42]. Note also that the alkalinity of seawater (pH measured between 8.0 and 8.2 at Naval Group Research lab), is mainly associated with the presence of hydrogencarbonate ions.

The present article describes the behavior of carbon steel and thus does not deal with the influence of alloying elements. More precisely, the article relates to studies performed with three main materials. The first one is the alloy typical of sea harbor steel sheet piles, i.e., carbon steel S355GP. The other alloys are S355NL and TU250B carbon steels. The composition of each steel grade is given in Table 1. The slight observed differences, e.g., less carbon and manganese and more copper for TU250B, did not lead to detectable effects [41] so that it can be considered that the composition of the corrosion product layer does not depend, or only marginally, on the considered carbon steel grade. This is also illustrated by the results obtained with different carbon steel alloys [26,30,36,41,42]. Experimental details about the preparation of the carbon steel samples can be found in the original articles [26,30,36,37,40–44]. In any case, the steel surface was shot blasted to remove any trace of previous rusting or mill scale and degreased with acetone.

Table 1. Composition (wt. %) of the three carbon steels used in this study (nominal composition for S355GP or as given by the manufacturer, i.e., Arcelor-Mittal for S355NL, and Vallourec for TU250B), rest = Fe.

Carbon Steel	C	Mn	P	S	Si	Al	Cr	Cu	Ni
S355GP	≤0.27	≤1.7	≤0.055	≤0.055	≤0.6	0.02	-	-	-
S355NL	0.17	1.4	0.015	0.005	0.21	0.02	0.02	0.01	0.02
TU250B	0.12	0.6	0.017	0.006	0.22	-	0.08	0.19	0.09

2.2. Characterization of the Corrosion Product Layers

The characterization of the corrosion product layers requires that samples are preserved from oxidation by air (or more exactly oxygen). This important point is developed in the article (Section 5). In the lab, the samples can be kept in a freezer (−24b °C), a simple procedure that hinders the oxidation of Fe(II)-based corrosion products that are reactive towards oxygen [40–42]. This frozen state moreover facilitates the sampling of small slices of corrosion product layers that can be easily handled. Once warmed up, these solid slices transform to sludge due to their high content of water.

The thorough and unambiguous characterization of the phases present in the corrosion product layers cannot be achieved using a single analytical tool. Moreover, the characterization techniques must be adapted to the particular nature of the layers that are thick, heterogeneous, and very porous. The pores are filled with water, microorganisms, and organic matter. When analyzing a corrosion product layer, the first objective is to identify all the crystalline solid phases that constitute this layer. This can be achieved using X-ray diffraction (XRD). Another technique must at least be used to characterize amorphous, nanocrystalline or poorly crystallized compounds. The second objective is to determine how the various compounds are distributed inside the corrosion product layer, in particular, to establish if there exists some kind of stratification from metal to seawater. A local method such as μ-Raman spectroscopy (μ-RS) is adequate for that purpose. μ-RS is moreover suitable for the analysis of wet (and even immersed) samples and the characterization of non-crystalline matter. The combination of μ-RS with XRD, used in various works [26,30,36,37,39–44], is then a minimal requirement.

μ-RS analysis was performed in any case at room temperature using a Jobin Yvon High Resolution Raman spectrometer (LabRam HR or LabRam HR evo) equipped with a confocal microscope and a Peltier-based cooled charge-coupled device (CCD) detector. Two laser sources were used indifferently, a solid-state diode-pumped green laser (532 nm) or a He-Ne laser (632.8 nm). The laser power must be reduced down to 10% (0.6 mW), and in some cases, even to 1% (0.06 mW), of the maximum to prevent the transformation of the analyzed compounds into hematite α-Fe$_2$O$_3$. This transformation can take place due to excessive heating, as observed for Fe(III)-oxyhydroxides [45] and magnetite Fe_3O_4 [45,46]. The acquisition time was variable and depended on the nature of the analyzed phase. It was generally equal to 60 s but could be increased up to 2 min to optimize the signal to noise ratio. At least 20 zones (diameter of 3–6 µm) of a given sample were analyzed through a 50× long working distance objective (50 × LWD). A motorized XY stage was used in order to focus on the different zones of the sample and determine the possible organization/stratification of the corrosion product layer. Most of the investigations could be performed without specific protection from air, which was possible because the time required for analysis was short and samples remained wet during analysis, which delays oxidation.

X-ray diffraction (XRD) analysis was performed with an Inel EQUINOX 6000 diffractometer, using a curved detector (CPS 590), with the Co-Kα radiation (λ = 0.17903 nm) at 40 kV and 40 mA. The CPS 590 detector is designed for the simultaneous detection of the diffracted photons on a 2θ range of 90°. The analysis was performed with a constant angle of incidence (5 degrees) for 30 to 60 min depending on the sample. To prevent the oxidation of Fe(II)-based compounds during preparation and analysis, the samples were mixed with a few drops of glycerol in a mortar before being crushed until a homogenous oily paste was obtained. With this procedure, the various particles that constitute the sample are coated with glycerol and thus sheltered from the oxidizing action of O_2 [47]. Glycerol may only give rise to a very broad "hump" visible on the XRD pattern between 2θ~25° and 2θ~35° [48].

XRD and µ-RS results are displayed throughout the present article to illustrate the complexity of the corrosion product layers formed on carbon steel in natural seawater.

3. General Features of the Corrosion Product Layer

3.1. Stratification of the Corrosion Product Layer

After several years in natural seawater, the surface of carbon steel structures permanently immersed in natural seawater is covered by a corrosion product layer itself covered by biofouling. Figure 1 displays the main features of the layers formed on carbon steel coupons after 6 to 11 years of immersion [36,41,42,44].

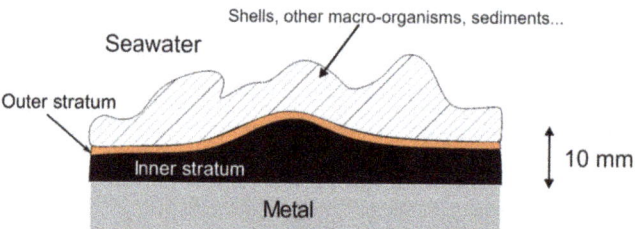

Figure 1. Schematic representation (cross-section) of the layer covering a carbon steel coupon after 6–11 years of permanent immersion in natural seawater.

The corrosion product layer can be several millimeters thick, reaching locally a thickness of 1 cm or more. Under the biofouling layer that may incorporate mineral compounds (sand, clay, fragments of shells) and macro-organisms, an orange-brown corrosion product layer is generally seen. This is the outer stratum, mainly composed of Fe(III)-oxyhydroxides (FeOOH). Lying underneath is found the inner stratum, characterized by a black color, that is in contact with the steel surface. This inner stratum is mainly composed of Fe(II)-based corrosion products, mixed with magnetite Fe_3O_4. The Fe(II)-based compounds are reactive towards oxygen and the Fe(III)-oxyhydroxides that constitute the orange-brown outer layer are the end products of their oxidation by dissolved O_2. After 6–11 years in natural seawater, the inner black stratum is much thicker than the orange-brown outer stratum. This demonstrates that the steel surface and the main part of the corrosion product layer are no more reached by dissolved oxygen, i.e., that anoxic conditions are met. This explains why anaerobe microorganisms develop systematically inside the corrosion product layers and influence necessarily the corrosion process, as perfectly illustrated by the phenomenological model proposed by Melchers et al. [17,22]. Dissolved oxygen cannot reach the inner part of the corrosion product layer because the aerobe microorganisms that colonize the biofouling layer and the orange-brown outer stratum consume it. The little oxygen that could possibly reach the dark inner stratum would then react with the Fe(II)-based corrosion products to form FeOOH phases and/or magnetite.

The stratification of the corrosion product layer can be clearly observed as soon as 1–2 months of immersion [30].

3.2. Composition of the Corrosion Product Layer

Figure 2 shows the XRD pattern of the corrosion product layer that covered a carbon steel coupon immersed 6 months in natural seawater (Les Minimes marina exposure site, La Rochelle, Atlantic Ocean). The main crystalline solid phases that constitute such corrosion product layers are all detected here. Two Fe(III)-oxyhydroxides, namely goethite α-FeOOH and lepidocrocite γ-FeOOH, are identified. They constitute the orange-brown outer stratum. Magnetite, the Fe(II,III) mixed-valence oxide Fe_3O_4, is identified too. It is present in the inner dark layer. Magnetite can form in anoxic conditions as shown by thermodynamic data e.g., [49,50]. The overall reaction involves the oxidation of three Fe(0) atoms to one Fe(II) cation and two Fe(III) cations, and the reduction of four water molecules to hydrogen. It can be written as:

$$3Fe + 4H_2O \rightarrow Fe_3O_4 + 4H_2 \tag{1}$$

This first process explains the formation of magnetite on or close to the metal surface. However, magnetite can also be obtained indirectly as an oxidation product of Fe(II)-based corrosion products at very low oxygen flow rates [51]. This second process explains why magnetite was in some cases mainly detected at the interface between the inner dark stratum and the outer orange stratum [41], where a small amount of oxygen can access after having crossed over the outer parts of the biofouling/corrosion product layer.

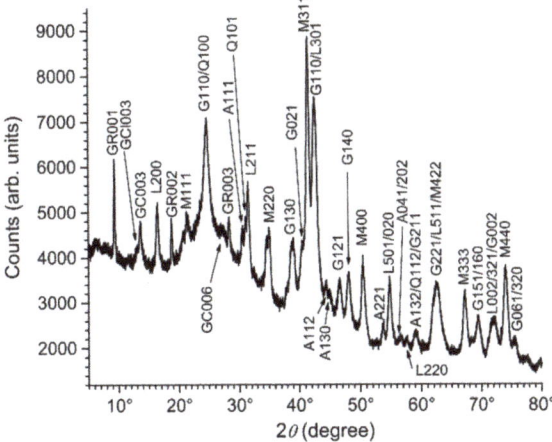

Figure 2. XRD analysis of the corrosion product layer covering a carbon steel coupon after 6 months in natural seawater (exposure site: Les Minimes marina, La Rochelle, Atlantic Ocean). A = aragonite, G = goethite, GR = sulfate green rust, GC = carbonate green rust, GCl = chloride green rust, L = lepidocrocite, M = magnetite, and Q = quartz. The diffraction lines are denoted with the corresponding Miller index.

The other corrosion products identified via the XRD pattern of Figure 2 are green rust (GR) compounds, namely the sulfate GR, the carbonate GR, and the chloride GR. Green rust compounds are mixed-valence Fe(II,III) layered double hydroxides (LDH) mainly containing Fe(II) cations (67% to 75%). In contrast, magnetite contains mostly Fe(III) cations (67% i.e., only 33% of Fe(II) cations). The crystal structure of GR compounds consists of the stacking of hydroxide layers, built on Fe(OH)$_6$ octahedra, carrying a positive charge because of the presence of the Fe(III) cations [52–55]. So-called interlayers constituted of anions and water molecules are intercalated between two adjacent hydroxide layers. Various GRs can be obtained depending on the anions present in the surrounding environment, and seawater, given its composition, can induce the formation of GR(Cl$^-$), GR(SO$_4^{2-}$), and GR(CO$_3^{2-}$).

The sulfate green rust is predominant as revealed by the numerous results accumulated in the past ten years [26,30,36,41,42]. This is also illustrated in Figure 2 by the higher relative intensity of the diffraction peaks of GR(SO$_4^{2-}$). The chloride GR, which could be expected to be the predominant GR variety because of the high chloride concentration of seawater, was rarely identified or as a very minor compound. However, the stability of the various GR compounds does not only depend on the intercalated anion concentration. It depends on the geometry, size, and number of charges of the anion, or more generally, on any parameter that influences the stability of the LDH crystal structure. The studies dealing with the affinity of this structure for various anions revealed that divalent anions were generally preferred to monovalent ones [56–58]. In solutions characterized by large [Cl$^-$]/[SO$_4^{2-}$] concentration ratios, it was demonstrated accordingly that GR(SO$_4^{2-}$) was obtained preferentially to GR(Cl$^-$) [58].

Other compounds can be identified with the XRD pattern of Figure 2. Quartz, i.e., sand, comes from the environment. Aragonite, a CaCO$_3$ phase, can also come from the environment as some shells are made of it. However, its formation may be associated with the corrosion process, as explained in Section 4.1.

Table 2 lists the compounds positively identified via XRD and/or μ-RS spectroscopy in the corrosion product layers formed on carbon steel permanently immersed in natural seawater, as reported in references [26,30,36,41,42,44]. Chukanovite Fe$_2$(OH)$_2$CO$_3$ and greigite Fe$_3$S$_4$ were rarely identified and do not appear among the products revealed by the XRD pattern of Figure 2. In contrast, mackinawite FeS is one of the main corrosion products, present in the inner dark layer, but it is not detected either via the XRD pattern. This is generally the case because mackinawite is mainly found in a nanocrystalline state and is rarely identified by XRD [26,30,36,41,42,44]. It can however be detected using μ-Raman spectroscopy [59].

Table 2. List of corrosion products/minerals identified on carbon steel coupons or structures permanently immersed in natural seawater according to references [26,30,36,41,42,44] and their possible link with the anodic/cathodic nature of the underlying metal surface.

Compound Name: Chemical Formula	Importance/Abundance and Localization	Link to Anodic and Cathodic Zones
Sulfate green rust, GR(SO$_4^{2-}$): Fe$^{II}_4$Fe$^{III}_2$(OH)$_{12}$SO$_4$·8H$_2$O	Main component of the inner layer	Favored in anodic zones
Carbonate green rust, GR(CO$_3^{2-}$): Fe$^{II}_4$Fe$^{III}_2$(OH)$_{12}$CO$_3$·2H$_2$O	Minor component of the inner layer	Favored in cathodic zones
Chloride green rust, GR(Cl$^-$): Fe$^{II}_3$Fe^{III}(OH)$_8$Cl·2H$_2$O	Minor component of the inner layer	Not known
Chukanovite: Fe$_2$(OH)$_2$CO$_3$	Rare	Cathodic zones
Mackinawite/Fe(III)-containing mackinawite: Fe^{II}S/ Fe$^{II}_{1-3x}$Fe$^{III}_{2x}$S	Main component of the inner layer, due to SRB	Not known
Greigite: Fe$_3$S$_4$	Rare	Not known
Magnetite: Fe$_3$O$_4$	Main component of the inner layer	Favored in cathodic zones
Lepidocrocite: γ-FeOOH	Main component of the outer layer	Anodic zones or uniform corrosion
Goethite: α-FeOOH	Main component of the outer layer	Anodic zones or uniform corrosion
Akaganeite: β-FeOOH	Rare, minor component	Not known
Aragonite: CaCO$_3$	Outer layer	Cathodic zones

Figure 3 shows a picture obtained with the optical microscope of the μ-RS apparatus. It was taken during the analysis of the inner dark layer covering a carbon steel coupon immersed 6 years in natural seawater (exposure site: Naval Group laboratory, Cherbourg, English Channel). At the center of the image, a stack of thin bluish platelets can be seen. Only a part of these platelets is visible, as the rest is covered with shiny black matter, but one can see that these particles seem to have the characteristic hexagonal shape of GR crystals [36,41,60–62]. μ-RS analysis of these platelets (not shown) confirmed that they corresponded to a GR compound. The shiny black matter that surrounds them was also analyzed by μ-RS (Figure 4). Figure 4a displays the most commonly found spectrum. It is composed of two Raman peaks, the main one located at 283 cm^{-1} and the other one at 207 cm^{-1}. This spectral signature is typical of nanocrystalline mackinawite [59]. This nanocrystalline compound, formerly considered as "amorphous FeS" or "poorly ordered FeS", is the solid that precipitates from dissolved Fe(II) and S(-II) species [63,64]. Aging in the solution can improve the crystallinity but this process is very slow in alkaline conditions and at ambient temperatures [59]. In seawater (pH~8.1), mackinawite then remains essentially in its nanocrystalline state.

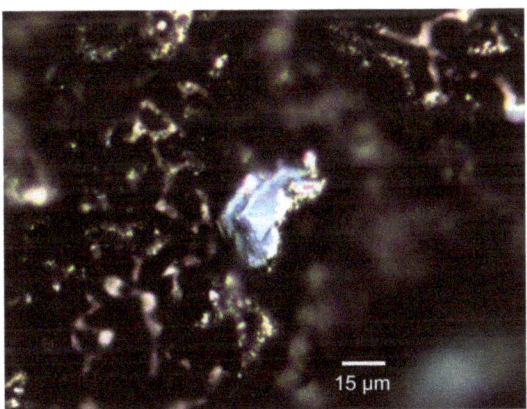

Figure 3. Photograph of a stack of hexagonal platelets of GR(SO$_4^{2-}$) surrounded by shiny black matter (Optical microscope of the Raman apparatus, ×50 objective), taken during the analysis of the inner stratum of the corrosion product layer covering cathodic zones of a carbon steel coupon after 6 years in natural seawater (exposure site: Naval Group laboratory, Cherbourg, English Channel).

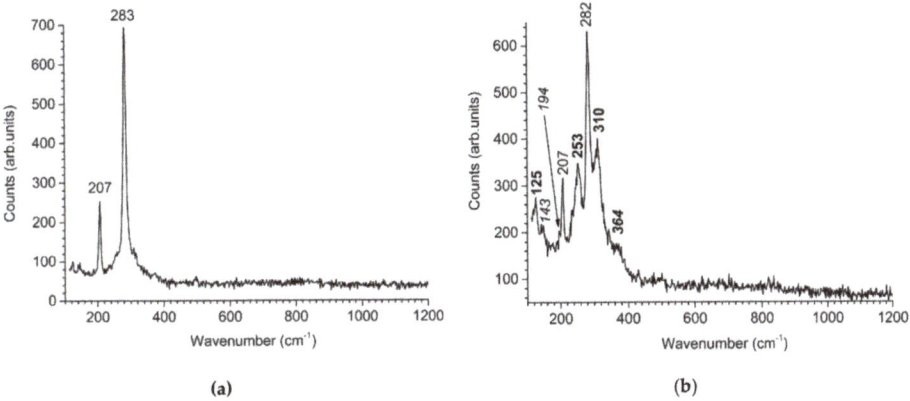

Figure 4. Raman spectra obtained from the shiny black matter shown in Figure 3: **(a)** nanocrystalline mackinawite, **(b)** mixture of various iron sulfides.

Figure 4b displays another Raman spectrum of the shiny black matter, less frequently observed than the previous one. First, the two Raman peaks of nanocrystalline mackinawite are seen, at 207 cm^{-1} and 282 cm^{-1}. Secondly, other intense peaks are seen at 125 cm^{-1}, 253 cm^{-1}, 310 cm^{-1}, and 364 cm^{-1}. They are attributed to the so-called "Fe(III)-containing mackinawite", a slightly oxidized form of mackinawite [59,65] that may contain up to 20% of Fe(III) [59,66]. The proposed chemical formula is then $Fe^{II}_{1-3x}Fe^{III}_{2x}S$ [59]. The in-situ oxidation of Fe(II) into Fe(III) can take place even in anoxic conditions, and this process finally leads to greigite Fe_3S_4 [65] via a solid-state transformation [67]. Actually, the spectrum of Figure 4b shows typical features of greigite that are the peaks at 143 cm^{-1}, 194 cm^{-1}, and 364 cm^{-1} (shared with Fe(III)-containing mackinawite) [65,68].

The iron sulfides are formed because of bacterial activity. Sulfide species are not present in seawater except in deaerated conditions that promote the growth and activity of sulfide-producing bacteria such as SRB. In studies combining a microbiological analysis of the bacterial flora with a characterization of the corrosion products, the presence of FeS was indeed associated with that of SRB (e.g., [18,26,27,30,69]). As explained above, anoxic conditions are established after some time in the inner black stratum of the corrosion product layers and consequently anaerobe microorganisms such as SRB can grow and be active. At the beginning of the corrosion process, only a few deaerated regions are found and FeS forms only locally. It could be detected in one particular zone of a steel coupon after 1 month of immersion, a finding associated with the concomitant detection of SRB [30]. After 6–12 months, FeS is scattered more or less homogeneously in the inner black layer [26], which indicates that anaerobic conditions prevail all over the steel surface. This corresponds to the moment when the influence of SRB becomes significant, as it affects the whole steel surface, a particular time of the corrosion process that is characterized by an increase of the corrosion rate [22,69].

Finally, maghémite (γ-Fe_2O_3) and ferrihydrite ($Fe_5HO_8 \cdot 4H_2O$) are two compounds possibly present as minor components of the orange outer stratum. They are difficult to characterize by XRD because: (i) the XRD pattern of maghémite is close to that of magnetite and (ii) ferrihydrite is a very poorly ordered and crystallized form of Fe(III)-(oxy)hydroxide. They could not be unambiguously identified in previous works [26,30,36,41,42,44] and for this reason, were omitted from Table 2.

3.3. GR(SO_4^{2-}), the First Compound Resulting of the Corrosion of Carbon Steel in Seawater

Analysis of the corrosion product layers formed on carbon steel coupons after 1 week of immersion in natural seawater (Les Minimes marina, La Rochelle, Atlantic Ocean) revealed only two phases, the sulfate GR and lepidocrocite γ-FeOOH [30]. Actually, lepidocrocite is the main oxidation product of GR(SO_4^{2-}) by dissolved O_2 [58], which indicates that GR(SO_4^{2-}) is the first solid phase to form from the dissolved species produced by the corrosion of the metal. This point was confirmed by anodic polarization experiments performed with carbon steel electrodes immersed in artificial seawater-like solutions deaerated with argon [36]. After 15 min, 3 h, and 26 h of polarization, only the sulfate GR (with traces of the carbonate GR) could be identified on the steel surface.

The formation of GR compounds in seawater-like conditions was also studied using another method. GR compounds could be precipitated by mixing a solution of NaCl, $Na_2SO_4 \cdot 10H_2O$, $FeCl_2 \cdot 4H_2O$, and $FeCl_3 \cdot 6H_2O$ with a solution of NaOH [70]. The obtained precipitates were analyzed immediately after their formation and after 1 week of aging in suspension at room temperature. The initial precipitate was in any case a mixture of GR(SO_4^{2-}) with GR(Cl$^-$). After one week of aging, only traces of GR(Cl$^-$) remained, GR(SO_4^{2-}) being the final result of the overall process. This study shows that GR(Cl$^-$) may be an intermediate transient compound, rapidly transformed to GR(SO_4^{2-}). This would explain why GR(Cl$^-$) could be sometimes identified, always in small amounts, in the dark inner stratum of the corrosion product layer.

From all these findings, it can be proposed that GR(SO_4^{2-}) forms from the dissolved Fe(II) species produced by the corrosion of steel through the process illustrated in Figure 5. This process may take place within a few minutes or a few seconds as GR(SO_4^{2-}) was the only product detected after 15 min of immersion [36].

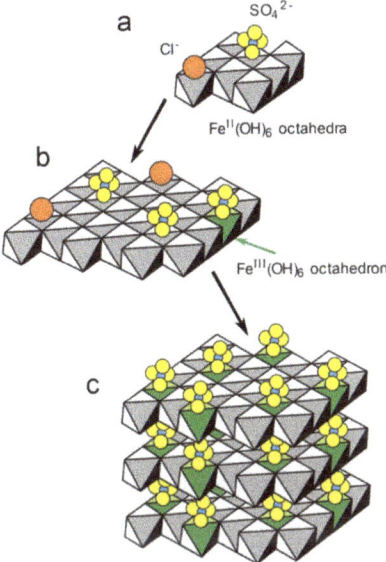

Figure 5. Schematic representation (a–c) of the possible mechanism for GR(SO$_4^{2-}$) formation as the first solid phase resulting from the corrosion of carbon steel in seawater. The water molecules present in the interlayer of the GR or adsorbed on the hydroxide sheets are omitted for clarity.

It is generally admitted that the Fe^{2+} cations produced by the anodic reaction precipitate first with the OH$^-$ ions produced by the cathodic reaction (see Section 4.1), which leads to the Fe(II) hydroxide Fe(OH)$_2$. The crystal structure of Fe(OH)$_2$ is based on sheets formed by hydroxide-bridged Fe^{II}(OH)$_6$ octahedra. The formation of the solid phase may then begin by the formation of small units of such sheets, where Cl$^-$ and SO$_4^{2-}$ anions (predominant in seawater) are adsorbed (Figure 5a). At the pH of seawater, Fe(OH)$_2$ can be oxidized to GR(SO$_4^{2-}$) even in anoxic conditions [50], so that in any case some Fe(II) cations are rapidly oxidized to Fe(III). A Fe^{III}(OH)$_6$ octahedron necessarily bears an excess positive charge and attracts anions, facilitating their adsorption on the forming hydroxide sheets, as illustrated in Figure 5b with the example of a sulfate anion. The stacking of such Fe(III)-containing hydroxide sheets, with intercalated anions and water molecules, actually corresponds to the structure of GR compounds [52–55], as explained in previous Section 3.1. Since divalent anions give greater stability to this structure, Cl$^-$ ions should be finally expelled from the interlayers, finally leading to GR(SO$_4^{2-}$) as illustrated in Figure 5c.

The overall process leading from the Fe(II) dissolved species to the first solid corrosion product, i.e., GR(SO$_4^{2-}$), can then be summarized by the reactions (2), if dissolved O$_2$ is present, or (3), if anoxic conditions are established:

$$6Fe^{2+} + 10OH^- + SO_4^{2-} + \frac{1}{2}O_2 + 9H_2O \rightarrow Fe^{II}_4Fe^{III}_2(OH)_{12}SO_4 \cdot 8H_2O \tag{2}$$

$$6Fe^{2+} + 10OH^- + SO_4^{2-} + 10H_2O \rightarrow Fe^{II}_4Fe^{III}_2(OH)_{12}SO_4 \cdot 8H_2O + H_2 \tag{3}$$

4. Characterization of Anodic and Cathodic Zones

4.1. Heterogeneity of the Corrosion Product Layer and Role of Interfacial pH

The use of sufficiently large carbon steel coupons, e.g., 10 cm × 10 cm sized coupons, generally enables us to observe the heterogeneity of the corrosion process [41]. An example is displayed in

Figure 6, showing the surface of a coupon after 6 years in seawater. Although the coupon surface is mostly covered by biofouling and macro-organisms (a shell is visible near the center) different zones can be easily distinguished. At the bottom (middle of the lower edge) of the coupon, a large orange tubercle has formed. This accumulation of corrosion products indicates that the corrosion process has been accelerated here. The underlying metal surface then more likely corresponds to an anodic zone. This can be rigorously demonstrated after the corrosion product layer is removed via the determination of the localized degradation depth [9,41,42]. Other zones are covered by a much thinner dark layer of corrosion products, as indicated in Figure 6, where the corrosion rate was less important. They correspond to cathodic zones of the underlying carbon steel surface.

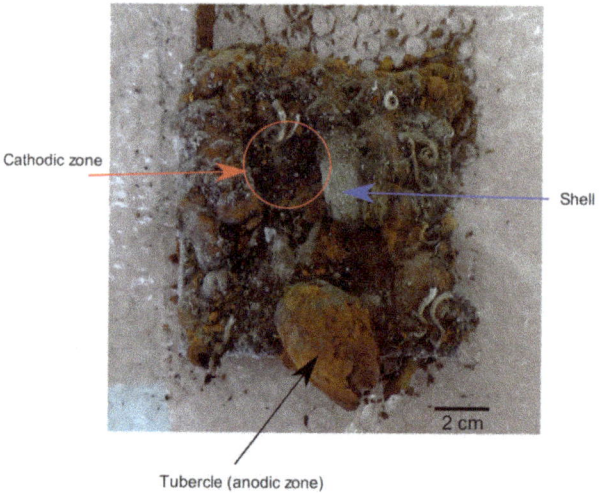

Figure 6. Picture showing the surface of a 10 cm × 10 cm carbon steel coupon after 6 years in natural seawater (exposure site: Naval Group laboratory, Cherbourg, English Channel).

The formation of cathodic zones, where the cathodic reaction rate is higher than that of the anodic reaction rate, and anodic zones, where the cathodic reaction rate is lower than the anodic reaction rate, may have various origins. The possible causes of localized corrosion are well established and include the inherent heterogeneity of the biofilm, the differential aeration cells (especially for vertical surfaces), and the heterogeneity of the steel surface (inclusions, welds, segregation ...). The possible persistence of the galvanic coupling between anodic and cathodic zone is however governed, for long immersion times, by the properties of the corrosion product layers that progressively grow on each zone. The detailed characterization of the products forming in cathodic and anodic zones was then a key point towards the understanding of the long-term localized corrosion processes of carbon steel in seawater [41,42].

The thick tubercles of corrosion products such as the one shown in Figure 6 are, like the corrosion product layers resulting from uniform corrosion, composed of two main strata, an inner dark stratum in contact with the metal surface and an outer orange-brown stratum [41,42]. Figure 7 displays the XRD pattern of a fragment of the inner dark stratum sampled close to the metal surface in an anodic zone. This pattern is mainly composed of the diffraction peaks of the sulfate GR that are very intense. Other GR compounds are not identified and only the main diffraction peak of magnetite (M311) is clearly seen.

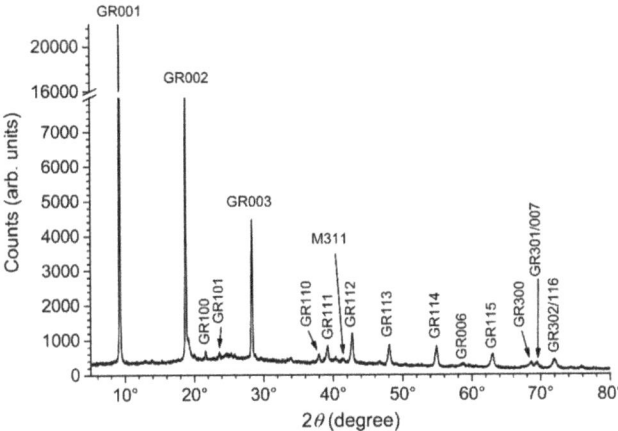

Figure 7. XRD analysis of the inner stratum of the corrosion product layer covering anodic zones of a carbon steel coupon after 6 years in natural seawater (exposure site: Naval Group laboratory, Cherbourg, English Channel). GR = sulfate green rust, and M = magnetite. The diffraction lines are denoted with the corresponding Miller index.

Figure 8 displays the XRD pattern of a fragment of the whole dark layer that covered a cathodic zone of the metal surface. The most intense peak is in this case the main diffraction peak M311 of magnetite. The diffraction peaks of other GR compounds are now clearly detected, and those of the carbonate GR are particularly intense. Aragonite is also clearly identified. The Fe(III)-oxyhydroxides are minor components of the corrosion product layer formed in this cathodic area, in agreement with its visual aspect (black color). This demonstrates that the composition of the corrosion product layer depends on the anodic/cathodic character of the underlying metal surface.

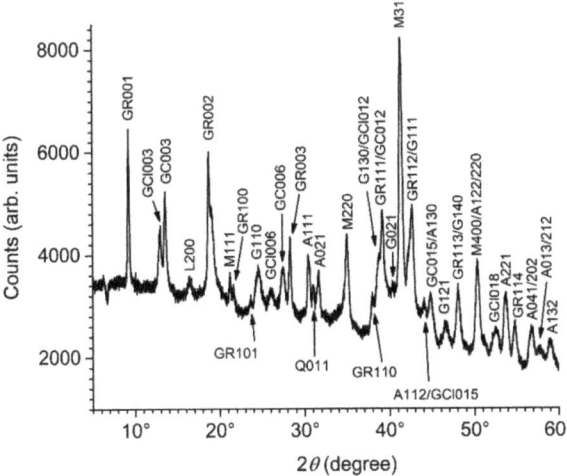

Figure 8. XRD analysis of the inner stratum of the corrosion product layer covering cathodic zones of a carbon steel coupon after 1 year in natural seawater (exposure site: cargo port of La Rochelle, Atlantic Ocean). A = aragonite, GR = sulfate green rust, GC = carbonate green rust, GCl = chloride green rust, L = lepidocrocite, M = magnetite, and Q = quartz. The diffraction lines are denoted with the corresponding Miller index.

Iron sulfides were identified in any case in both anodic and cathodic zones using μ-RS. Actually, the Raman spectra of mackinawite displayed in Figure 4 were obtained from the corrosion product layer covering a cathodic zone. This demonstrates that SRB can grow and be active in both kinds of zones, which mainly indicates that oxygen does not reach the inner dark layer neither in the anodic nor in the cathodic zones (this point is further discussed in Section 4.4). At the present time, it is not possible to state whether SRB develops first or preferentially in one kind of zone.

In Table 2 that lists the products positively identified inside the corrosion product layers [26,30,36,41,42,44], it is indicated whether a compound is preferentially formed in anodic or cathodic zones. The compounds favored in cathodic zones are carbonate-containing phases, namely aragonite, chukanovite, and GR(CO_3^{2-}), and magnetite. In each case, the main reason is the increase of the interfacial pH induced by the difference between the cathodic and the anodic reaction rates. On a metal surface where uniform corrosion takes place, the cathodic and anodic reaction rates are equal. The anodic reaction produces Fe^{2+} cations:

$$Fe \rightarrow Fe^{2+} + 2e^- \tag{4}$$

The corresponding cathodic reaction, whether it is O_2 or H_2O reduction, produces 2 OH^- ions per 2 consumed electrons:

$$\frac{1}{2}O_2 + H_2O + 2e^- \rightarrow 2OH^- \tag{5}$$

$$2H_2O + 2e^- \rightarrow H_2 + 2OH^- \tag{6}$$

Fe^{2+} and OH^- ions are then produced in a ratio $OH^-/Fe^{2+} = 2$ and subsequently incorporated inside the solid phases that constitute the corrosion product layer. This is commonly summarized by the following reaction:

$$Fe^{2+} + 2OH^- \rightarrow Fe(OH)_2 \tag{7}$$

As discussed in Section 3.3, the formation of Fe(II)-hydroxide sheets is rapidly followed in seawater by that of GR(SO_4^{2-}) but, as shown by reaction (7), the pH at the steel/seawater interface is unaffected. The same conclusion is drawn if another corrosion product is considered, as illustrated by the example of magnetite formation (reaction 1). In cathodic zones, the rates of reduction reactions (5) and/or (6) is higher than that of the anodic reaction (4) because part of the consumed electrons comes from the anodic zones. The ratio between the produced OH^- and Fe^{2+} ions in a cathodic zone is then higher than 2, which induces an increase of the interfacial pH. This effect is well-known for steel structures subjected to cathodic protection and leads to the formation of the so-called calcareous deposit, mainly composed of aragonite $CaCO_3$ e.g., [44,71–74]. The presence of aragonite among the corrosion products of steel inside the layers forming on a cathodic zone is then clearly associated with the increase of the interfacial pH. This assumption was confirmed via experiments performed in artificial seawater where shells, the other possible origin of aragonite, were of course not present [9]. The increase of the interfacial pH leads to changes in the inorganic carbonic equilibrium at the metallic interface and favor CO_3^{2-} over HCO_3^-:

$$HCO_3^- + OH^- \rightarrow CO_3^{2-} + H_2O \tag{8}$$

This explains why the formation of aragonite is not the only process favored in the cathodic zones: the formation of other CO_3^{2-} based compounds, i.e., chukanovite and GR(CO_3^{2-}), is also facilitated.

In contrast, the ratio between the produced OH^- and Fe^{2+} ions is lower than 2 in the anodic zones. As Fe^{2+} is a Lewis acid, the interfacial pH may tend to decrease in the anodic areas. This well-known phenomenon is the origin of various localized corrosion phenomena such as crevice corrosion, pitting corrosion, etc...

4.2. Equilibrium Conditions between GR(SO_4^{2-}) and GR(CO_3^{2-})

The formation of GR(CO_3^{2-}) in the cathodic zones was thoroughly studied. The first results were obtained with a study of the impact of cathodic protection on aged (5 years of immersion) and already strongly corroded carbon steel coupons. It was observed after one year of cathodic polarization that the whole amount of GR(SO_4^{2-}) initially present in the corrosion product layers had been entirely transformed to GR(CO_3^{2-}) [44]. The reaction induced by the polarization can be written as follows:

$$Fe^{II}_4Fe^{III}_2(OH)_{12}SO_4 \cdot 8H_2O + CO_3^{2-} \rightarrow Fe^{II}_4Fe^{III}_2(OH)_{12}CO_3 \cdot 2H_2O + 6H_2O + SO_4^{2-} \quad (9)$$

This writing clearly shows that the transformation is favored by the increase of the CO_3^{2-} concentration. More precisely, the equilibrium conditions between GR(SO_4^{2-}) and GR(CO_3^{2-}) are governed by the pH, the sulfate concentration, and the carbonate species concentration [9,44]. This can be highlighted by writing the reaction with the main carbonate dissolved species present in seawater, i.e., HCO_3^-:

$$Fe^{II}_4Fe^{III}_2(OH)_{12}SO_4 \cdot 8H_2O + HCO_3^- \leftrightarrows Fe^{II}_4Fe^{III}_2(OH)_{12}CO_3 \cdot 2H_2O + 6H_2O + SO_4^{2-} + H^+ \quad (10)$$

The equilibrium conditions are then expressed by:

$$pK = pH - \log[a(SO_4^{2-})/a(HCO_3^-)] \quad (11)$$

In Equation (11), $a(SO_4^{2-})$ and $a(HCO_3^-)$ are the activities of SO_4^{2-} and HCO_3^- in the solution where GR(SO_4^{2-}) and GR(CO_3^{2-}) coexist. A study of these equilibrium conditions was achieved to obtain a numerical value of the equilibrium constant [9], which led to:

$$pK = 7.85 \pm 0.35 \quad (12)$$

Using the typical values of $a(SO_4^{2-})$ and $a(HCO_3^-)$ for seawater [9], it was computed that the pH corresponding to the equilibrium conditions between both GR compounds was equal to 8.24 ± 0.35 [9], which is similar or only slightly higher than the average pH of seawater. This explains why a slight increase of the interfacial pH can indeed favor the formation of GR(CO_3^{2-}) at the detriment of that of GR(SO_4^{2-}).

4.3. Variations of the Magnetite Content

The key point is however the competitive formation processes of magnetite and GR(SO_4^{2-}) because magnetite is a well-known electronic conductor [32–34] whereas, to our better knowledge, GR(SO_4^{2-}) is an insulator. According to thermodynamic data, magnetite would indeed be favored by an increase of pH, and conversely, GR(SO_4^{2-}) by a decrease of pH [50]. This was also confirmed by experimental results [50].

As for the corrosion product layers formed on carbon steel permanently immersed in seawater, the relative proportions of magnetite and GR(SO_4^{2-}) proved to depend on the anodic/cathodic character of the underlying metal surface. This is illustrated by the results synthesized in Table 3.

This table gathers data from [41,42] and unpublished results coming from the same research program. To quantify the changes in the relative proportions of magnetite and GR(SO_4^{2-}) revealed by XRD analysis, the M_{311} diffraction line intensity to GR_{003} diffraction line intensity ratio, i.e., $I(M_{311})/I(GR_{003})$, was used. This ratio was chosen because, in the considered experimental conditions (Naval Group laboratory, Cherbourg, English Channel), it was close to 1 (actually between 0.77 and 1.05) for the corrosion product layers generated by uniform corrosion [41]. After 6 years of immersion in these conditions, the average corrosion rate was determined, from thickness loss measurements, to be equal to 0.07 ± 0.02 mm/year for uniform corrosion. Therefore, a zone of the metal surface where the local corrosion rate is significantly higher than 0.07 mm/year should be considered as an anodic zone. Conversely, if the local corrosion rate is significantly lower than 0.07 mm/year then this part of the steel surface should be considered as a cathodic zone.

Table 3. Determined values of the $I(M311)/I(GR003)$ ratio for inner layers on anodic and cathodic zones and associated corrosion rates from References [41,42] and previously unpublished results from the same research program. The corrosion rates were determined by local thickness measurements as described in [40–42] on carbon steel coupons permanently immersed 6–8 years in natural seawater (exposure site: Naval Group laboratory, Cherbourg, English Channel).

Zone	Corrosion Rate (mm/Year)	$I(M311)/I(GR003)$
Strongly anodic	0.50 ± 0.01	0
Anodic	0.20 ± 0.03	0.12
Anodic	0.15 ± 0.04	0.5
Anodic	0.14 ± 0.02	0.03
Slightly anodic	0.08 ± 0.02	0.51
Uniform corrosion	**0.07 ± 0.02**	**0.77 to 1.05**
Slightly cathodic	0.06 ± 0.01	0.5
Slightly cathodic	0.06 ± 0.01	3.1
Cathodic	0.05 ± 0.01	1.9 to 3.3
Cathodic	0.040 ± 0.005	8.9 to 10
Cathodic	0.035 ± 0.005	2.6
Strongly cathodic	~0.01	4.9 to 14.3

The data related to the various analyzed zones are listed in Table 3 with decreasing measured local corrosion rates or, in other words, from the most severely degraded areas (0.5 mm/year, top line) to less degraded regions (0.01 mm/year, bottom line). In each case, the corrosion product layer covering the metal was analyzed by XRD so that the composition of the layer could be associated with the corrosion rate. It can then be seen in Table 3 that the $I(M_{311})/I(GR_{003})$ ratio determined from the XRD analysis increases with decreasing corrosion rate. This trend is more clearly illustrated by the diagram displayed in Figure 9, where the average value of the $I(M_{311})/I(GR_{003})$ ratio is plotted against the average local corrosion rate. This clearly shows that the magnetite content of the corrosion product layer increases with the cathodic character of the metal surface.

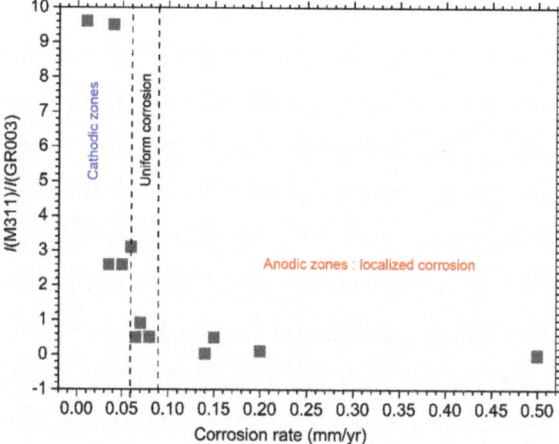

Figure 9. Variation of the $I(M311)/I(GR003)$ ratio (determined from XRD analysis of the dark inner layers covering steel surfaces after 6 years of immersion in natural seawater), with the corresponding average local corrosion rate (determined by local thickness measurements as described in [40–42]).

4.4. The Localized Corrosion Mechanism Associated with Magnetite-Rich Cathodic Zones

As described in the previous section, the corrosion product layers covering cathodic zones are significantly enriched with magnetite that is an electronic conductor. Moreover, these layers do not contain, or only in small amounts, FeOOH compounds [41,42], in agreement with visual observations: these layers are black, i.e., the orange-brown outer stratum composed of FeOOH compounds did not form. This is only possible if dissolved O_2 does not react with the Fe(II)-based compounds present in the corrosion product layer. This means that the O_2 molecules reaching the outer surface of the corrosion product layer are consuming, through the reduction process, electrons coming from another part of the system, i.e., necessarily the anodic zones of the steel surface. This process is possible only if an electronic pathway exists inside the corrosion product layer. This pathway can be constituted by a network of interconnected magnetite particles also connected to the metal surface. Such a network may indeed be present in the magnetite-rich layers covering the cathodic zones, that are moreover much thinner than those covering the anodic zones ([41,42], see also Figure 6). A schematic representation of the mechanism is presented in Figure 10.

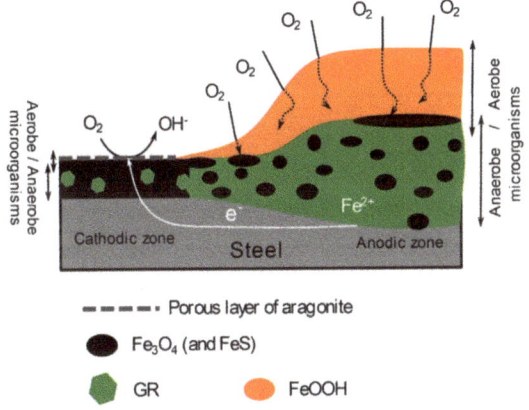

Figure 10. Schematic representation of the localized corrosion process associated with the heterogeneity of the corrosion product layer.

Actually, the anodic zones of the metal surface are thick and enriched with insulating compounds such as Fe(III)-oxyhydroxides and GR(SO_4^{2-}). Dissolved O_2 cannot reach the metal surface underneath or be reduced at the surface of magnetite particles that are present in small amounts and consequently not interconnected. It is consumed in the outer stratum by aerobe microorganisms or in the dark inner stratum through its reaction with Fe(II)-based corrosion products.

Localized corrosion can only lead to severe issues if it can persist for a long time. The particular mechanism associated with the heterogeneity of the corrosion product layers is based on the assumption that dissolved O_2 participates in the corrosion process because it can be reduced on the outer surface of the magnetite-rich layer covering the cathodic zones [41]. Since the formation of magnetite is favored by an increase of pH, this mechanism has then a self-sustaining ability as it creates conditions that favor its own persistence: If oxygen reduction takes place preferentially in the cathodic zones, then the interfacial pH increases and consequently the formation of magnetite is favored. Conversely, the process favors a decrease of pH in the anodic zones that hinders the formation of magnetite and leads to an accumulation of insulating compounds on the metal surface.

The role of SRB in this mechanism may also be important. Anaerobic conditions are met at the metal surface in both anodic and cathodic zones and SRB can grow and be active in both zones, as demonstrated by the identification of FeS in any case. In the cathodic zones, dissolved O_2 may reach the outer part of the dark corrosion product layer covering the metal before to be reduced at the surface of magnetite particles so that SRB would preferentially develop closer to the metal. In any case, iron sulfides are electronic conductors [16,23] and their formation in the cathodic zones would then favor the persistence of the corrosion cell described above thus reinforcing the corresponding localized corrosion mechanism. Electroactive SRB [23–25] could also take advantage of the interconnected Fe$_3$O$_4$/FeS particles network linked to the metal surface in the cathodic zones, and favor its persistence. Finally, this process would stop once:

(i) the layer covering the cathodic zone is covered by a thick layer of biofouling/calcareous deposit that prevents (as in anodic zones) dissolved O$_2$ to reach magnetite particles,
(ii) the corrosion product layer covering the cathodic zone becomes so thick that the network of interconnected Fe$_3$O$_4$/FeS particles is no more linked to the metal surface.

5. Oxidation of the Corrosion Product Layers and FORMATION of Akaganeite

Akaganeite (β-FeOOH) is a common corrosion product of steel in marine atmospheres e.g., [38,39,75,76]. Though it is considered as the β-phase of Fe(III) oxyhydroxides, it necessarily contains Cl$^-$ ions and its chemical formula is more exactly FeO$_{1-x}$(OH)$_{1+x}$Cl$_x$ [77]. However, akaganeite was rarely observed, and always as a minor component, in the studies [26,30,36,41,42,44] dealing with corrosion products formed on carbon steel permanently immersed in seawater.

The corrosion processes involved during permanent immersion are of course different from those of atmospheric corrosion. The wet/dry cycles typical of atmospheric corrosion induce oxidation and transformation of Fe(II) species/compounds that do not take place during permanent immersion and may be responsible for the formation of akaganeite. To clarify this point a coupon was, after 1 year of immersion in natural seawater, exposed 11 days to the atmosphere of the laboratory so that the corrosion product layer dried completely. This layer was then analyzed by XRD and the result is shown in Figure 11.

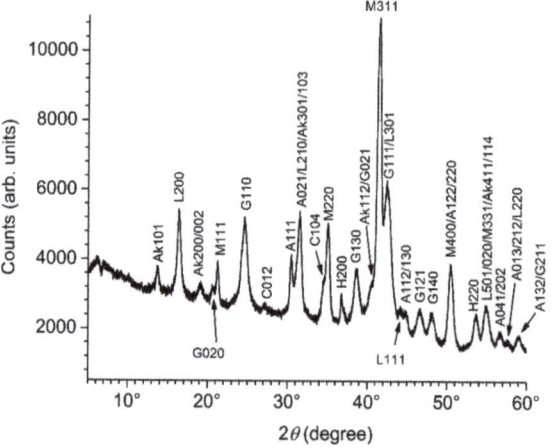

Figure 11. XRD analysis of the inner stratum of the corrosion product layer of a carbon steel coupon after 1 year in natural seawater (exposure site: cargo port of La Rochelle, Atlantic Ocean) and subsequent exposure of 11 days to the atmosphere (laboratory). Ak = akaganeite, A = aragonite, C = calcite, G = goethite, H = halite (NaCl), L = lepidocrocite, and M = magnetite. The diffraction lines are denoted with the corresponding Miller index.

The Fe(III)-based compounds already formed during the immersion period, i.e., magnetite, lepidocrocite, and goethite, are the main components of the dried oxidized layer. The Fe(II)-based corrosion products, i.e., GR compounds, are not detected anymore. They were integrally oxidized and transformed into Fe(III) compounds during the 11 days of exposure to air. Aragonite is also identified, it comes more likely from a cathodic zone of the metal. Calcite is also present but may come from fragments of shells. The drying of the sample has moreover induced the formation of NaCl (halite).

Finally, another Fe(III) compound, not initially present in the corrosion product layer, is now identified: it is akaganeite.

This shows that akaganeite may mostly result from the oxidation and drying of corrosion product layers previously formed during the immersion of steel in seawater. Actually, the formation of akaganeite requires a high concentration of both dissolved Fe(II) and Cl$^-$ species [78]. During the drying of a corrosion product layer, water evaporates progressively and dissolved species (Cl$^-$, Fe^{2+}) concentrations increase continuously until the formation of akaganeite is possible. The overall reaction could be written as:

$$4Fe^{2+} + O_2 + 6H_2O + 4HCl \rightarrow 4FeO_{1-x}(OH)_{1+x}Cl_x + 8H^+ \quad (13)$$

Note that intermediate compounds/species could be involved, e.g., GR(Cl$^-$) [78,79], Fe(II)-hydroxychloride [78,79] or dissolved Fe(III) species.

This result points out the necessity to shelter the corrosion product layer from the air before and during analysis. First, the Fe(II)-based compounds must be preserved because, as they form directly from the dissolution of steel, they convey the most important information about the corrosion mechanism. Secondly, the formation of compounds not really associated with the corrosion process must be avoided because it may lead to a misinterpretation of the mechanisms involved.

A final corrosion product must be discussed, the Fe(II)-hydroxychloride β-Fe$_2$(OH)$_3$Cl known as a precursor of akaganeite [78,79]. Though it has not been clearly identified on "modern" steel structures, it has however been reported as the main component of the corrosion product layer covering Gallo-Roman iron ingots that remained at the bottom of the Mediterranean Sea for 2000 years [43]. This finding shows that the corrosion mechanisms may change drastically after a (very) long time of permanent immersion in seawater. However, no information is currently available about this possible transition period and the associated mechanisms and processes.

6. Conclusions

The thorough characterization of the corrosion product layers provided important information that led to a better understanding of the mechanisms of marine corrosion of carbon steel permanently immersed in natural seawater. Important facts must be pointed out:

- It is generally admitted that, after a sufficiently long immersion time (variable but at least 6 months), anoxic conditions are met at the steel surface and inside the inner part of the corrosion product layer. Anaerobe microorganisms such as SRB can develop and be active, which leads to the formation of iron sulfides. Marine corrosion is then intrinsically a biologically influenced process and its complexity cannot be entirely mimicked by laboratory experiments and/or using artificial seawater. The biofilm itself and other microorganisms are also known to have an influence on corrosion processes.
- The first solid phase to form on the steel surface is the sulfate GR. It is favored, with respect to magnetite and carbonate GR, in the anodic zones, by the decrease of the interfacial pH. A mechanism is proposed, involving (i) the adsorption of anions (mainly Cl$^-$ and SO$_4^{2-}$) on the nuclei of Fe(OH)$_2$ hydroxide sheets, (ii) the oxidation of part of the Fe(II) cations to Fe(III) in the hydroxide sheets and (iii) the stacking of the sheets leading to GR(SO$_4^{2-}$) after the release of Cl$^-$ ions.
- A particular mechanism can however involve oxygen even if anoxic conditions are established at the steel surface. The pH proved to have a significant influence on the composition of the corrosion product layer and this composition then depends on the anodic/cathodic character of the underlying metal surface. In cathodic zones, the increased interfacial pH favors the formation of magnetite (among other compounds) that is an electronic conductor. Associated with a low corrosion rate, the process then leads to a magnetite-rich layer that remains moderately thick (~2–5 mm) even after 6–8 years [41,42]. The reduction of dissolved O$_2$ can then take place at the

- outer surface of this layer as long as the magnetite particles remain interconnected and connected to the steel surface.
- The activity of SRB (and other sulfide-producing bacteria) in the cathodic zones may favor and reinforce this mechanism because it generates additional conductive corrosion products, i.e., iron sulfides.
- Finally, it is necessary to preserve the samples from air to avoid the transformation of Fe(II)-based compounds. The nature of these compounds is directly related to the corrosion mechanisms. Moreover, the oxidation/transformation of the Fe(II)-based corrosion products may produce other compounds (e.g., akaganeite), thus possibly leading to an erroneous interpretation of the mechanisms.

Author Contributions: Conceptualization, all authors; methodology, all authors; validation, all authors; formal analysis, all authors; investigation, all authors; writing—original draft preparation, P.R.; writing—review and editing, all authors; visualization, P.R., M.J.; supervision, A.-M.G., P.R.; project administration, A.-M.G., P.R.; funding acquisition, A.-M.G., P.R. All authors have read and agreed to the published version of the manuscript.

Funding: The various studies synthesized here were funded by: the Technological Research Institute Jules Verne of Nantes (France) through the ADUSCOR research program; the European Regional Development Fund (ERDF) and the CDA of La Rochelle through the DYPOMAR research project; the seaports (GPM) of La Rochelle, Nantes-Saint-Nazaire, Marseille, and Le Havre.

Conflicts of Interest: The authors declare no conflict of interest.

References

1. Chaves, I.A.; Melchers, R.E. Long term localised corrosion of marine steel piling welds. *Corros. Eng. Sci. Technol.* **2013**, *48*, 469–474. [CrossRef]
2. Melchers, R.E.; Jeffrey, R.J.; Usher, K.M. Localized corrosion of steel sheet piling. *Corros. Sci.* **2014**, *79*, 139–147. [CrossRef]
3. Melchers, R.E.; Chaves, I.A.; Jeffrey, R. A conceptual model for the interaction between carbon content and manganese sulphide inclusions in the short-term seawater corrosion of low carbon steel. *Metals* **2016**, *6*, 132. [CrossRef]
4. Liu, C.; Cheng, X.; Dai, Z.; Liu, R.; Li, Z.; Cui, L.; Chen, M.; Ke, L. Synergistic effect of Al_2O_3 inclusion and pearlite on the localized corrosion evolution process of carbon steel in marine environment. *Materials* **2018**, *11*, 2277. [CrossRef] [PubMed]
5. Liduino, V.S.; Lutterbach, M.T.S.; Sérvulo, E.F.C. Biofilm activity on corrosion of API 5L X65 steel weld bead. *Colloids Surf. B Biointerfaces* **2018**, *172*, 43–50. [CrossRef]
6. Evans, U.R. *The Corrosion and Oxidation of Metals: Scientific Principles and Practical Applications*; Edward Arnold (Publishers) Ltd.: London, UK, 1960.
7. Jeffrey, R.; Melchers, R.E. Corrosion of vertical mild steel strips in seawater. *Corros. Sci.* **2009**, *51*, 2291–2297. [CrossRef]
8. Zou, Y.; Wang, J.; Bai, Q.; Zhang, L.L.; Peng, X.; Kong, X.F. Potential distribution characteristics of mild steel in seawater. *Corros. Sci.* **2012**, *57*, 202–208. [CrossRef]
9. Duboscq, J.; Sabot, R.; Jeannin, M.; Refait, P.H. Localized corrosion of carbon steel in seawater: Processes occurring in cathodic zones. *Mater. Corros.* **2019**, *70*, 973–984. [CrossRef]
10. Lee, W.; Lewandowski, Z.; Nielsen, P.H.; Hamilton, W.A. Role of sulfate-reducing bacteria in corrosion of mild steel: A review. *Biofouling* **1995**, *8*, 165–194. [CrossRef]
11. Starovetsky, D.; Khaselev, O.; Starovetsky, J.; Armon, R.; Yahalom, J. Effect of iron exposure in SRB media on pitting initiation. *Corros. Sci.* **2000**, *42*, 345–359.
12. Huang, Y.; Duan, J.; Ma, S. Effects of anaerobe in sea bottom sediment on the corrosion of carbon steel. *Mater. Corros.* **2004**, *55*, 46–48. [CrossRef]
13. Jeffrey, R.; Melchers, R.E. The changing topography of corroding mild steel surfaces in seawater. *Corros. Sci.* **2007**, *49*, 2270–2288. [CrossRef]
14. Castaneda, H.; Benetton, X.D. SRB-biofilm influence in active corrosion sites formed at the steel-electrolyte interface when exposed to artificial seawater conditions. *Corros. Sci.* **2008**, *50*, 1169–1183. [CrossRef]

15. Beech, I.B.; Campbell, S.A. Accelerated low water corrosion of carbon steel in the presence of a biofilm harbouring sulphate-reducing and sulphur-oxidising bacteria recovered from a marine sediment. *Electrochim. Acta* **2008**, *54*, 14–21. [CrossRef]
16. Dong, Z.H.; Shi, W.; Ruan, H.M.; An Zhang, G. Heterogeneous corrosion of mild steel under SRB-biofilm characterised by electrochemical mapping technique. *Corros. Sci.* **2011**, *53*, 2978–2987. [CrossRef]
17. Melchers, R.E.; Jeffrey, R. Corrosion of long vertical steel strips in the marine tidal zone and implications for ALWC. *Corros. Sci.* **2012**, *65*, 26–36. [CrossRef]
18. Stipaničev, M.; Turcu, F.; Esnault, L.; Schweitzer, E.W.; Kilian, R.; Basseguy, R. Corrosion behavior of carbon steel in presence of sulfate-reducing bacteria in seawater environment. *Electrochim. Acta* **2013**, *113*, 390–406. [CrossRef]
19. World Steel Association. "World Steel in Figures 2019", Brussels, Belgium. 2019. Available online: https://www.worldsteel.org/publications/infographics.html (accessed on 2 June 2020).
20. Melchers, R.E. Mathematical modelling of the diffusion controlled phase in marine immersion corrosion of mild steel. *Corros. Sci.* **2003**, *45*, 923–940. [CrossRef]
21. Melchers, R.E.; Jeffrey, R. Early corrosion of mild steel in seawater. *Corros. Sci.* **2005**, *47*, 1678–1693. [CrossRef]
22. Melchers, R.E.; Wells, T. Models for the anaerobic phases of marine immersion corrosion. *Corros. Sci.* **2006**, *48*, 1791–1811. [CrossRef]
23. Enning, D.; Venzlaff, H.; Garrelfs, J.; Dinh, H.T.; Meyer, V.; Mayrhofer, K.J.J.; Hassel, A.W.; Stratmann, M.; Widdel, F. Marine sulfate-reducing bacteria cause serious corrosion of iron under electroconductive biogenic mineral crust. *Environ. Microbiol.* **2012**, *14*, 1772–1787. [CrossRef] [PubMed]
24. Venzlaff, H.; Enning, D.; Srinivasan, J.; Mayrhofer, K.J.J.; Hassel, A.W.; Widdel, F.; Stratmann, M. Accelerated cathodic reaction in microbial corrosion of iron due to direct electron uptake by sulfate-reducing bacteria. *Corros. Sci.* **2013**, *66*, 88–96. [CrossRef]
25. Yu, L.; Duan, J.; Du, X.; Huang, Y.; Hou, B. Accelerated anaerobic corrosion of electroactive sulfate-reducing bacteria by electrochemical impedance spectroscopy and chronoamperometry. *Electrochem. Commun.* **2013**, *26*, 101–104. [CrossRef]
26. Pineau, S.; Sabot, R.; Quillet, L.; Jeannin, M.; Caplat, C.H.; Dupont-Morral, I.; Refait, P.H. Formation of the Fe(II-III) hydroxysulphate green rust during marine corrosion of steel associated to molecular detection of dissimilatory sulphite-reductase. *Corros. Sci.* **2008**, *50*, 1099–1111. [CrossRef]
27. Duan, J.; Wu, S.; Zhang, X.; Huang, G.; Du, M.; Hou, B. Corrosion of carbon steel influenced by anaerobic biofilm in natural seawater. *Electrochim. Acta* **2008**, *54*, 22–28. [CrossRef]
28. Boudaud, N.; Coton, M.; Coton, E.; Pineau, S.; Travert, J.; Amiel, C. Biodiversity analysis by polyphasic study of marine bacteria associated with biocorrosion phenomena. *J. Appl. Microbiol.* **2010**, *109*, 166–179. [CrossRef]
29. Usher, K.M.; Kaksonen, A.H.; MacLeod, I.D. Marine rust tubercles harbour iron corroding archaea and sulphate reducing bacteria. *Corros. Sci.* **2014**, *83*, 189–197. [CrossRef]
30. Lanneluc, I.; Langumier, M.; Sabot, R.; Jeannin, M.; Refait, P.; Sablé, S. On the bacterial communities associated with the corrosion product layer during the early stages of marine corrosion of carbon steel. *Int. Biodeter. Biodegrad.* **2015**, *99*, 55–65. [CrossRef]
31. Stipanicev, M.; Turcu, F.; Esnault, L.; Rosas, O.; Basseguy, R.; Sztyler, M.; Beech, I.B. Corrosion of carbon steel by bacteria from North Sea offshore seawater injection systems: Laboratory investigation. *Bioelectrochemistry* **2014**, *97*, 76–88. [CrossRef]
32. Rlinger, M.I.; Samokhvalov, A.A. Electron conduction in magnetite and ferrites. *Phys. Stat. Sol. B* **1977**, *79*, 9–48. [CrossRef]
33. Ihle, D.; Lorenz, B. Small-polaron band versus hopping conduction in Fe_3O_4. *J. Phys. C Solid State Phys.* **1985**, *18*, L647–L650. [CrossRef]
34. Lopez Maldonado, K.L.; de la Presa, P.; de la Rubia, M.A.; Crespo, P.; de Frutos, J.; Hernando, A.; Matutes-Aquino, J.A.; Elizalde Galindo, J. Effects of grain boundary width and crystallite size on conductivity and magnetic properties of magnetite nanoparticles. *J. Nanopart. Res.* **2014**, *16*, 2482. [CrossRef]
35. Li, Y.; Hou, B.; Li, H.; Zhang, J. Corrosion behavior of steel in Chengdao offshore oil exploitation area. *Mater. Corros.* **2004**, *55*, 305–309. [CrossRef]
36. Refait, P.; Nguyen, D.D.; Jeannin, M.; Sablé, S.; Langumier, M.; Sabot, R. Electrochemical formation of green rusts in deaerated seawater-like solutions. *Electrochim. Acta* **2011**, *56*, 6481–6488. [CrossRef]

37. Refait, P.; Jeannin, M.; Sabot, R.; Antony, H.; Pineau, S. Corrosion and cathodic protection of carbon steel in the tidal zone: Products, mechanisms and kinetics. *Corros. Sci.* **2015**, *90*, 375–382. [CrossRef]
38. Morcillo, M.; Chico, B.; Alcantara, J.; Diaz, I.; Wolthuis, R.; de la Fuente, D. SEM/Micro-Raman characterization of the morphologies of marine atmospheric corrosion products formed on mild steel. *J. Electrochem. Soc.* **2016**, *163*, C426–C439. [CrossRef]
39. Morcillo, M.; Chico, B.; de la Fuente, D.; Alcantara, J.; Odnevall Wallinder, I.; Leygraf, C. On the Mechanism of Rust Exfoliation in Marine Environment. *J. Electrochem. Soc.* **2017**, *164*, C8–C16. [CrossRef]
40. Refait, P.; Grolleau, A.-M.; Jeannin, M.; François, E.; Sabot, R. Corrosion of carbon steel at the mud zone/seawater interface: Mechanisms and kinetics. *Corros. Sci.* **2018**, *130*, 76–84. [CrossRef]
41. Refait, P.; Grolleau, A.-M.; Jeannin, M.; François, E.; Sabot, R. Localized corrosion of carbon steel in marine media: Galvanic coupling and heterogeneity of the corrosion product layer. *Corros. Sci.* **2016**, *111*, 583–595. [CrossRef]
42. Refait, P.; Jeannin, M.; François, E.; Sabot, R.; Grolleau, A.-M. Galvanic corrosion in marine environments: Effects associated with the inversion of polarity of Zn/carbon steel couples. *Mater. Corros.* **2019**, *70*, 950–961. [CrossRef]
43. Rémazeilles, C.; Neff, D.; Kergourlay, F.; Foy, E.; Conforto, E.; Guilminot, E.; Reguer, S.; Refait, P.; Dillmann, P. Mechanisms of long-term anaerobic corrosion of iron archaeological artefacts in seawater. *Corros. Sci.* **2009**, *51*, 2932–2941. [CrossRef]
44. Refait, P.; Jeannin, M.; Sabot, R.; Antony, H.; Pineau, S. Electrochemical formation and transformation of corrosion products on carbon steel under cathodic protection in seawater. *Corros. Sci.* **2013**, *71*, 32–36. [CrossRef]
45. De Faria, D.L.A.; Silva, S.V.; Oliveira, M.T.D. Raman micro spectroscopy study of some iron oxides and oxyhydroxides. *J. Raman Spectrosc.* **1997**, *28*, 873–878. [CrossRef]
46. Shebanova, O.N.; Lazor, P. Raman study of magnetite (Fe_3O_4): Laser-induced thermal effects and oxidation. *J. Raman Spectrosc.* **2003**, *34*, 845–852. [CrossRef]
47. Hansen, H.C.B. Composition, stabilisation, and light absorption of Fe(II)-Fe(III) hydroxycarbonate (green rust). *Clay Miner.* **1989**, *24*, 663–669. [CrossRef]
48. Duboscq, J.; Abdelmoula, M.; Jeannin, M.; Sabot, R.; Refait, P. On the formation and transformation of Fe(III)-containing chukanovite, $Fe^{II}_{2-x}Fe^{III}_x(OH)_{2-x}O_xCO_3$. *J. Phys. Chem. Solids* **2020**, *138*, 109310. [CrossRef]
49. Pourbaix, M. Thermodynamics and corrosion. *Corros. Sci.* **1990**, *30*, 963–988. [CrossRef]
50. Refait, P.; Géhin, A.; Abdelmoula, M.; Génin, J.-M.R. Coprecipitation thermodynamics of iron(II-III) hydroxysulphate green rust from Fe(II) and Fe(III) salts. *Corros. Sci.* **2003**, *45*, 656–676. [CrossRef]
51. Ruby, C.; Abdelmoula, M.; Naille, S.; Renard, A.; Khare, V.; Ona-Nguema, G.; Morin, G.; Génin, J.-M.R. Oxidation modes and thermodynamics of Fe^{II-III} oxyhydroxycarbonate green rust: Dissolution-precipitation versus in situ deprotonation. *Geochim. Cosmochim. Acta* **2010**, *74*, 953–966. [CrossRef]
52. Stampfl, P.P. Ein basisches eisen-II-III-karbonat in rost. *Corros. Sci.* **1969**, *9*, 185–187. [CrossRef]
53. Allmann, R. Doppelschichtstrukturen mit brucitähnlichen Schichtionen $[Me(II)_{1-x}Me(III)_x(OH)_2]^{x+}$. *Chimia* **1970**, *24*, 99–108.
54. Refait, P.; Abdelmoula, M.; Génin, J.-M.R. Mechanisms of formation and structure of green rust one in aqueous corrosion of iron in the presence of chloride ions. *Corros. Sci.* **1998**, *40*, 1547–1560. [CrossRef]
55. Simon, L.; François, M.; Refait, P.; Renaudin, G.; Lelaurain, M.; Génin, J.-M.R. Structure of the Fe(II-III) layered double hydroxysulphate green rust two from Rietveld analysis. *Sol. State Sci.* **2003**, *5*, 327–334. [CrossRef]
56. Miyata, S. Anion-exchange properties of hydrotalcite-like compounds. *Clays Clay Miner.* **1983**, *31*, 305–311. [CrossRef]
57. Mendiboure, A.; Schöllhorn, R. Formation and anion exchange reactions of layered transition metal hydroxides $[Ni_{1-x}M_x](OH)_2(CO_3)_{x/2}(H_2O)_z$ (M=Fe,Co). *Rev. Chim. Miner.* **1986**, *23*, 819–827. [CrossRef]
58. Refait, P.; Memet, J.B.; Bon, C.; Sabot, S.; Génin, J.-M.R. Formation of the Fe(II)-Fe(III) hydroxysulphate green rust during marine corrosion of steel. *Corros. Sci.* **2003**, *45*, 833–845. [CrossRef]
59. Bourdoiseau, J.A.; Jeannin, M.; Sabot, R.; Rémazeilles, C.; Refait, P. Characterisation of mackinawite by Raman spectroscopy: Effects of crystallisation, drying and oxidation. *Corros. Sci.* **2008**, *50*, 3247–3255. [CrossRef]
60. Bocher, F.; Géhin, A.; Ruby, C.; Ghanbaja, J.; Abdelmoula, M.; Génin, J.-M.R. Coprecipitation of Fe(II-III) hydroxycarbonate green rust stabilised by phosphate adsorption. *Sol. State Sci.* **2004**, *6*, 117–124. [CrossRef]
61. Barthélémy, K.; Naille, S.; Despas, C.; Ruby, C.; Mallet, M. Carbonated ferric green rust as a new material for efficient phosphate removal. *J. Colloid Interface Sci.* **2012**, *384*, 121–127. [CrossRef]

62. Usman, M.; Hanna, K.; Abdelmoula, M.; Zegeye, A.; Faure, P.; Ruby, C. Formation of green rust via mineralogical transformation of ferric oxides (ferrihydrite, goethite and hematite). *Appl. Clay Sci.* **2012**, *64*, 38–43. [CrossRef]
63. Jeong, H.Y.; Lee, J.H.; Hayes, K.F. Characterization of nanocrystalline mackinawite: Crystal structure, particle size, and specific surface area. *Geochim. Cosmochim. Acta* **2008**, *72*, 493–505. [CrossRef] [PubMed]
64. Ohfuji, H.; Rickard, D. High resolution transmission electron microscopic study of synthetic nanocrystalline mackinawite. *Earth Planet. Sci. Lett.* **2006**, *241*, 227–233. [CrossRef]
65. Bourdoiseau, J.-A.; Jeannin, M.; Rémazeilles, C.; Sabot, R.; Refait, P. The transformation of mackinawite into greigite studied by Raman spectroscopy. *J. Raman Spectrosc.* **2011**, *42*, 496–504. [CrossRef]
66. Mullet, M.; Boursiquot, S.; Abdelmoula, M.; Génin, J.-M.R.; Ehrhardt, J.J. Surface chemistry and structural properties of mackinawite prepared by reaction of sulfide with metallic iron. *Geochim. Cosmochim. Acta* **2002**, *66*, 829–836. [CrossRef]
67. Lennie, A.R.; Redfern, S.A.T.; Champness, P.E.; Stoddart, C.P.; Schofield, P.F.; Vaughan, D.J. Transformation of mackinawite to greigite: An in situ X-ray powder diffraction and transmission electron microscope study. *Am. Min.* **1997**, *82*, 302–309. [CrossRef]
68. Rémazeilles, C.; Saheb, M.; Neff, D.; Guilminot, E.; Tran, K.; Bourdoiseau, J.-A.; Sabot, R.; Jeannin, M.; Matthiesen, H.; Dillmann, P.; et al. Microbiologically influenced corrosion of archaeological artefacts; characterisation of iron(II) sulphides by Raman spectroscopy. *J. Raman Spectrosc.* **2010**, *41*, 1135–1143. [CrossRef]
69. Malard, E.; Kervadec, D.; Gil, O.; Lefevre, Y.; Malard, S. Interactions between steels and sulphide-producing bacteria-Corrosion of carbon steels and low-alloy steels in natural seawater. *Electrochim. Acta* **2008**, *54*, 8–13. [CrossRef]
70. Refait, P.; Sabot, R.; Jeannin, M. Role of Al(III) and Cr(III) on the formation and oxidation of the Fe(II-III) hydroxysulfate Green Rust. *Colloids Surf. A Phys. Eng. Asp.* **2017**, *531*, 203–212. [CrossRef]
71. Hartt, W.H.; Culberson, C.H.; Smith, S.W. Calcareous deposits on metal surfaces in seawater- A critical review. *Corrosion* **1984**, *40*, 609–618. [CrossRef]
72. Lee, R.U.; Ambrose, J.R. Influence of cathodic protection parameters on calcareous deposit formation. *Corrosion* **1986**, *44*, 887–891. [CrossRef]
73. Yan, J.F.; White, R.E.; Griffin, R.B. Parametric studies of the formation of calcareous deposits on cathodically protected steel in seawater. *J. Electrochem. Soc.* **1993**, *141*, 1275–1280. [CrossRef]
74. Barchiche, C.; Deslouis, C.; Festy, D.; Gil, O.; Refait, P.; Touzain, S.; Tribollet, B. Characterization of calcareous deposits in artificial sea water by impedance techniques. 3- Deposit of $CaCO_3$ in the presence of Mg(II). *Electrochim. Acta* **2003**, *48*, 1645–1654. [CrossRef]
75. Cook, D.C.; van Orden, A.C.; Carpio, J.J.; Oh, S.J. Atmospheric corrosion in the Gulf of Mexico. *Hyperf. Interact.* **1998**, *113*, 319–329. [CrossRef]
76. Li, S.; Hihara, L.H. A micro-Raman spectroscopic study of marine atmospheric corrosion of carbon steel: The effect of akaganeite. *J. Electrochem. Soc.* **2015**, *162*, C495–C502. [CrossRef]
77. Stahl, K.; Nielsen, K.; Jiang, J.; Lebech, B.; Hanson, J.C.; Norby, P.; van Lanschot, J. On the akaganeite crystal structure, phase transformations and possible role in post-excavational corrosion of iron artifacts. *Corros. Sci.* **2003**, *45*, 2563–2575. [CrossRef]
78. Rémazeilles, C.; Refait, P. On the formation of β-FeOOH (Akaganéite) in chloride-containing environments. *Corros. Sci.* **2007**, *49*, 844–857. [CrossRef]
79. Refait, P.; Génin, J.-M.R. The mechanisms of oxidation of ferrous hydroxychloride β-$Fe_2(OH)_3Cl$ in chloride-containing aqueous solution: The formation of β-FeOOH akaganeite; an X-ray diffraction, Mössbauer spectroscopy and electrochemical study. *Corros. Sci.* **1997**, *39*, 539–553. [CrossRef]

© 2020 by the authors. Licensee MDPI, Basel, Switzerland. This article is an open access article distributed under the terms and conditions of the Creative Commons Attribution (CC BY) license (http://creativecommons.org/licenses/by/4.0/).

Experience-Based Physico-Chemical Models for Long-Term Reinforcement Corrosion

Robert E. Melchers

Centre for Infrastructure Performance and Reliability, The University of Newcastle, Callaghan 2308, Australia; rob.melchers@newcastle.edu.au

Abstract: The long-term corrosion progression of steel reinforcement is important for estimating the life of reinforced concrete infrastructure. Reviews of field experience and results from recent controlled long-term experiments show that the development of reinforcement corrosion is much more complex than the classical empirical Tuutti model. A new, comprehensive model is proposed, referencing observations and inferences from many field and laboratory observations and built on the bi-modal model for the corrosion of steel. It includes the critical roles of air-voids in the concrete at the concrete-steel interface and the effect of long-term alkali leaching as accelerated by the presence of chlorides. Both are affected by compaction and concrete permeability. The role of chlorides in the early stages is confined to pitting within air-voids. These are critical for allowing initiation to occur, while their size influences the severity of early corrosion. Empirical data show that for seawater with an average water temperature in the range of 10–20 °C, the corresponding rate of long-term corrosion r_a is in the range of 0.012–0.015 mm/y.

Keywords: reinforcement; corrosion; chlorides; progression; alkalinity; cracking

1. Introduction

Reinforcement corrosion of marine structures can be a major problem for structural safety and serviceability. Despite much research attention over many years, the causes of such corrosion remain unclear: "after more than half a century of research on the issue of steel corrosion in concrete, many questions remain open" [1]. Seldom mentioned in these overviews is that a considerable amount of practical experience over many years has shown, repeatedly, that for many reinforced concrete structures exposed for decades in high chloride environments, reinforcement corrosion has not occurred or is negligible despite very high concentrations of chlorides at the reinforcement bars [2–4]. One example of this type of behavior is the set of some 900 driven reinforced concrete piles, constructed during the 1930s that were found, on extraction from their foundations in 2012, to show almost no evidence of reinforcement corrosion. During that time they had been exposed, continuously, to the immersion, tidal, splash, and atmospheric zones of the coastal Pacific Ocean [5]. The state of the reinforcement and the lack of corrosion for most of the surfaces of the reinforcing bars were verified by breaking open randomly selected piles. This showed the high density of the concrete, the lack of air-voids within the concrete and at the steel interface surfaces, and that the rusts that were present were very thin and of the type generated under very low oxygen conditions. Very high chloride concentrations were observed inside the concrete, including immediately adjacent to the steel reinforcement bars. The concrete cross-sections yielded pH values everywhere around 12 other than the pH around 7 in the 2–3 mm outer edges. This indicated that much of the concrete cross-sections still contained high levels of calcium hydroxide ($Ca(OH)_2$) even after about 80 years of exposure. There were some exceptions to this trend. The most notable was for one pile that showed very severe localized corrosion at the point where it was inferred that the pile had been deeply cracked in flexure at the time of construction [5].

Similar findings have been made recently [6] for the massive reinforced concrete Phoenix caissons hastily produced during WW2 and now lying abandoned along the coast of Normandy (F). Those that could be inspected directly or through aerial photography showed little or no obvious corrosion of reinforcement [7], despite having been exposed in the chloride-rich immersion, tidal, splash, and marine atmospheric zones since 1944. Reinforcement corrosion was evident mostly only where early structural damage (through a major storm event in 1944) had occurred or at poor construction joints. Parallel findings are available for a range of other practical reinforced concrete structures [3,8–11]. However, the conventional wisdom, based largely on laboratory research, appears focused primarily on cases of reinforced (and prestressed) concrete structures and laboratory samples that showed early initiation of reinforcement corrosion and relatively fast development of some type of structural damage.

The reasons for the poor performance of some practical reinforced concretes have become clearer as a result of recent long-term experimental findings [12,13]. These are reviewed briefly in the next section. They provide a background for critical aspects of initiation of reinforcement corrosion in marine conditions and for the subsequent rate of its progression. A new model for reinforcement corrosion progression as a function of longer-term exposures is then introduced, extended from an earlier empirical model [14] that was based on the empirical analysis of data from actual structures and on modern understanding of the development of corrosion of bare steel in seawater. The extended model proposed herein accounts for the current and new understanding of the relevant physico-chemical mechanisms and criteria.

As a first step in the calibration of the proposed model, it is compared with data from experiments conducted during the 1950s on a range of model reinforced concretes covering different water-cement and aggregate-cement ratios. Comments are made about the principal factors that govern the model, including the important aspect of the interfacial zone between the steel and concrete. The roles of the depth of concrete cracking, of fractures, and of poor construction joints are discussed, including the likely rates of very localized corrosion. Some comments about practical implications are made throughout the paper.

At this point it is noted that apart from empirical and physico-chemical modeling approaches, the literature, as reviewed by Raupach [15], also contains some models based on interpretations from electro-chemical testing. Although they have been advocated for many years (cf. [16]), these are not used herein. The reason is that the results obtained are known to be problematic when compared to physical observations of corrosion and pitting in actual seawater conditions (e.g., [17]) and this also has been noted repeatedly for reinforcement corrosion (e.g., [18–23]). This confirms the need for calibration and validation of electrochemical test results against empirical field data [24]. The approach herein is to work directly with the available empirical data for calibration of the physico-chemical model.

2. Background

The classical, entirely empirical model for the initiation and progression of reinforcement corrosion is attributed to Tuutti [25] although there was a similar antecedent [26]. The Tuutti model provides a period during which chlorides permeate through the concrete cover to eventually reach the reinforcement. Then, it is assumed that when a sufficiently high concentration of chlorides is reached at reinforcement, corrosion initiates, followed by a steady increase in corrosion with time (Figure 1a).

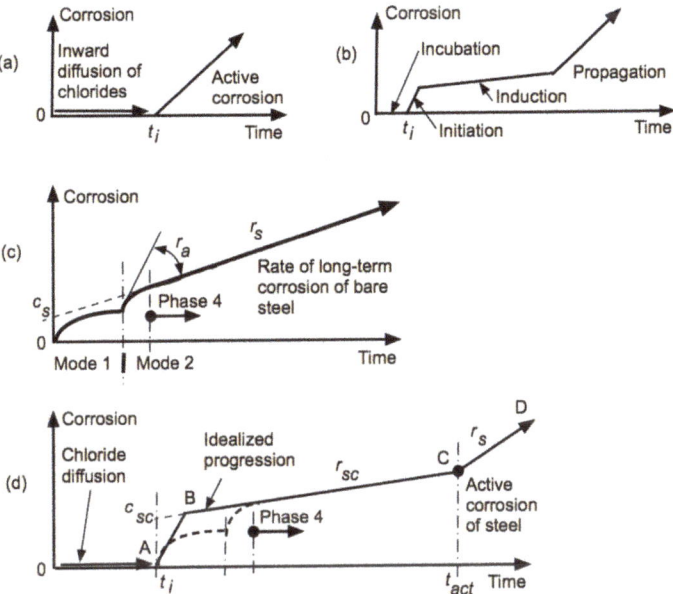

Figure 1. (a) Traditional Tuutti model showing corrosion "initiation" and an immediate serious corrosion for high chloride concentration conditions at the reinforcing bars, (b) phenomenological model proposed by François et al., 1994 [27] to consider the effect of cracking in facilitating the transportation of chloride ions to the reinforcement, (c) bi-modal model for the corrosion of steel in marine (and other) conditions, (d) corrosion loss model proposed by Melchers and Li, 2006 [14] with parameter c_{sc} related to void size, long-term rate r_{sc} related to concrete permeability, and the time t_{act} to commencement of active corrosion.

To allow for cracking of the concrete as might be caused by tensile flexural stresses in the concrete François et al. [27] proposed, on the basis of their own test results, the phenomenological model shown in Figure 1b (cf. [28]). However, a practical assessment of test conditions indicates that the constant load applied for the beams is considerably more than the beams of the proportions used, which would sustain in normal service. While the stated maximum moment that was applied to the beams is realistic for the nominal working load capacity used in design, typically the "sustained" loading, that is the loading applicable for most of the operational life of a beam, is some 10–20% of the design load [29]. It follows that the crack sizes in the experiments are some 5–10 times greater than those that would be expected under normal service conditions. In fact, most beams in practice show no signs of flexural cracking. In practice, if severe cracking does occur, it almost invariably is the result of overloading or poor design. It follows that the model is unduly conservative for realistic structures (Figure 1b).

In the model of Figure 1b, t_i is the time at which initiation occurs. Due to the large crack sizes in the experimental work, the initiation of corrosion will occur relatively early in the life of the structure, presumably as a result of chlorides (and likely oxygen) being able to reach the reinforcement relatively quickly. The model assumes that shortly after t_i the rate of corrosion drops to a very low value, attributed to the rate-controlling reaction stated as then being cathodic oxygen reduction. The reason for the large reduction in the corrosion rate is considered to be a build-up of corrosion products. Eventually, at t_{prop}, the model enters the "propagation" phase that has a damaging rate of reinforcement corrosion.

One difficulty with both the Tuutti and the François et al. models is in the role they assign to chlorides. As noted already by Foley [30] and as evident in results from carefully controlled experiments by Heyn and Bauer [31] and Mercer and Lumbard [32] in zero velocity conditions (such as inside concretes), the chloride concentration has very little effect on the rate of corrosion (although it can affect the propensity for pitting). Potentially, this is the reason the much-studied critical chloride concentration, at which t_i is assumed to occur, has proved so elusive, with very wide variability in the experimental results (e.g., [33]). It also may explain why some actual reinforced concretes have very high chloride concentrations inside the concrete but little or no evidence of reinforcement corrosion (cf. [4]). Despite these observations, the concept of chloride as the critical factor for initiating reinforcement corrosion in marine environments appears still firmly entrenched [34–36], although the modern terminology has become "chloride-induced" corrosion. However, the precise meaning of this term remains uncertain.

A second difficulty with these models is the assumption that the oxidization of the steel in the presence of water is always through the cathodic oxygen reduction reaction (ORR) $O_2 + 2H_2O + 4e^- \rightarrow 4OH^-$. This has also been assumed as the case for extended exposure periods, such as over decades (e.g., [37]), even though the reinforcement has already corroded significantly and there has been a considerable build-up of rusts. According to the bi-modal model for the corrosion of steel [38], a considerable build-up of rusts should produce predominantly anaerobic corrosion conditions after only a few years of exposure. For reinforcement corrosion in concrete, direct evidence of anaerobic corrosion is available for concrete structures exposed in marine conditions since WW2 (see Figure 1c in [39]).

A model that accounts for these factors and which is consistent with long-term corrosion behavior for steel in marine environments, was proposed by Melchers and Li [14]. Rather than assuming the corrosion of steel is a linear function of time as in the models of Tuutti and François et al., it was built on the more accurate bi-modal model for the corrosion of steel (Figure 1c). That model has been verified for a wide range of environments including soils and also a variety of steels and other alloys [40]. Hence, it can be expected to be valid also in concrete. The model for reinforcement corrosion is shown in Figure 1d. As in the Tuutti and François et al. models, it has a period of initiation ($0–t_i$) during which inward diffusion of chloride ions is likely to occur. Reinforcement corrosion commences at t_i (Figure 1d). However, as described further below, the conditions under which this occurs are more complex than the mere achievement of a "critical chloride concentration". After initiation, corrosion progresses initially in Mode 1, governed, as explained further below, by the availability of oxygen (and water) in air-voids in the concrete at the concrete-steel interface. As the oxygen is depleted and corrosion products build-up, the corrosion of the reinforcement transitions into Mode 2 with a corresponding relatively fast increase in reinforcement corrosion loss (Figure 1c). The overall effect is shown as (A-B) in Figure 1d. Thereafter, the reinforcement corrosion is the rate controlled predominantly by the cathodic hydrogen evolution reaction. This relatively slow reaction accounts for the plateau-effect (B-C) in Figure 1d. Eventually, at t_{act}, new conditions arise that permit a relatively fast and damaging corrosion (C-D)—these also are considered further below.

In wet oxygenated environments, the corrosion of steels in Mode 1 is predominantly under aerobic conditions (Figure 1c). The corresponding cathodic oxygen reduction reaction (ORR) is rate-controlled by oxygen diffusion from the external environment. As rusts build-up, the environment at the steel-rust interface changes predominantly under anaerobic conditions, for which corrosion occurs essentially by pitting under very low pH values [41]. The usual anodic reaction $Fe \rightarrow Fe^{2+} + 2e^-$ still applies but the process is now rate-controlled by the cathodic hydrogen evolution reaction (HER): $2H_2O + 2e^- \rightarrow H_2 \uparrow + 2OH^-$. The dissolution of water provides the hydroxide ions necessary to form rusts.

While oxygen is not directly involved in the HER, oxygen is not entirely excluded from the overall longer-term corrosion behavior. For atmospheric and for immersion corrosion, that is without the presence of concrete, oxidation may occur at the external rust layers [42], releasing ferrous ions and thus diminishing the overall rust layer [43]. The net effect of this is that the long-term corrosion rate r_s depends both on the rate of loss of external rust by oxidation and on the build-up of rusts by anaerobic processes at the metal-rust interface. In effect, oxygen is still the ultimate electron acceptor but the process is more convoluted.

It is clear that the concentration or availability of oxygen at the external rust surface can exert some influence over the rate of long-term corrosion r_s and also that both oxygen availability and r_s can be affected by encasing the bar in concrete.

Before proceeding, it is noted that field data [44] show that, closely enough for practical purposes, the longer-term part of the process, denoted as phase 4 in Mode 2 (Figure 1c), can be considered a linear function in time. It may be represented in a simplified manner by parameters c_s and r_s. It is certainly not the usual "corrosion rate"—this is a linear function passing through the origin and driven at the metal-rust interface region by the oxygen reduction reaction and the availability of oxygen.

It is reasonable to assume that for steel bars inside concrete, the progression of the corrosion process will follow a pattern of behavior similar to that for the corrosion of steels in other environments. It is likely that for the steel encased in concrete relatively impermeable to oxygen Mode 1 will be rather short in duration. In this sense, encasement in concrete would have an effect essentially similar to a lower oxygen concentration in the external environment [45]. Therefore, encasement would also tend to depress the rate of oxidation of the external rust layers in phase 4 and reduce the net value of r_s (Figure 1c). Let this reduced value, due to concrete encasement, be denoted r_{sc} noting that it is likely to also depend on factors such as concrete over thickness, concrete compaction, and the permeability of that concrete potentially as affected by the wetness of the concrete. When the cover concrete is very dense and of very low permeability oxygen diffusion to the external rust layers will be much inhibited and, in the limit, $r_{sc} \rightarrow 0$. This is consistent with observations of essentially no corrosion in very dense, low permeability concretes even after more than 80 years of exposure [5]. However, such a scenario is unlikely to continue ad-infinitum. Other mechanisms are likely to intervene, shown in Figure 1d as commencing at t_{act}. Originally proposed purely empirically [14], this has recently been shown to be caused by the gradual, long-term loss of concrete alkali such that at t_{act} the concrete will have lost so much material that it has greatly increased pore spaces and much greater pore connectivity. This permits a high level of local oxygen diffusion and thus much increased corrosion by direct oxidation [13].

On the other hand, for concretes with high permeability (and cracking) access of oxygen from the external environment and thus the exterior oxidation of the rust layers is likely to be somewhat easier. The result will be that rusts are permeated into the concrete pre spaces immediately surrounding corroded steel bars. However, the effect of greater oxygen permeation through the concrete cover on the rate of corrosion r_s is likely to be slight, since r_s depends mainly on the rate of the cathodic HER at the metal-rust interface. The situation changes dramatically, however, if there is significant damage, such as from widespread concrete cover cracking and spalling.

The corrosion behavior denoted schematically by (A-B) in Figure 1d arises from two aspects. One, as noted above, is the transition from Mode 1 to Mode 2 for the corrosion of steel (Figure 1c). The other, as will be seen, and the more important effect is that from air-voids and similar imperfections in the concrete matrix at the concrete-steel interface. The overall corrosion loss is shown idealized by parameter c_{sc} and occurs in the relative short-time period immediately after initiation at t_i.

Corrosion initiation for general (or uniform) corrosion at the usual potentials for iron in water is possible only for a local pH below about 9, dictated by thermodynamic conditions (Gibbs free energy or Pourbaix). This is irrespective of chloride concentration. It tends to rule out the initiation of general corrosion inside concretes and concrete pore waters with their usually high pH. The situation for pitting corrosion is rather different. Pitting corrosion involves a higher (more active) potential, and is thermodynamically possible, even at elevated pH environments when the chloride concentration is sufficiently high [46]. This possibility directly permits the initiation of reinforcement corrosion at chloride-rich wet air-voids in the concrete adjacent to the reinforcement steel.

The severity of corrosion associated with an air-void depends on the amount of oxygen in the air-void and the local availability of pore water. It has been shown to commence as differential aeration, localized at the edges of the air-voids [1] that then causes localized (pitting) corrosion of the adjacent steel [12]. Once initiated, such localized corrosion is only very mildly inhibited by diffusion considerations and will increase rapidly until eventually limited by the availability of oxygen or water. The net result is an almost step-wise increase in corrosion loss just after t_i, idealized as (A-B) and c_{sc} in Figure 1d. This type of behavior also has been observed for near-full-scale beams, for example, by Yu et al. [28] who attributed it purely to corrosion products inhibiting oxygen diffusion.

The size and distribution of the air-voids in the concrete matrix at the steel surface reflect the degree of concrete compaction achieved prior to concrete setting. Moreover, they are likely to be functions of the composition of the concrete and properties such as water-cement ratio and aggregate-cement ratios. All these tend to have a degree of statistical uncertainty and this is likely reflected in the amount of corrosion at the air-voids, i.e., in c_{sc} in Figure 1d.

Collecting together the various factors noted above provides the overall schematic model shown in Figure 2. It can be seen that after t_i the amount of corrosion is governed not just by the rate of progression of corrosion, that is by r_{sc}, but also by the volume of the air-voids, as these govern c_{sc}. Estimates for the values of the parameters (c_{sc} and r_{sc}) for a longer-term corrosion, based on physical tests, are given in the next section.

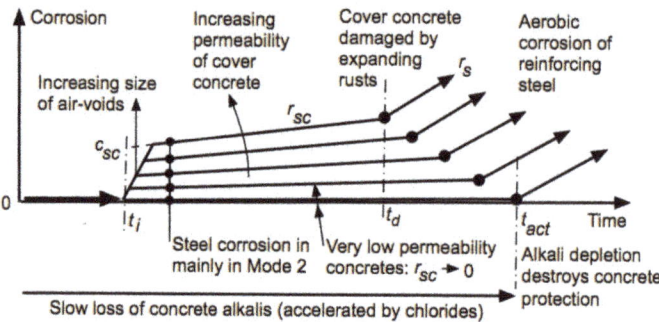

Figure 2. Model generalized from Figure 1d for the development of corrosion loss as a function of exposure time, concrete permeability (and wetness), and concrete compaction.

3. Reinforcement Corrosion after Initiation (Parameters c_{sc} and r_{sc})

There are few longer-term experimental programs covering a range of concrete mixes for which both reinforcement corrosion initiation and progression were observed and which were sufficiently detailed to observe the bi-modal corrosion behavior of the steel. One of these is the program reported by Shalon and Raphael [47]. It used multiple model concrete specimens each 40 mm × 40 mm × 140 mm long made from local (limestone) aggregates and commercial cement without additives. Each specimen was provided with a longitudinal, centrally-placed 6 mm diam. mild steel bar. The mixing water consisted of local natural seawater. As a result, chlorides were present at a high concentration in the concrete matrix from the outset. Thus, the initiation period (0–t_i) can be considered negligible. This is a valid experimental technique to accelerate the overall process [48]. A range of aggregate-cement and water-cement ratios was used for the concrete specimens. They were cast horizontally in steel molds.

There is no information on concrete compaction other than the fact that the bars were "inserted" and "embedded" in the concrete of each specimen, apparently after the molds were filled with concrete [49]. All the specimens were stored in a laboratory fog-room at about RH 98% and average air temperature about 25 °C until required for examination. At 3, 6, 12, 24, and 48 months, one or two specimens from each concrete mix was broken up and the surface condition of the bars examined. Any rusts on the bars were removed using a protocol generally similar to that currently specified for reinforcement bar cleaning. The cleaned bars were then weighed and the masses compared with the original masses. The original paper only provides percentage mass losses. For the present analysis, these were converted to corrosion losses (in mm) using the reported nominal diameter and the typical density of steel (7800 kg/m^3).

A parallel project using specimens of the same size and with comparable water-cement and aggregate-cement ratios with exposures extending over more than 12 years has been reported recently [12,13]. This program used low-heat as well as blended commercial cement. Some mixes were made with calcareous aggregates. Unlike the Shalon and Raphael [47] experiments, the parallel project found no or negligible corrosion losses even over the 12 years of exposure, which was insufficient to obtain accurate quantitative results. The major difference was that a high degree of compaction had been carried out. Only microscopic voids were visible in the concrete at the steel interface.

A completely different project has yielded information on reinforcement corrosion and its progression over some 28 years of exposure. Three-meter long reinforced concrete beams (36 in total) were exposed to artificial chloride-rich wet and dry cyclic laboratory conditions in ambient temperatures between about 5 and 20 °C [50]. After only a few months of exposure, some initial corrosion was reported but after the first few years (about 4–5) the corrosion rate declined significantly [28]. This was assumed to be due to rust products and calcite blocking oxygen access through the cracks. More severe general and pitting corrosion of the steel reinforcement bars was observed after 14, 23, 26, and 28 years of exposure, together with concrete cracking and damage [34]. However, even after 28 years of exposure, the corrosion of the reinforcement was considered very mild and highly erratic along the steel bars, with some longitudinal sections still showing no obvious corrosion, despite the low concrete cover in some beams (10 mm). In all cases, most of the corrosion occurred along the bottom of the bars (of the horizontally-cast beams).

The corrosion losses were reported as a loss of the cross-sectional area of the main reinforcement bars, sampled at numerous locations along the bars. The cross-section area loss for the two 12 mm diam. main reinforcement bars in each beam tested show considerable variation but mostly in a range of 20–25 mm^2. Converting this to corrosion loss produces an average (radial) corrosion loss of 0.32 mm after 28 years. The corrosion losses at the shorter exposure periods show a linear trend from negligible corrosion at 4.5 years [34] which, taken together, are equivalent to a long-term rate of reinforcement corrosion r_{sc} = 0.014 mm/y (Figure 14). Since only one concrete mix was used throughout, with a similar workmanship for all the specimen beams, there is no information on the potential effects of concrete mix design or concrete compaction.

Returning now to the experimental results reported by Shalon and Raphael [47], Figure 3 summarizes their observations of corrosion losses as functions of water-cement (w/c) and aggregate-cement (a/c) ratios. For the combinations shown, trends have been added through the data points, in most cases fitted using the Stineman [51] non-linear "best-fit" function. In a few cases interpreted trends are shown. These are based on the data points but in between build on the expected overall consistency with the majority of the best-fit trends.

Remarkably, throughout all the plots, the bi-modal corrosion loss trend for the corrosion of the reinforcement steel is clearly evident (Figure 3). It occurs within the first 1–2 years of exposure. Remarkably also, the rate of longer-term corrosion r_{sc} is highly consistent, in all cases around 0.015 mm/y for the whole of the (wide) ranges of aggregate-cement and water-cement ratios. Due to the inverse relationship between the concrete strength and concrete permeability [52], this result can be interpreted immediately since showing r_{sc} is not strongly dependent on concrete permeability.

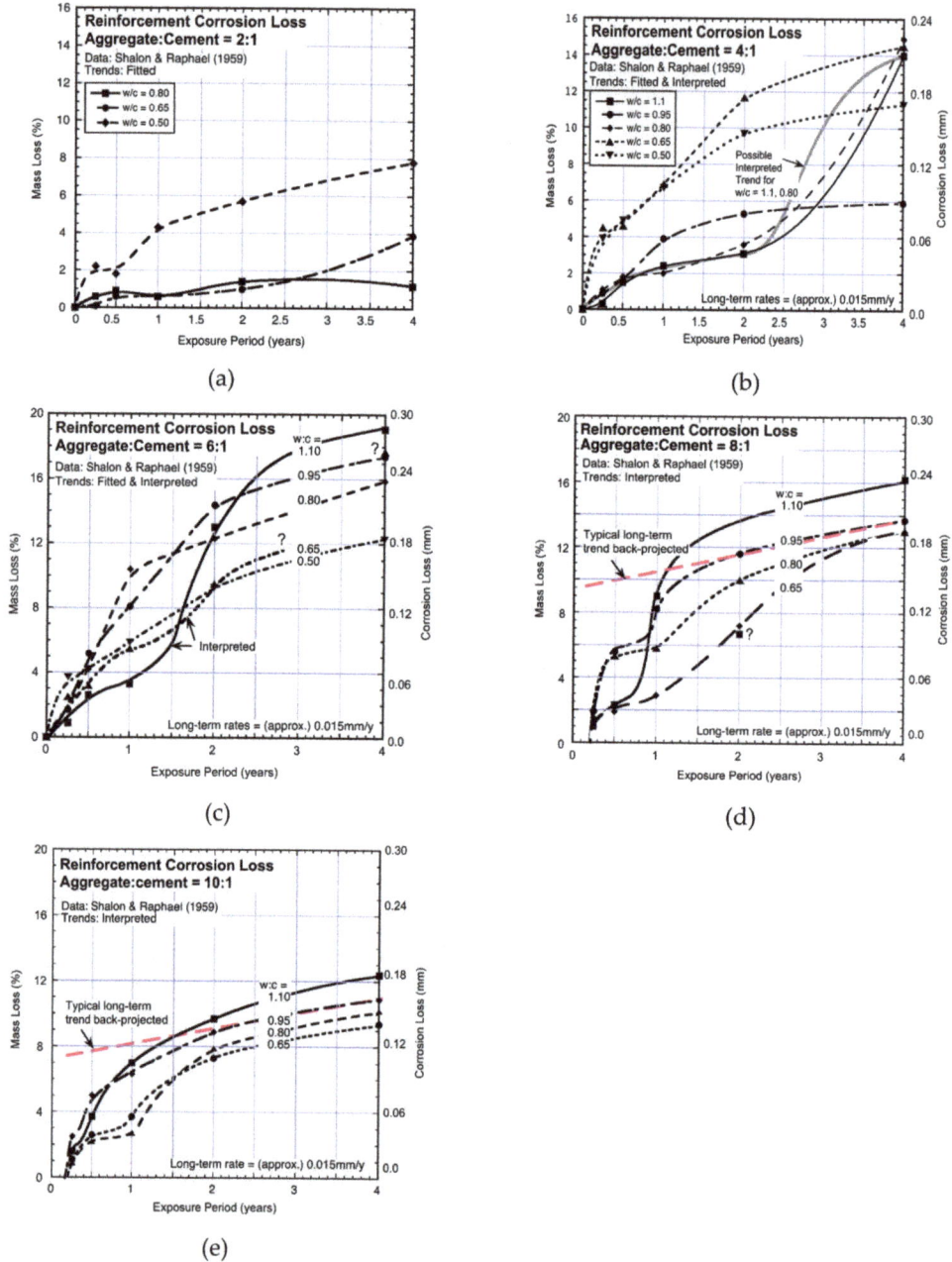

Figure 3. (a–e) Data and trends for mass losses derived from data reported by Shalon and Raphael, 1959 [47] showing dependence on the exposure period for different aggregate-cement and water-cement ratios. Most of the trends are best-fit, some are interpreted. Where shown, the long-term tangent line can be used to estimate c_{sc} and r_{sc}. In all cases, r_{sc} is about 0.015 mm/y, across all aggregate-cement (a/c) and water-cement (w/c) ratios. Note the bi-modal corrosion loss trend within the 1–2 year period of exposure for most trends.

The majority of the trends in Figure 3 show that after the first 2–3 years the trends tend to be linear at a rate $r_{sc} \approx 0.015$ mm/y. This appears almost independent of the precise proportions of the concrete mixes. In Figure 3a,b, some concrete mixes with high water-cement ratios (i.e., very wet mixes) show very little corrosion, at least for the first 3–4 years, followed by corrosion losses that are more consistent with the other data sets. Although the exposure periods are not sufficiently long to confirm, the data trends do suggest that for these cases, too, the pattern is the same as the others, albeit delayed in time.

For the cement-rich trends in Figure 3a, it is seen that corrosion losses are relatively low, and one case, at least, shows r_{sc} approaching zero. This is consistent with the expectations noted above for high impermeability concretes. It also is consistent with practical observations even after periods of marine exposure exceeding 80 years [5].

The plots in Figure 3 allow the parameter c_{sc} to be extracted. The results are summarized in Figure 4 as functions of the w/c and a/c ratios. Evidently, c_{sc} increases with increased water-cement (w/c) ratio and then declines for the further increase in w/c. This trending for c_{sc} is slightly later and also higher for concretes with high w/c ratios.

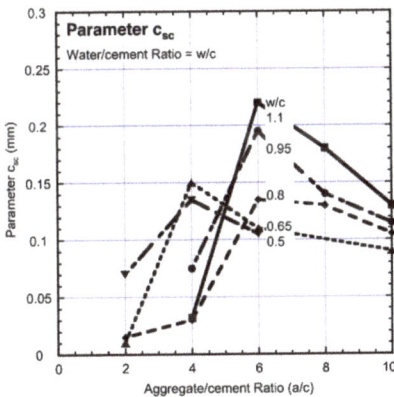

Figure 4. Parameter c_{sc} as a function of a/c and w/c ratios.

In interpreting the results in Figure 4, it is reasonable to assume that the lack of compaction of the concrete once the bars had been "placed" in them [47] would have left air-voids at the steel-concrete interface. These air-voids can be expected, after concrete setting, to be greater for the concretes with higher w/c ratios as a result of greater shrinkage with higher water content. This is likely the reason for greater values of c_{sc} with increased w/c ratio. As shown in Figure 4, initially the higher a/c ratio produced higher values of c_{sc} but this is not the case for a/c greater than about 4–6, for which c_s declines with a/c. As noted, this behavior is likely to be a result of the permeability of the concrete. Comparisons with other scenarios are of interest. For example, for sand particle-steel interfaces, permeability and voids are known to influence localized corrosion [53]. This holds also for spherical glass beads on metal surfaces [54] and for poorly compacted clays [55]. In each case, the observations can be attributed to the effect of voids on differential aeration at the void-space and also on their size (volume).

For the more permeable concretes, i.e., those with a/c ratios > 2, the overall long-term corrosion loss $c(t)$ as a function of continuous exposure time t is given by:

$$c(t) = c_{sc} + 0.015(t - t_i) \quad \text{for } t > t_i \tag{1}$$

where t is the actual elapsed time, $t_i > 0$ is the (estimated) time to initiation, and c_{sc} is obtained from Figure 4. For a/c ratios < 2 Equation (1) overestimates $c(t)$. There is insufficient data to be quantitatively more definitive but it is clear from Figure 2 and the above discussion that for such cases $c_{sc} \to 0$ and $r_s \to 0$.

While there appears to be no data in the literature for the parameter c_{sc} (rightly, since this parameter has only recently been identified), some information is available from which to make estimates for r_s. Beaton et al. [56] reported that typical corrosion rates equivalent to about 0.012 mm/y for RC piles above the mudline, and up to 0.18 mm/y elsewhere, both for 37 years of exposure, are sufficiently long to be taken as estimating r_s. Stewart and Rosowsky [57] proposed long-term corrosion rates in the range 0.011–0.23 mm/y, based on current density measurements for superficially sound concretes [58,59]. Moreover, Andrade and Alonso [60] and Sagüés et al. [10] reported longer-term rates around 0.01 mm/y based on electrochemical measurements. These estimates bracket the rate derived from the experiments in Figure 3.

4. Commencement of Corrosion at t_{act}

The lower trend line shown in Figure 2 represents the practical observations that high quality, very low permeability concretes with no discernable air-voids show almost no corrosion. Figure 5 shows an example, for marine concretes recovered from bridge piles more than 80 years old [5]. As noted, for these there was no observable corrosion so that both c_{sc} and $r_s \to 0$. Under these conditions, some other mechanism must come into play if reinforcement corrosion is to become possible. Recent experimental observations have shown that this involves the loss, through dissolution, of calcium hydroxide ($Ca(OH)_2$) in the concrete surrounding the reinforcing bars, leaving behind a concrete matrix with pH around 7–8 [13]. Usually the dissolution process is very slow but it is accelerated proportionally to the concentration of chlorides in the solution [61]. The experimental results showed that the dissolution process leaves behind a permeable concrete matrix and clear evidence that oxygen can readily permeate through it to oxidize the reinforcement [13]. Thus, not only the lowering of concrete pH at the reinforcement bars but also the greater permeability for oxygen leads to the severe rate of corrosion after t_{act} (Figure 2).

Figure 5. Example of high quality, well-compacted concrete broken open after more than 80 years of continuous marine exposure, showing the void-free surface that interfaced with the steel 32 mm diameter reinforcing bar (removed for clarity). No corrosion was detected along the whole 5–7 m of reinforcement bar (photograph © RE Melchers, 2020).

The rate of loss of alkalis for high quality concrete may be estimated, to a first approximation, by assuming a constant rate of loss of $Ca(OH)_2$ from first exposure onwards. Figure 6 shows a summary of the loss of $Ca(OH)_2$ over a period of 10 years for different water-cement and aggregate-cement ratios and for concretes made with seawater and with freshwater [13].

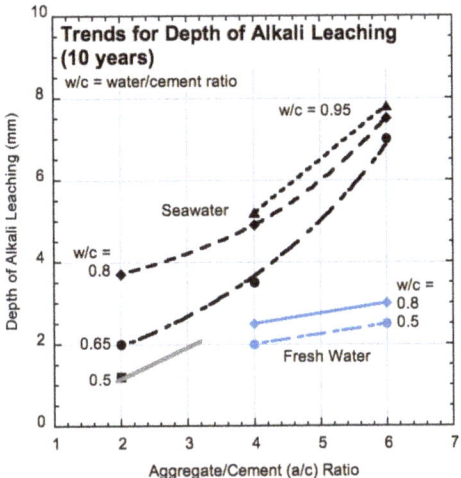

Figure 6. Depth as measured from the exterior concrete surface of loss of concrete alkali as measured by pH on the concrete cross-sections (based on data in Melchers and Chaves, 2020 [13]).

The trends in Figure 6 may be used to estimate the expected time before a complete loss of alkali material and thus, the local development of a concrete matrix permeable to oxygen. By way of example, for an (uncracked) concrete made with freshwater with (moderate) water cement ratio of 0.5 and an aggregate cement ratio of 4:1, Figure 6 indicates that the depth of alkali dissolution is about 2 mm in 10 years or 0.2 mm/y. For a concrete made with seawater the corresponding rate is about 0.3 mm/y. Thus, a concrete structure with a cover of 50 mm would commence with active corrosion caused by the loss of alkalis after t_{act} = 250 and 165 years, respectively. For a leaner concrete, say a/c = 5:1, the respective depths of loss of $Ca(OH)_2$ are greater and the rates are higher (about 0.23 and 0.45 mm/y), and the expected times shorter, 220 and 110 years, respectively. These comparative times demonstrate the significant effect of chlorides on the rate of alkali dissolution. They also demonstrate the importance of aggregate-cement ratio, which is the importance of cement content relative to the aggregate content.

In both examples the time estimates appear to be high, but they are not unrealistic when compared, for example, with observations for reinforced concrete piles exposed to Pacific Ocean immersion, tidal and splash conditions for over 80 years [5]. Full-sized (380 mm × 460 mm) cross-sectional samples of these showed concrete cross-section pH readings around 12, except for the outer 2–3 mm, despite very high chloride concentrations—around ten times the normally accepted threshold. Importantly, the cement content for these piles was high, with a/c ratios estimated around 4.5:1. Similar to the specimens in the Shalon and Raphael [47] experiments, these piles were all uncracked. The effect of cracks is considered in the next section.

5. Corrosion at Deep (Hairline) Cracks and Other Imperfections

The conventional wisdom, for example, as codified in standard specifications, is that cracks in the concrete of less than about 0.3 mm across are of negligible importance. Tracing the origin of this criterion shows that it is derived from short-term laboratory experiments [62]. However, other reports have discounted crack width as the important parameter in favor of the "existence" of a crack [50,63], while several practical reports have noted severe localized reinforcement corrosion for very narrow (i.e., hairline) cracks that extended to the reinforcement [10,64,65]. Similar observations have been made in other practical cases.

Figure 7 shows an example of very severe, so-called "tunneling" corrosion of a 6 mm diam. steel bar after 65 years of exposure in marine atmospheric conditions along the North Sea [66,67]. In this case, corrosion had penetrated along the bar axis for about 6–8 mm but had left a "sleeve" at the outer surface of the bar. Figure 8 shows the very considerable localized reinforcement corrosion of two of four 32 mm diam. steel bars together with watery-looking rust stains located at a cracked cross-section exposed in Pacific Ocean tidal conditions for about 85 years [5]. In both cases the cracks were "hairline" in width. In neither case were rust deposits or rust stains visible on the exterior surfaces of the concretes including at or near the hairline cracks.

Figure 7. Corroded end of 6 mm diam. reinforcement bar extracted from 65-year-old concrete exposed to a severe marine atmosphere. Note the tunneling corrosion extending inwards about 6–8 mm.

Figure 8. End view of remains of 32 mm diam. reinforcement bar, with watery-looking rust stains on a cracked concrete cross-section, after 85 years of exposure to seawater in tidal conditions (photograph courtesy of Clayton Smith).

Two questions arise immediately from these cases: (a) What were the corrosion mechanism(s) and (b) where did the corroded steel go, and how? In both cases, a critical observation is that there was a (hairline) crack from the exterior concrete surface into the concrete and deeper than the location of the reinforcing bars. While initially the hairline crack could permit some access of atmospheric or dissolved oxygen to the reinforcement bar, any build-up of rusts would soon convert local conditions to predominantly anaerobic, and move the local steel corrosion process into Mode 2, governed by the cathodic HER with the generation of pits and acidic ferric chlorides (Figure 1c) [41]. Being water-soluble, the ferrous chlorides are able to leach easily from the corrosion site via the hairline cracking. Taken together, these aspects allow a mechanism to be postulated to explain the observations in Figures 7 and 8.

At the commencement of Mode 2, corrosion will be, as noted, predominantly by pitting under anaerobic conditions at a corrosion rate r_a (Figure 1c). As corrosion develops it will be through successive pitting, each pit depth step restrained in depth, followed by sideways growth of the pits with amalgamation of adjacent pits, followed by further pitting [68]. This pattern leads to the sequential development of corrosion by pitting as shown in Figure 9. In the earlier stages, corrosion of the steel bar is from the concrete crack inwards with, through radial amalgamations of pits, the development of (for a circular bar) an annular ring of corrosion. Eventually, the center of the steel bar will be reached, leaving an annular grove around the bar. One-half of this forms the sharp-pointed bar geometry shown in Figure 8. Further corrosion is possible only for the remaining exposed metal of the annular ring, attacking each side independently. Corrosion will progress along the axis of the bar, more severely along the center as this is the predominant source of iron and also the location of impurities that arise from hot-rolling, which are known to increase the rate of corrosion slightly (Figure 9) [69,70]. The eventual effect is to produce the tunneling corrosion seen in Figure 7. The fact that the tunneling shown in Figure 7 occurred at about 65 years of exposure, some 20 years earlier than that shown in Figure 8, is largely due to the considerable difference in bar diameter—6 mm compared with 32 mm.

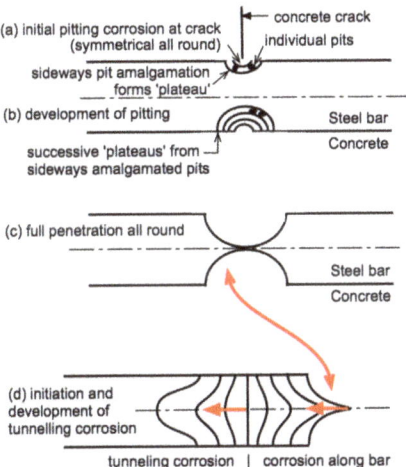

Figure 9. Schematic representation of the progression of initially very localized corrosion at a hairline crack, development of corrosion further into the bar, and the eventual development of tunneling corrosion along the centerline of the bar.

Much of the corrosion development shown in Figure 9 is governed by anaerobic, high chloride conditions, producing, as noted, highly soluble $FeCl_2$ that can move easily through the (hairline) crack to the external environment, leaving little or no trace of rust deposits such as red-brown rust spots (Figure 8). On reaching the external environment the $FeCl_2$ will be oxidized to FeOOH or to essentially similar insoluble rusts [71]. These may leave characteristic rust stains on the concrete or more likely are washed away, by rainwater or seawater, and thus leaving little or no trace.

Since there is no deposition of corrosion products within the crack to inhibit the rate of the corrosion reaction, r_a remains the governing corrosion rate. It can be considered to act perpendicular to all corroding surfaces, including at the deepest penetration (i.e., perpendicular to the longitudinal axis of the reinforcing bar) (Figure 9). It follows that a first estimate of the loss of bar radius, Δr, over a time interval $(t - t_0)$ is given by:

$$\Delta r(t) = r_a (t - t_0) \qquad (2)$$

Here, t_0 represents the time period prior to the commencement of the above corrosion process. In most cases, this will be approximately $t_0 = 0$. The rate r_a may be extracted from earlier work that considered the commencement of anaerobic corrosion for steel in seawater [38]. Figure 10 shows the results of field observations at different average seawater temperatures for exposure sites in different parts of the world.

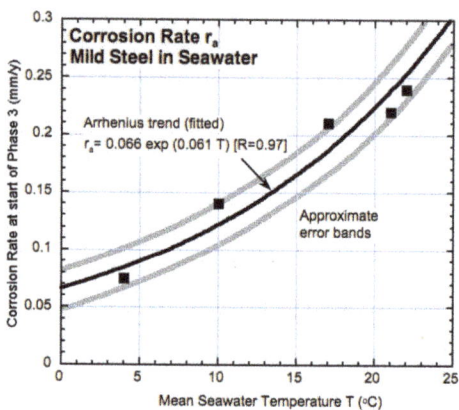

Figure 10. Initial corrosion rate (r_a) at the start of Mode 2 of the corrosion trend model for steel (Figure 1c) with a pitting corrosion uninhibited by the rust build-up (based on data in Melchers, 2003 [38]).

Assuming no obstructions or rust products develop to inhibit the free movement of $FeCl_2$, Equation (2) can be expected to also apply to the rate of tunneling corrosion once the centerline of the reinforcing bar is reached and the only available steel for corrosion is the steel bar cross-section. In essence, this means that Relationship (2) also applies for corrosion that goes "around corners".

Predictions from Equation (2) may be compared with observations, such as for reinforced concrete handrail elements along the North Sea at Arbroath, Scotland [66,67]. According to Figure 10, r_a = 0.11–0.14 mm/y for an annual average temperature of 10 °C. For 60 years of exposure and with r_a = 0.11 mm/y it would take 27 years to fully penetrate a 3 mm radius bar, leaving 33 years to produce a tunneling depth of 3.6 mm. If the rate r_a = 0.14 mm/y, these figures become 21 years and 5.4 mm of tunneling depth. The latter is consistent with physical observations (Figure 7).

For the Hornibrook bridge case (Figure 8), the average water temperature is about 22 °C, so that r_a is in the range of 0.22–0.27 mm/y, with, over 85 years, an estimated penetration of about 19–23 mm. This estimate is somewhat greater than the physical radius of the bars (16 mm) but is not inconsistent with the observations. Better consistency can be derived if t_0 is in the range of 15–25 years—not unreasonable for a concrete that had maintained pH around 12 after 85 years of exposure [5].

Corrosion in hairline flexural cracks in beams is likely to follow a pattern similar to that outlined above. Unfortunately, the flexural cracks observed in the full-scale laboratory tests of François et al. [27] and Zhu et al. [34] are much wider than the hairline cracks and thus, more exposed to the environment. As noted, such cracks are not typical of those in actual structures under normal service ("sustained") load conditions.

6. Discussion

The analyses presented above permits the development of practical models for initiation and for progression of corrosion of steel reinforcement in concretes—models based on practical observations and on interpretations from fundamental theory. Most empirical models in the literature do not reference back, even partially, to fundamentals. Predominantly, they rely on the assumption that chlorides are the main driver for reinforcement corrosion. This is despite many practical cases having provided empirical contrary evidence. Recent detailed experimental observations, coupled with observations discounting the role of chlorides for general corrosion in quiescent conditions have demonstrated the important role of wet air-voids in the concrete matrix at the steel surface as reservoirs for oxygen and water and that only when these are present do chlorides advance the potential for corrosion, and specifically for pitting corrosion [39]. The experimental work showed that it is pitting corrosion that ultimately drives the initiation process, even in high pH environments. Remarkably, the role of chlorides in relation to the pitting corrosion potential has been known for many years [46]. However, it has almost always been ignored in discussions of the initiation of corrosion in concretes with their normally high alkalinity. Moreover, while some studies have reported the presence of air-voids, very few have mentioned any related pitting and its possible influence on initiation [33]. Failure to account for these effects may explain (a) the wide variation in the "critical chloride concentration" derived from various studies and (b) the apparent conundrum in establishing a relationship between the concentration of chlorides and corrosion initiation in practical structures. Without considering air-voids to supply oxygen (and water) there can be no corrosion, irrespective of the concentration of chlorides at the reinforcement steel.

The size of the air-voids where the concrete interfaces with the steel is important. As follows directly from theoretical considerations, and as shown in practical cases, it governs the amount of corrosion that can occur after initiation and is reflected in the parameter c_{sc} (Figure 2). Void-size is related to permeability but more importantly to the degree of concrete compaction. Where compaction of the concrete around the bars is poor, larger air-voids are likely both throughout the concrete matrix and against the formwork and at reinforcement bars. For the latter, air-voids are often observed on the underside of horizontal reinforcement bars, irrespective of whether they are "top" bars or bars in the lower parts of a beam or slab [72]. This has been observed also for laboratory specimens [12,73,74]. These observations provide one explanation to why spalling is mostly from the underside of beams rather than from the sides, even for a similar concrete cover and when exposure to seawater or seawater spray is similar.

For well-made, well-compacted, impermeable concretes air-voids tend to be very small or negligible (Figure 5). Practical experience shows that the initiation of reinforcement corrosion, if it occurs at all, is then not a significant issue. Experiments that observe reinforcement corrosion over extended time periods show that for such concretes reinforcement corrosion effectively ceases after a relatively short time (cf. AB in Figure 2). There is little corrosion product and negligible or no structural damage. As the size of air-voids increases, corrosion at initiation tends to increase also (Figure 2). Moreover, in this scenario concrete permeability (for oxygen) is likely to be greater. This will result in a somewhat greater longer-term rate r_s and an earlier time t_{act} to serious damaging corrosion (Figure 2).

Shalon and Raphael [47] interpreted their data along conventional lines at the time. Their observations led others to make a conclusion that concretes should not be made with seawaters due to their high chloride content (even though the use of seawater had been a standard practice in many locations for many years). The analysis given herein shows that the Shalon and Raphael data can be interpreted in an entirely different way when using the recently exposed importance of air-voids in the concrete, as well as when the bi-modal model for the progression of steel corrosion are taken into account. The analysis given above shows that air-voids, with chlorides present to allow pitting corrosion to be thermodynamically feasible, drive the amount of corrosion soon after initiation. Reinforcement corrosion then follows the bi-modal trending and, since the rate of oxygen availability is inhibited by the concrete cover, reinforcement corrosion soon reaches linear trending, idealized as BC in Figure 2. It corresponds to the predominantly anaerobic (Mode 2) part of the bi-modal model.

The data reported by Shalon and Raphael [47] clearly show the effect of concrete mix design, with concretes having higher a/c ratios (i.e., lean concretes) and having linear trending for much longer than for the other concrete mixes (Figure 3c,d). However, the effect on r_s is small, while the effect on c_{sc} is consistent with the parameter (aggregate-cement ratio) having the most effect on permeability (Figure 4). The overall consistency seen in Figures 3 and 4 adds a degree of confidence to the proposed model.

The concrete cover thickness was not considered by Shalon and Raphael [47] but from experience it is known to be an important parameter, even in non-chloride environments [52]. Depending on the exposure environment, it normally is assigned the role of inhibiting diffusion of oxygen, chloride, and carbon dioxide to the reinforcement. However in terms of the present exposition, the concrete envelope around reinforcement also is important. It retards the loss of ferrous ions from the external oxidation of the rusts [43] and, in the earlier stages, it tends to keep the rusts in place, protecting them from oxidation, abrasion, erosion or velocity effects as typical in some other environments. Again, this interpretation provides a completely different perspective to the conventional view that the rate of supply of oxygen governs the rate of oxidation at the concrete-steel interface. Such behavior would not result in the long-term linear trends seen in the experimental data.

The oxidation of the external rust layers likely accounts also for the practical observations of concrete cover spalling and high degrees of reinforcement corrosion when the cover thickness is small. Usually, such spalling is attributed to an overall expansion of the rust envelope around a reinforcement bar. This may be exacerbated by excessive (e.g., atmospheric) temperature variations damaging the concrete material. However, the concept of oxidation of the external rust layers is likely to be just as damaging, but has the advantage of being consistent with theory.

One question about the data from the Shalon and Raphael [47] experiments is whether their use of calcareous aggregates had an effect on the results obtained. This should be seen in the context of the pH for concrete in many actual structures being still around 12 after many years of exposure [13]. Where serious alkali leaching has begun to occur the presence of calcareous material may delay the drop in pH of the concrete, keeping it at around 9 by virtue of the calcareous material. This was noted, for example, for 65 year-old concretes exposed to marine atmospheres [66]. In addition, a survey of many reinforced concrete structures and the likely aggregates used for their concretes indicated that those made with calcareous aggregates tended to have longer effective lives [75]. For these, the time to initiation was much more difficult to estimate *ex post*. However, the practical experience suggested that the use of calcareous aggregates is not a critical issue for initiation or for r_{sc}. Instead, it appears to have the effect of maintaining somewhat longer the alkalinity necessary to maintain a concrete pH above 9. It is an area that has had little investigation and could benefit from further research.

A second question about the data from the Shalon and Raphael [47] experiments is whether the permeability of the concrete, and the air-voids at the steel-concrete interface, is reflected properly by their aggregate-cement and water-cement ratios. Both are often associated with permeability (e.g., [52]) but whether these parameters provide realistic representations of permeability for actual concretes is an open question.

As noted, in the Shalon and Raphael experiments the steel bars were placed into the concretes. There is no information on compaction. Air-voids were not mentioned in the published paper. It is reasonable to assume that if they had been observed they were unlikely to have been considered important and therefore were not measured. More broadly, it appears that since the availability of vibrators from the 1940s onwards it has been assumed that the vibrators produce adequate concrete compaction. It is unclear whether this was ever assessed in terms of air-voids around, and in particular under, reinforcement bars. It is also unclear whether mechanical vibration has been assessed relative to the earlier techniques of hand-rodding and hand-tamping. In view of the above discussion about the importance of air-voids, a further investigation appears warranted, preferably using realistic concretes and realistic compaction techniques.

Overall, more regard should be paid to the actual experience of actual structures, particularly those with highly repetitive, but individually made elements since these could be considered examples of very large experiments. They certainly are realistic, much more so than any laboratory concretes and even more so than electrochemical tests. There are also related experiences, not for reinforced concrete but for systems with a steel-particle contact, and without high pH conditions, and in some cases with seawater present. Despite these apparently adverse conditions, the evidence is compelling. It has long been recognized that bare steel piles driven into sands and muds in seawater conditions show essentially no corrosion even after many decades of exposure except at the sand/mud-seawater interface zone [76]. Similarly, ferrous iron pipes buried in extremely well compacted clay soils with acidic soil pH around 5–7 show almost no corrosion, again over many decades, simply since oxygen is excluded, particularly from the external rust surfaces [77]. The lessons from these observations are obvious and it is clear that the pre-occupation with chloride-induced corrosion for already high pH concretes does not sit well with these observations. It also does not sit well with the practice over many years of permitting concretes to be made with seawater [8], even though this practice was banned in many countries in the 1960s. The ban has been attributed by some authors (e.g., [52]) in part to the very paper used herein—namely Shalon and Raphael [47]. The present paper, and the earlier work [4,44], indicate that the problems with many reinforced concrete structures are not so much with chlorides but with the conditions (poor concrete permeability, permeable perhaps thin concrete cover, deep and possibly hairline cracking, and damage to the concrete matrix and cover from material issues such as alkali-aggregate reactions [78]) that permit such corrosion to progress after initiation. For atmospheric exposures, there is also the issue of thin concrete covers deteriorating under high temperature fluctuations. These are all potentially important matters of detail that may affect the progression of reinforcement corrosion.

Finally, the present developments show that provided the above matters of detail are properly considered, reinforced concrete structures can have extended service lives, even in high chloride environments. This is provided the concretes are well-made and have no or negligible air-voids in the concrete at the steel-concrete interface. The model proposed herein allows for some degree of corrosion after initiation, as caused by the volume of air-voids and as subsequently increased at a rate of about 0.015 mm/y, influenced only a small amount by the permeability of the exterior (cover) concrete. In this model it is not necessary to consider the imposition of concrete flexural cracking since in practice most concrete structures show little or negligible degrees of concrete flexural or other cracking. Therefore, it is inappropriate to use data for the corrosion of reinforcement in concrete structures when these have induced crack sizes much larger than occur under normal sustained loadings. The present results and proposed model show that it is possible to design and make reinforced concretes that avoid reinforcement corrosion or reduce it to negligible levels, as has been shown by experience to be feasible in practical concrete structures. As noted, this requires low permeability and very well compacted concretes to ensure there are minimal air-voids at the reinforcement, in particular under horizontal bars. Such concretes will also delay the loss of concrete alkalinity, a rate that can also be reduced through an adequate cement content so as to add the acid-buffering capacity and thereby delay the long-term development of alkali-leached concrete permeability, which will permit a much greater rate of oxidation commencing at t_{act}. The fundamentals for achieving good quality durable concretes have been well-known in the industry for many decades, but the experiences have not been placed in quite the context outlined herein. The present analyses provide the theoretical support for such experiences. It also makes clear where attention must be focused to achieve long-term durable reinforced concrete structures.

7. Conclusions

The following conclusions may be drawn from the material presented herein.

1. The development of reinforcement corrosion after initiation can be represented by a relatively fast increase to c_{sc} followed by a slow rate of corrosion defined by r_{sc}, where c_{sc} depends on the aggregate-cement ratio and water-cement ratio of the concrete and r_{sc} is about 0.015 mm/y of the general corrosion for most concrete mixes. Concretes with extremely low permeability are associated with very low values of r_{sc}.

2. For aggregate-cement ratios less than about 4, the parameter c_{sc} increases with the aggregate-cement ratio but decreases for higher aggregate/cement ratios, in both cases more so for higher water-cement ratios. This is attributed to the effect of the water-cement ratio on concrete permeability.

3. The slow loss of the low solubility concrete alkali calcium hydroxide leads, for extended exposures, to a much reduced concrete pH and to a porous concrete matrix that permits the entry of atmospheric or dissolved oxygen, which then causes oxidation of the revealed reinforcement bars. Since the solubility of calcium hydroxide is accelerated in the presence of seawater, corrosion will likely occur earlier for otherwise similar conditions.

4. In some conditions, severe corrosion may occur at deep hairline cracks intersecting the reinforcement and in the presence of chlorides, without necessarily leaving "tell-tale" rust stains. The process involves water-soluble ferrous chloride formed in corrosion pits leaching to the external environment, eventually leaving a characteristic tunneling type of corroded bar. Empirical data show that for seawater with an average water temperature in the range of 10–20 °C, the corresponding rate of corrosion r_a is in the range of 0.22–0.27 mm/y.

Funding: This research received no external funding.

Data Availability Statement: All data used and described in this study are in the public domain and in the open literature.

Acknowledgments: The author acknowledges his colleagues at The University of Newcastle in supporting this research and, through laboratory facilities and technical staff, permitting the continued exploration of the practical aspects of the nature of reinforcement corrosion. The author acknowledges the encouragement of Joost Gulikers, Rijkswaterstaat (Ministry of Infrastructure and Water Management), the Netherlands, that led to an earlier (2019) version of this manuscript.

Conflicts of Interest: There are no conflicts of interest.

References

1. Angst, U.M.; Geiker, M.R.; Alonso, M.C.; Polder, R.; Isgor, O.B.; Elsener, B.; Wong, H.; Michel, A.; Hornbostel, K.; Gehlen, C.; et al. The effect of the steel-concrete interface on chloride-induced corrosion initiation in concrete: A critical review by RILEM TC 262-SCI. *Mater. Corros.* **2019**, *52*, 88. [CrossRef]
2. Melchers, R.E. Observations about the time to commencement of reinforcement corrosion for concrete structures in marine environments. In *Consec'10, Concrete under Severe Conditions*; Castro-Borges, P., Moreno, E., Sakai, K., Gjorv, O.E., Banthia, N., Eds.; CRC Press: Boca Raton, FL, USA, 2010; pp. 617–624.
3. Melchers, R.E. Carbonates, carbonation and the durability of reinforced concrete marine structures. *Aust. J. Struct. Eng.* **2010**, *10*, 215–226. [CrossRef]
4. Melchers, R.E.; Chaves, I.A. A comparative study of chlorides and longer-term reinforcement corrosion. *Mater. Corros.* **2017**, *68*, 613–621. [CrossRef]
5. Melchers, R.E.; Pape, T.M.; Chaves, I.A.; Heywood, R. Long-term durability of reinforced concrete piles from the Hornibrook Highway bridge. *Aust. J. Struct. Eng.* **2017**, *18*, 41–57. [CrossRef]
6. Anteagroup. *Etude sur L'etat de Conservation du Port Artificiel Winston Churchill, Phase 0 Analyse Documentaire (Rapport A78136/B), Phase 1–Diagnostic des Vestiges (Rapport A79782/B)*; Anteagroup: Paris, France, September 2015.
7. Melchers, R.E.; Howlett, C.H. Reinforcement corrosion of the Phoenix caissons after 75 years of marine exposure. *Proc. Inst. Civ. Engrs. Marit. Eng.* **2020**. [CrossRef]
8. Wakeman, C.M.; Dockweiler, E.V.; Stover, H.E.; Whiteneck, L.L. Use of concrete in marine environments. *Proc. ACI* **1958**, *54*, 841–856, Discussion 54, 1309–1346.
9. Lukas, W. Relationship between chloride content in concrete and corrosion in untensioned reinforcement on Austrian bridges and concrete road surfacings. *Betonw. Fert. Tech.* **1985**, *51*, 730–734.
10. Sagüés, A.A.; Kranc, S.C.; Presuel-Moreno, F.; Rey, D.; Torres-Costa, A.; Yao, L. *Corrosion Forecasting for 75-Year Durability Design of Reinforced Concrete, Final Report to Florida Department of Transport*; University of South Florida: Tampa, FL, USA, 2001.
11. Lau, K.; Sagüés, A.A.; Yao, L.; Powers, R.G. Corrosion performance of concrete cylinder piles. *Corrosion* **2007**, *63*, 366–378. [CrossRef]
12. Melchers, R.E.; Chaves, I.A. Reinforcement corrosion in marine concretes–1. Initiation. *ACI Mater. J.* **2019**, *116*, 57–66. [CrossRef]
13. Melchers, R.E.; Chaves, I.A. Reinforcement Corrosion in Marine Concretes–2. Long-Term Effects. *ACI Mater. J.* **2020**, *117*, 217–228. [CrossRef]
14. Melchers, R.E.; Li, C.Q. Phenomenological modelling of corrosion loss of steel reinforcement in marine environments. *ACI Mater. J.* **2006**, *103*, 25–32.
15. Raupach, M. Models for the propagation phase of reinforcement corrosion–An overview. *Mater. Corros.* **2006**, *57*, 605–613. [CrossRef]
16. Andrade, C. Propagation of reinforcement corrosion: Principles, testing and modelling. *Mater. Struct.* **2019**, *52*, 1–26. [CrossRef]

7. Stockert, L.; Haas, M.; Jeffrey, R.J.; Melchers, R.E. Electrochemical measurements and short-term-in-situ exposure testing. In Proceedings of the Corrosion & Prevention, Melbourne, Australia, 11–14 November 2012. CD ROM, Paper No. 100.
8. Burkowsky, B.; Englot, J. Analyzing good deck performance on Port Authority bridges. *Concr. Int.* **1988**, *10*, 25–33.
9. Wallbank, E.J. *The Performance of Concrete in Bridges. A Survey of 200 Highway Bridges*; Department of Transport, HMSO: London, UK, 1989.
10. Gjorv, O.E. Steel corrosion in concrete structures exposed to Norwegian marine environment. *Concr. Int.* **1994**, *16*, 35–39.
11. Gulikers, J.; Raupach, M. Modelling o reinforcement corrosion in concrete. *Mater. Corros.* **2006**, *57*, 603–604. [CrossRef]
12. Hornbostel, K.; Angst, U.M.; Elsener, B.; Larsen, C.K.; Geiker, M.R. Influence of mortar resistivity on the rate-limiting step of chloride0induced macro-cell corrosion of reinforcing steel. *Corros. Sci.* **2016**, *110*, 46–56. [CrossRef]
13. Sassine, E.; Laurens, S.; François, R.; Ringot, E. A critical discussion on rebar electrical continuity and usual interpretation thresholds in the field of half-cell potential measurements in steel reinforced concrete. *Mater. Struct.* **2018**, *51*, 93. [CrossRef]
14. Kelly, R.G.; Scully, J.R.; Shoesmith, D.W.; Buchheit, R.G. *Electrochemical Techniques in Corrosion Science and Engineering*; Marcel Dekker: New York, NY, USA, 2002.
15. Tuutti, K. Corrosion of steel in concrete, Swedish Cement and Concrete Research Institute, Stockholm, Research Report No. 4. 1982. See also Service life of structures with regard to corrosion of embedded steel. In *Performance of Concrete in Marine Environment*; ACI SP-65; American Concrete Institute: Detroit, MI, USA, 1984; pp. 223–236.
16. Clear, K.C. *Time-to-Corrosion for Reinforcing Steel in Concrete Slabs, V. 3: Performance after 830 Daily Salt Applications, FHWA-RD-76-70*; Federal Highway Administration: Washington, DC, USA, 1976; 64p.
17. François, R.; Arliguie, G.; Maso, J.-C. Durabilité du béton armé soumis à l'action des chlorures. *Ann. l'Institut Tech. Batim. Trav. Publics* **1994**, *529*, 1–48.
18. Yu, L.; François, R.; Dang, V.H.; l'Hostis, V.; Gagné, R. Development of chloride-induced corrosion in pre-cracked RC beams under sustained loading: Effect of load-induced cracks, concrete cover, and exposure conditions. *Cem. Concr. Res.* **2015**, *67*, 246–258. [CrossRef]
19. *CIB Commission W81 Actions on Structures: Live Loads in Buildings*; CIB Report No. 116; International Council for Research and Innovation in Building and Construction, AIBC: Ottawa, ON, Canada, 1989.
20. Foley, T.R. The role of the chloride ion in iron corrosion. *Corrosion* **1970**, *26*, 58–70. [CrossRef]
21. Heyn, E.; Bauer, O. Ueber den Angriff des Eisens durch Wasser und wässerige Losungen. *Stahl Eisen* **1908**, *28*, 1564–1573.
22. Mercer, A.D.; Lumbard, E.A. Corrosion of mild steel in water. *Br. Corros. J.* **1995**, *30*, 43–55. [CrossRef]
23. Angst, U.M.; Elsener, B.; Larsen, C.K.; Vennesland, O. Critical chloride content in reinforced concrete–A review. *Cem. Concr. Res.* **2009**, *39*, 1122–1138. [CrossRef]
24. Zhu, W.; François, R.; Liu, Y. Propagation of corrosion and corrosion patterns of bars embedded in RC beams stored in chloride environment for various periods. *Constr. Build. Mater.* **2017**, *145*, 147–156. [CrossRef]
25. Loreto, G.; di Benedetti, M.; De Luca, A.; Nanni, A. Assessment of reinforced concrete structures in marine environment: A case study. *Corros. Rev.* **2018**, *37*. [CrossRef]
26. Chalhoub, C.; François, R.; Carcasses, M. Critical chloride threshold values as a function of cement type and steel surface conditions. *Cem. Concr. Res.* **2020**, *134*, 106086. [CrossRef]
27. Chitty, W.-J.; Dillmann, P.; L'Hostis, V.; Millard, A. Long-term corrosion of rebars embedded in aerial and hydraulic binders–Parametric study and first step of modelling. *Corros. Sci.* **2008**, *50*, 3047–3055. [CrossRef]
28. Melchers, R.E. Modeling of marine immersion corrosion for mild and low alloy steels–Part 1: Phenomenological model. *Corrosion* **2003**, *59*, 319–334. [CrossRef]
29. Melchers, R.E. Modelling durability of reinforced concrete structures. *Corros. Eng. Sci. Technol.* **2020**, *55*, 171–181. [CrossRef]
30. Melchers, R.E. A review of trends for corrosion loss and pit depth in longer-term exposures. *Corros. Mater. Degrad.* **2018**, *1*, 4. [CrossRef]
31. Wranglen, G. Pitting and Sulphide Inclusions in Steel. *Corros. Sci.* **1974**, *14*, 331–349. [CrossRef]
32. Evans, U.R.; Taylor, C.A.J. Mechanism of atmospheric rusting. *Corros. Sci.* **1972**, *12*, 227–246. [CrossRef]
33. Stratmann, M.; Bohnenkamp, K.; Engell, H.J. An electrochemical study of phase-transitions in rust layers. *Corros. Sci.* **1983**, *23*, 969–985. [CrossRef]
34. Southwell, C.R.; Bultman, J.D.; Alexander, A.L. Corrosion of metals in Tropical environmwnts–Final report of 16 years exposures. *Mater. Perform.* **1976**, *15*, 9–25.
35. Melchers, R.E.; Chernov, B.B. Corrosion loss of mild steel in high temperature hard freshwater. *Corros. Sci.* **2010**, *52*, 449–454. [CrossRef]
36. Pourbaix, M. Significance of protection potential in pitting and intergranular corrosion. *Corrosion* **1970**, *6*, 431–438. [CrossRef]
37. Shalon, R.; Raphael, M. Influence of seawater of corrosion of reinforcement. *J. ACI* **1959**, *30*, 1251–1268.
38. Poursaee, A.; Hansen, C.M. Potential pitfalls in assessing chloride-induced corrosion of steel in concrete. *Cem. Concr. Res.* **2009**, *39*, 391–400. [CrossRef]
39. Friedland, R. Influence of the quality of mortar and concrete upon corrosion of reinforcement. *J. ACI* **1950**, *22*, 125–139.
40. François, R.; Arliguie, G. Effect of microcracking and cracking on the development of corrosion in reinforced concrete members. *Mag. Concr. Res.* **1999**, *51*, 143–150. [CrossRef]
41. Stineman, R.W. A consistently well-behaved method of interpolation. *Creat. Comput.* **1980**, *6*, 54–57.

52. Richardson, M.G. *Fundamentals of Durable Reinforced Concrete*; SponPress: London, UK, 2002.
53. Gupta, K.; Gupta, B.K. The critical soil moisture content in the underground corrosion of mild steel. *Corros Sci.* **1979**, *19*, 171–178. [CrossRef]
54. Burns, M.; Salley, D.J. Particle size as a factor in the corrosion of lead by soils. *Ind. Eng. Chem.* **1930**, *22*, 293–297. [CrossRef]
55. Petersen, R.B.; Melchers, R.E. Effect of moisture content and compaction on the corrosion of mild steel buried in clay soils. *Corros Eng. Sci. Technol.* **2019**, *54*, 587–600. [CrossRef]
56. Beaton, J.L.; Spellman, D.L.; Stratfull, R.P. Corrosion of steel in continuously submerged reinforced concrete piling. *Highw. Res Rec.* **1967**, *204*, 11–21.
57. Stewart, M.G.; Rosowsky, D.V. Structural safety and serviceability of concrete bridges subject to corrosion. *J. Infrastruct. Syst* **1998**, *4*, 146–155. [CrossRef]
58. Dhir, R.K.; Jones, M.R. McCarthy, M.J. PFA concrete: Chloride-induced reinforcement corrosion. *Mag. Conc. Res.* **1994**, *46*, 269–277. [CrossRef]
59. Thoft-Christensen, P.; Jensen, F.M.; Middleton, C.; Blackmore, A. Revised rules for concrete bridges. In *Safety of Bridges*; Highway Agency: London, UK, 1996; pp. 1–12.
60. Andrade, C.; Alonso, M.C. Values of corrosion rate of steel in concrete to predict service life of concrete structures. In *Application of Accelerated Corrosion Tests to Service Life Prediction of Materials, ASTM STP 1194*; Cragnolino, G., Sridhar, N., Eds.; ASTM Philadelphia, PA, USA, 1994; 282p.
61. Johnston, J.; Grove, C. The solubility of calcium hydroxide in aqueous salt solutions. *J. Am. Chem. Soc.* **1931**, *53*, 3976–3991 [CrossRef]
62. Beeby, A.W. Corrosion of reinforcing steel in concrete and its relation to cracking. *Struct. Eng.* **1978**, *56*, 77–80.
63. Schiessl, P.; Raupach, M. Laboratory studies and calculations on the influence of crack width on chloride-induced corrosion of steel in concrete. *ACI Mater. J.* **1997**, *94*, 56–61.
64. Makita, M.; Mori, Y.; Katawaki, K. *Marine Corrosion Behavior of Reinfroced Concrete Exposed in Tokyo Bay*; SP 65-16; American Concrete Institute: Indianapolis, IN, USA, 1980; pp. 271–289.
65. Lewis, D.A.; Copenhagen, W.J. The corrosion of reinforcing steel in concrete in marine atmospheres. *S. Afr. Ind. Chem.* **1957**, *15*, 207–219. [CrossRef]
66. Melchers, R.E.; Li, C.Q.; Davison, M.A. Observations and analysis of a 63-year old reinforced concrete promenade railing exposed to the North Sea. *Mag. Concr. Res.* **2009**, *61*, 233–243. [CrossRef]
67. Melchers, R.E.; Li, C.Q. Reinforcement corrosion in concrete exposed to the North Sea for more than 60 years. *Corrosion* **2009**, *65*, 554–566. [CrossRef]
68. Jeffrey, R.; Melchers, R.E. The changing topography of corroding mild steel surfaces in seawater. *Corros. Sci.* **2007**, *49*, 2270–2288. [CrossRef]
69. Reger, M.; Vero, B.; Kardos, I.; Varga, P. The effect of alloying elements on the stability of centreline segregation. *Defect Diffus. Forum* **2010**, *297–301*, 148–153. [CrossRef]
70. Melchers, R.E.; Jeffrey, R.J.; Usher, K.M. Localized corrosion of steel sheet piling. *Corros. Sci.* **2014**, *79*, 139–147. [CrossRef]
71. Cornell, R.M.; Schwertmann, U. *The Iron Oxides: Structure, Properties, Reactions, Occurences and Uses*, 7th ed.; VCH Publishers: Weinheim, Germany, 1996.
72. Nawy, E.G. *Concrete Construction Engineering Handbook*; CRC Press: Boca Raton, FL, USA, 2008; pp. 30–57.
73. Horne, A.T.; Richardson, I.G.; Brydson, R.M.D. Quantitative analysis of the microstructure of interfaces in steel reinforced concrete. *Cem. Conc. Res.* **2007**, *37*, 1613–1623. [CrossRef]
74. Zhang, W.; Yu, L.; François, R. Inluence of top-casting-induced defects on the corrosion of the compressive reinforcement of naturally corroded beams under sustained loading. *Constr. Build. Mater.* **2019**, *229*, 116912. [CrossRef]
75. Melchers, R.E.; Li, C.Q. Reinforcement corrosion initiation and activation times in concrete structures exposed to severe marine environments. *Cem. Concr. Res.* **2009**, *39*, 1068–1076. [CrossRef]
76. Morley, J. The corrosion and protection of steel-piled structures. *Struct. Surv.* **1993**, *7*, 138–151. [CrossRef]
77. Wichers, C.M. Korrosion asphaltierter eiserner Rohre. *Das Gas Und Wasserfach* **1934**, *77*, 131–132.
78. Melchers, R.E. Long-term durability of marine reinforced concrete structures. *J. Mar. Sci. Eng.* **2020**, *8*, 290. [CrossRef]

Article

Effect of Multispecies Microbial Consortia on Microbially Influenced Corrosion of Carbon Steel

Hoang C. Phan [1,*], Linda L. Blackall [2] and Scott A. Wade [1,*]

1. Faculty of Science, Engineering and Technology, Swinburne University of Technology, Hawthorn, VIC 3122, Australia
2. School of BioSciences, University of Melbourne, Melbourne, VIC 3052, Australia; linda.blackall@unimelb.edu.au
* Correspondence: hoang.phan@virbac.vn (H.C.P.); swade@swin.edu.au (S.A.W.); Tel.: + 84-9-3396-8650 (H.C.P.); Tel.: +61-3-9214-4339 (S.A.W.)

Abstract: Microbially influenced corrosion (MIC) is responsible for significant damage to major marine infrastructure worldwide. While the microbes responsible for MIC typically exist in the environment in a synergistic combination of different species, the vast majority of laboratory-based MIC experiments are performed with single microbial pure cultures. In this work, marine grade steel was exposed to a single sulfate reducing bacterium (SRB, *Desulfovibrio desulfuricans*) and various combinations of bacteria (both pure cultures and mixed communities), and the steel corrosion studied. Differences in the microbial biofilm composition and succession, steel weight loss and pitting attack were observed for the various test configurations studied. The sulfate reduction phenotype was successfully shown in half-strength marine broth for both single and mixed communities. The highest corrosion according to steel weight loss and pitting, was recorded in the tests with *D. desulfuricans* alone when incubated in a nominally aerobic environment. The multispecies microbial consortia yielded lower general corrosion rates compared to *D. desulfuricans* or for the uninoculated control.

Keywords: corrosion; metabarcoding; MIC; multispecies; SRB

1. Introduction

The corrosion of metals immersed in the marine environment is a well-known issue, which can also be accelerated by the activities of a diverse range of microbes in a process known as microbially influenced corrosion (MIC) [1,2]. An example of MIC is the rapid failure of metal structures, typically in the form of localised corrosion around low tide level, which is referred to as accelerated low water corrosion (ALWC) [3,4]. ALWC has been reported on steel structures (e.g., sheet piling) in many ports and harbors worldwide. The reduction in thickness of steel pilings, which can be up to several mm/yr, is one of indications used to diagnose ALWC [5]. In the UK alone, the short-term cost of ALWC was estimated to be ~£250 million [3]. Methods to minimise ALWC include the application of cathodic protection, coatings and the use of special grade steel. Further work, however, is required in order to obtain a better understanding of the complex roles and types of microbes involved in ALWC.

Biofilm formation is a biological process that takes place in humid/aquatic environments, and involves a series of developmental steps. In the natural environment these intricate microbial structures typically contain a diverse range of microbes with complex interactions [6,7]. In relation to ALWC, different microbes are likely to play specific roles in the overall corrosion process, including the development of a biofilm on the steel structure, nutrient provision/cycling, as well as producing and maintaining an anaerobic environment. Specific microbes are responsible for the corrosion process [5,8,9]. Sulfate reducing bacteria (SRB), e.g., *Desulfovibrio desulfuricans*, are probably the best known bacterial group in relation to microbial corrosion and ALWC [10,11]. The SRB corrosion process is typically

associated with a black biofilm on the surface and localised corrosion underneath [12]. In addition to SRB, sulfur oxidising bacteria (SOB) have been implicated in ALWC, and the presence of both microbial groups will contribute to the sulfur cycle and subsequent corrosion [5,8,13–17].

While there have been a large number of studies of MIC in the laboratory, most of these have failed to replicate the accelerated corrosion rates observed in the field. One of the possible reasons for this is that most tests use single microbial strains in the experiments (most commonly SRB), which is very different to the mixed microbial species found in biofilms associated with accelerated corrosion in the real environment [9,18–21]. Another possible reason is that the testing arrangement used (e.g., test media, oxygen levels and test duration) may influence the processes (abiotic and biotic) required for the accelerated corrosion.

Various laboratory MIC experiments have employed undefined combinations of microbes (e.g., [5,22–25]). While this might be more similar to the field situation, the largely unknown combinations of microbes make replication impossible and the roles of any individual species cannot be determined. Several MIC studies using known microbial strains in various combinations have been undertaken, and have showed how the strains collectively affect corrosion (see Table S1). In some of the latter, the multispecies combinations have led to increased corrosion and in others decreased corrosion.

The mechanisms responsible for MIC are hypothesised to include transfer of electrons from the metal of interest to the SRB and/or biochemicals produced as part of metabolic processes [26,27]. Both require the presence of a suitable electron donor and electron acceptor. The artificial supply of electron donors could facilitate increased corrosion compared to real world scenarios since seawater typically has relatively low levels of organic carbon sources [28]. Several researchers have shown that different test media used in MIC studies can affect the level of corrosion [29–31]. Additionally, specific strains of SRB (e.g., *Desulfopila corrodens* and *Desulfovibrio* sp. strain IS7) have been linked to the direct electron uptake from an iron/steel surface [26] and have also been detected in field studies of ALWC [9,17,23].

This research adds to the understanding of the effect of multispecies microbial consortia in relation to microbial corrosion such as ALWC. The work involved exposing marine grade steel to an ALWC biofilm from a marine environment, pure culture isolates from an ALWC biofilm or adjacent locations, *D. desulfuricans* or to an uninoculated media (i.e., negative control). The hypothesis of this work was that the combinations of microbes would produce increased rates of corrosion compared to tests with single isolates or an uninoculated control.

2. Materials and Methods

2.1. Steel Test Coupons

Marine grade DH36 carbon steel [32] was chosen for this research due to its widespread use in shipbuilding and offshore structures. The metal coupon dimensions were 25 mm × 25 mm with a thickness of 6 mm. A 2 mm diameter hole was drilled in each coupon near the middle of one of the edges (3–4 mm from the edge) to allow coupons to be hung vertically in the aqueous corrosion tests. The top and bottom surfaces of the coupons were prepared using an automatic grinding machine (Tegramin-25, Struers, Cleveland, OH, USA) with a series of silicon carbide papers i.e., 320, 500 and 1200 grit. The edges were manually ground with 320 grit silicon carbide paper. After grinding, the coupons were cleaned by immersing in acetone in an ultrasonic bath for 8 min, rinsed thoroughly with distilled water and then absolute ethanol, and finally dried under warm air. Coupon weight was measured using a high accuracy analytical balance (MS205DU, readability 0.01 mg, Mettler Toledo, Mississauga, Canada) shortly before the corrosion tests.

2.2. Bacteria and Test Media Details

The four marine isolates were *Halomonas korlensis*, *Bacillus aquimaris*, *Prolixibacter bellariivorans* and *Sulfitobacter pontiacus*. Their presumed phenotypes cover a range of potential functions relevant to oxygen tolerance, biofilm formation and MIC. *H. korlensis* is a denitrifier/nitrate–nitrite reducing facultative aerobe [33] that was isolated from marine sediment [21]. *B. aquimaris* is a nitrate/metal ion utilising aerobe found in marine environments/biofilms/corrosion steel surfaces [34,35] that was isolated from an orange ALWC tubercle [36]. *P. bellariivorans* is an iron-relating/sugar-fermenting anaerobe [37,38] that was isolated from an orange ALWC tubercle [36]. Finally, *S. pontiacus* is an obligate aerobe which belongs to the Rhodobacteraceae group which are noted for surface association [6] and can reduce thiosulfate and sulfite to sulfate [39]. It was isolated from a microbial test kit (used to determine ALWC microbial functions) that was inoculated with an ALWC tubercle sample.

SRB are frequently associated with microbial corrosion (including ALWC), and a culture collection strain of an SRB, *D. desulfuricans* (ATCC 27774), was used. In addition to the five pure cultures, a consortium of microbes found in an orange tubercle from a suspected case of ALWC on steel sheet piling was also used in one of the test conditions [36].

Pure cultures of the marine isolates (stored at −80 °C in glycerol), were freshly grown on marine agar (MA, BD Difco, Franklin Lakes, NJ, USA) plates in an aerobic environment. These cultures were then used to inoculate 45 mL volumes of autoclaved (121 °C, 1 atm for 16 min) half-strength marine broth ($MB_{\frac{1}{2}}$, BD Difco, Franklin Lakes, NJ, USA), which were incubated aerobically at 25 °C for 72 h. Since MB has relatively high nutrient levels, it was diluted to better mimic seawater. Plate counts on MA were performed for all culture broths using serial ten-fold dilution and triplicate plating.

D. desulfuricans was grown in tryptic soy broth (Merck, Darmstadt, Germany), supplemented with sodium lactate (4 mL/L) and magnesium sulfate (2 g/L). The medium was adjusted to pH 7.5 prior to autoclaving (121 °C, 1 atm for 16 min), after which 0.5 g/L of 0.22 µm filter-sterilised ferrous ammonium sulfate was added. *D. desulfuricans* was incubated in an anaerobic jar at 25 °C for 72 h, with a gas generator sachet (AnaeroGen, Oxoid, Basingstoke, Hampshire, UK) and indicator (Resazurin, Oxoid, Basingstoke, Hampshire, UK). A black culture broth was obtained after incubation, which is an indication of ferrous sulfide production due to sulfate reduction. Plate counts of bacteria were performed anaerobically on modified tryptic soy agar (TSA+, prepared as for the modified tryptic soy broth but with 1.5% agar included prior to autoclaving).

An orange tubercle from an ALWC setting had been sampled previously and was stored frozen at −80 °C. The ALWC consortia inoculum for this work was prepared by adding 2.5 mL of thawed inner and 2.5 mL of outer tubercle layers to 45 mL of sterile MB1/2 broth (10% inoculation), and incubated aerobically at 25 °C for 72 h. All liquid cultures were gently shaken (~10 s) once a day to help stimulate microbial growth.

Corrosion tests were performed in 500 mL sterilised glass bottles containing 350 mL of sterile $MB_{\frac{1}{2}}$ with each containing a sterile test coupon suspended by Nylon string. The coupons and Nylon string were sterilised by immersion in absolute ethanol (Merck) for 30 min, then drying in a sterile dry bottle. Subsequently coupons were placed close to a Bunsen flame (10–15 cm) for 1–2 min to remove any excess ethanol prior to starting the corrosion tests. For the tests with pure cultures or orange tubercle material, 2 mL volumes of their 72 h culture broths were added and mixed thoroughly. A total of six test configurations were carried out (Table 1). All of the tests were nominally aerobic except for the test with *D. desulfuricans* (T6) for which anaerobic conditions were obtained by placing the test bottles in individual anaerobic bags together with AnaeroGen sachets and resazurin indicators (Oxoid). All the treatments were conducted statically with 6 individual replicates. All bottles were incubated in a Thermoline Scientific, Australia incubator at 25 °C.

Table 1. Relative abundance of microbes inoculated in laboratory corrosion tests and incubated at 25 °C in aerobic or anaerobic conditions.

Test (Incubation Atmosphere)	Brief Description	Microbes (cfu/mL)
T1 (aerobic)	No inoculum	None
T2 (aerobic)	Orange tubercle	Diverse consortium from an orange tubercle on steel sheet piling ($\geq 10^5$)
T3 (aerobic)	D. desulfuricans	D. desulfuricans (2×10^6)
T4 (aerobic)	Four marine isolates	B. aquimaris (9×10^5), H. korlensis (3×10^5), S. pontiacus (3×10^6), P. bellariivorans (7×10^4)
T5 (aerobic)	Four marine isolates + D. desulfuricans	D. desulfuricans (2×10^6), B. aquimaris (9×10^5), H. korlensis (3×10^5), S. pontiacus (3×10^6), P. bellariivorans (7×10^4)
T6 (anaerobic)	D. desulfuricans	D. desulfuricans (2×10^6)

2.3. Corrosion Tests

The duration of the corrosion tests was 8 weeks. Testing six replicates allowed a range of different studies to be performed using separate coupons. Periodic photos of test bottles were taken to record any changes in media or test coupons.

A total of 40 mL of fresh sterile MB$\frac{1}{2}$ was added into each test bottle fortnightly (i.e., fed-batch culture) to help maintain microbial viability during the study period. Sampling of the corrosion tests for bacterial counts was undertaken at weeks 2, 4 and 8 of incubation. This involved pipetting 10 mL of the culture (5 mL from near the test medium surface and 5 mL near the bottom of the medium) before nutrient amendment. Bacterial counts were obtained using MA plates for aerobic (T1–T5, Table 1) and TSA$^+$ plates for anaerobic (T6, Table 1) corrosion test using serial ten-fold dilutions and duplicate plating. pH was measured at the beginning of incubation and then after 4 and 8 weeks of incubation. Aliquots of liquid cultures were stored in −20 °C for use in subsequent microbial analyses.

After 8 weeks of incubation, all coupons were removed from the test bottles. Shortly after removal, a gentle stream of sterile MilliQ water was run over the coupons to remove any planktonic cells from the biofilms. Pictures of the coupons were then taken. Certain coupon biofilms/corrosion products were examined by scanning electron microscopy (SEM), X-ray powder diffractometery (XRD) or energy-dispersive X-ray spectroscopy (EDS). For these, sterile MilliQ water was used to gently wash the coupons three times, after which they were air-dried in a physical containment level 2 cabinet for 2 h and then stored in a desiccator. For SEM of coupon biofilms, one coupon per test condition underwent fixation with glutaraldehyde solution (2.5 vol %) then dehydration via an ethanol series (25%, 50%, 75%, 90% and 100%) [40]. Coupons for steel surface analyses (SEM, 3D optical profilometery, weight loss), underwent a series of cleaning steps to remove biofilms/corrosion products. Individual coupons were sonicated in sterile MilliQ water for 2 min (the fluid from this was stored at −20 °C for subsequent microbial analyses), then in Clark's solution [41] for 2 min, followed by rinsing with water and ethanol, drying under warm airflow and then storing in a desiccator.

2.4. Metal Coupon and Microbial Analysis

SEM (Zeiss SUPRA 40VP-25-38, Oberkochen, Germany) images were taken of the biofilms and the steel surfaces of separate fixed and cleaned coupons for each of the test conditions using a range of magnifications (150× to 12,000×). One coupon from each of the six treatments was used for EDS (Zeiss SUPRA 40VP-25-38, Oberkochen, Germany and INCA software, version 4.13) to provide semi-quantitative analyses of the biofilms/corrosion products formed on the coupons. Intensity with weight percent (wt.%) of chemical elements was averaged from two locations of the biofilm on each coupon surface. One uncleaned coupon from each of the six treatments was used for XRD (Bruker D8 Advance, Germany). Compositions of surface products were scanned with a copper line focus X-ray tube producing Kα radiation from a generator operating at 40 kV and 30 mA.

Three cleaned coupons from each of the six treatments were used to obtain average corrosion rates via weight loss using the method described in reference [41]. These coupons were also used for 3D optical profiling (Bruker Contour GT-K1) to obtain quantitative information about corrosion morphology. Each metal coupon was scanned over the entire surface (using 5× objective), and any areas with localised pitting were subjected to more detailed, higher resolution analysis. Pit depth (µm) and relative volume (µm^3) for pits greater than 10 µm depth were recorded and analysed using VISION 64™ software (Bruker), and then pit density was calculated by number of pit/mm^2.

For corrosion morphology analysis, the pitting depth, relative pit volume and corrosion rates were statistically compared for significant differences among the treatments ($p < 0.05$) using one-way ANOVA and Tukey's post-hoc test (IBM® SPSS™ Statistics version 25).

Microbial analysis was based on agar plate counts and 16S rRNA gene metabarcoding. Relative abundance was determined by a sum of the aerobic and anaerobic counts from the liquid in each treatment. For metabarcoding, liquid samples of T2, T4 and T5 experiments (Table 1) obtained at week 2 and week 8, together with biofilm suspensions obtained following sonication from week 8 samples were sent to a commercial test facility (AGRF, Adelaide, Australia) for DNA extraction and Illumina MiSeq sequencing. The primer pairs used were 341F (CCTAYGGGRBGCASCAG) and 806R (GGACTACNNGGGTATCTAAT) for amplification of the V3–V4 region of 16S rRNA genes. Data were processed using Quantitative Insights Into Microbial Ecology (QIIME 1.9) followed by METAGENassist [36].

3. Results

3.1. Surfaces of Coupons Before Cleaning

Photos of coupon surfaces were taken directly after removal from the test solution at the end of the 8-week immersion period (see Figure 1). The T1 and T3 test condition coupons had orange and black corrosion products on the surfaces, the T4 coupons were relatively free of corrosion products and black films were present on the remaining treatments (i.e., T2, T5 and T6).

The surfaces of individual test replicate coupons set aside for biofilm/corrosion product examination were observed using the SEM (Figure 2), and analysed by EDS (Figure 3). The coupon from the uninoculated test (T1) was covered in a general corrosion product, which appears to be a form of iron oxide. The images of the coupon exposed to the orange tubercle biofilm inoculum (T2) showed a biofilm present, with phosphorus and a relatively high sulfur content detected by the EDS (Figure 3). The coupon from the *D. desulfuricans* inoculum under aerobic incubation (T3) visually had a thicker biofilm on the surface compared to the coupon from the *D. desulfuricans* inoculum under anaerobic incubation (T6) (Figure 2). The biofilm of the T3 coupon also had much more sulfur present (relative to iron) compared to the T6 coupon, according to EDS (Figure 3). For the defined mixed microbial species tests (T4 and T5), the consortia including *D. desulfuricans* (T5) visually had a much denser biofilm compared to the sparse biofilm of T4 (Figure 2). No sulfur was detected in the mixed consortia without *D. desulfuricans* (T4) (Figure 3), while a small amount of sulfur in the biofilm was detected in the defined microbial consortia when the *D. desulfuricans* were present (T5) (Figure 3). The iron:sulfur ratios shown in Figure 3 provide an indication of the relative amount of sulfur detected in surface product analysis.

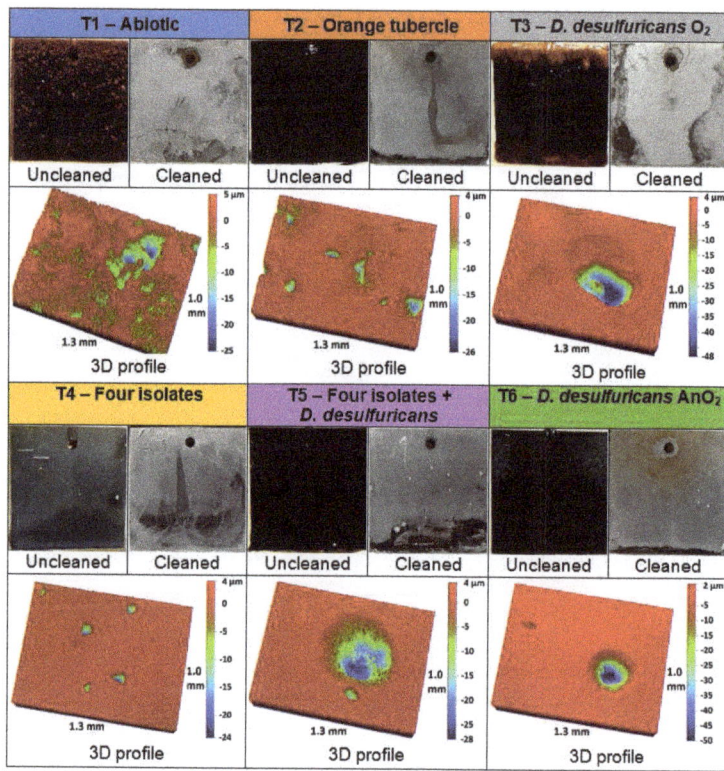

Figure 1. Photos of uncleaned and cleaned coupons after 8-weeks immersion in half-strength marine broth for the six different microbial testing conditions, together with 3D surface profiles of cleaned coupons showing surface and pit morphology.

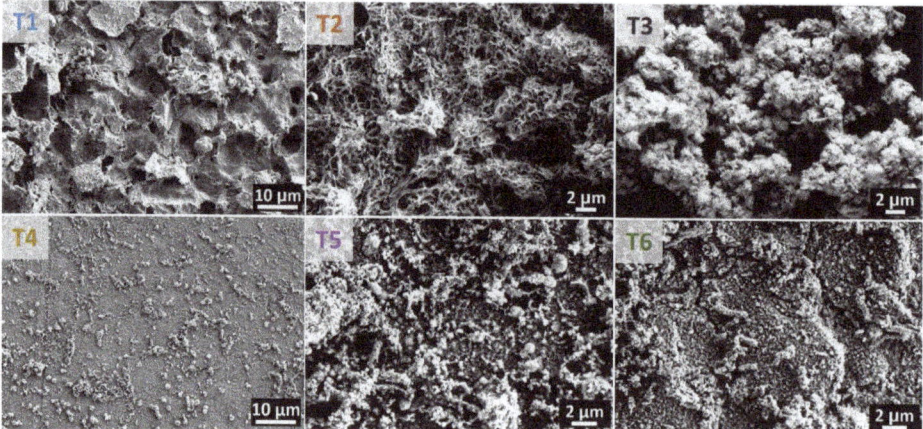

Figure 2. SEM images of biofilms/corrosion products on surfaces of coupons tested after 8-weeks incubation in the six different test conditions. All of these coupons underwent fixation (dehydration and 2% glutaraldehyde immersion) prior to imaging.

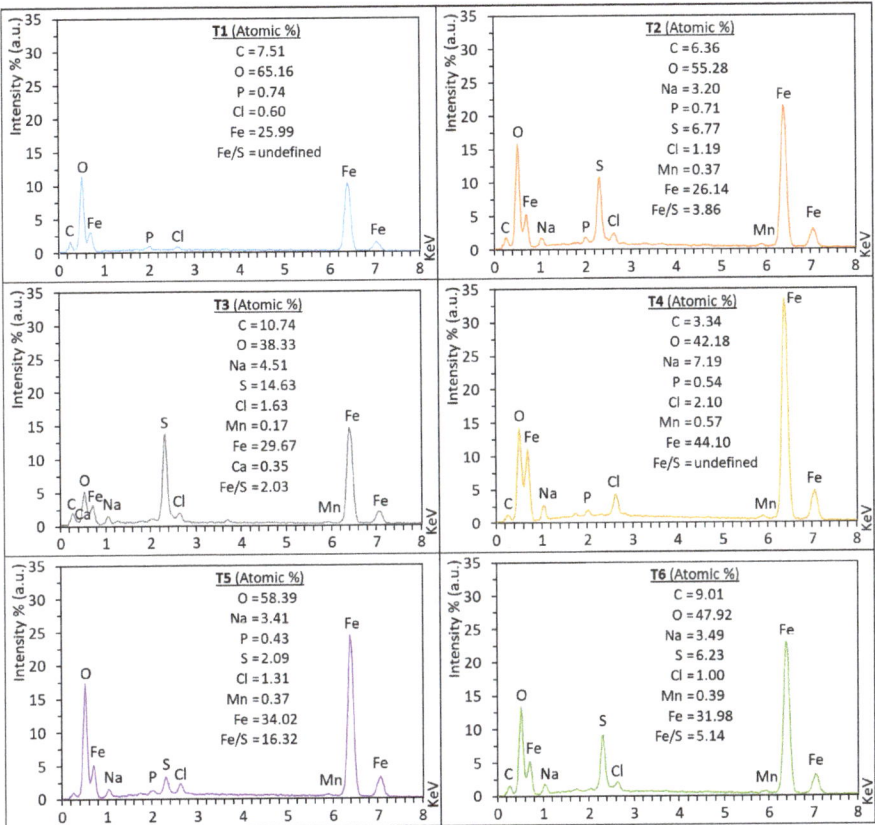

Figure 3. EDS spectra of the biofilms/corrosion products formed on DH36 coupons tested in the six different treatments after 8-week incubation.

The XRD spectra peaks of individual phases were identified by comparing the diffraction pattern to a known standard from the diffraction powder database of the International Centre for Diffraction Data (Diffrac Eva software, version 4.1). Unfortunately, the XRD spectra obtained (Figure S1) provided ambiguous results for corrosion products/biofilms, with the three peaks observed in each scan matching that of iron. It is possible that the intensity of iron dominated the scanning spectra or that the corrosion products formed were amorphous in nature.

3.2. Corrosion Evaluation

Example photos of the surfaces of test coupons after removing any biofilms and corrosion products are shown in Figure 1. The key difference observed was that while the uninoculated samples (T1) showed signs of general corrosion and occasional pitting, most of the coupons tested in different biotic conditions appeared to suffer from sparse localised pitting corrosion. SEM scans of the cleaned surfaces (Figure 4) confirmed that the uninoculated control (T1) coupon had a combination of general and localised corrosion attack, as did the coupon tested in the *D. desulfuricans* aerobic test configuration (T3). The coupons from the other tests (T2, T4, T5 and T6) contained localised corrosion but were relatively free from general corrosion. Finally, the diameters of pits observed for coupons tested with *D. desulfuricans* (T3, T5 and T6) appeared to be generally larger than those for the other test configurations.

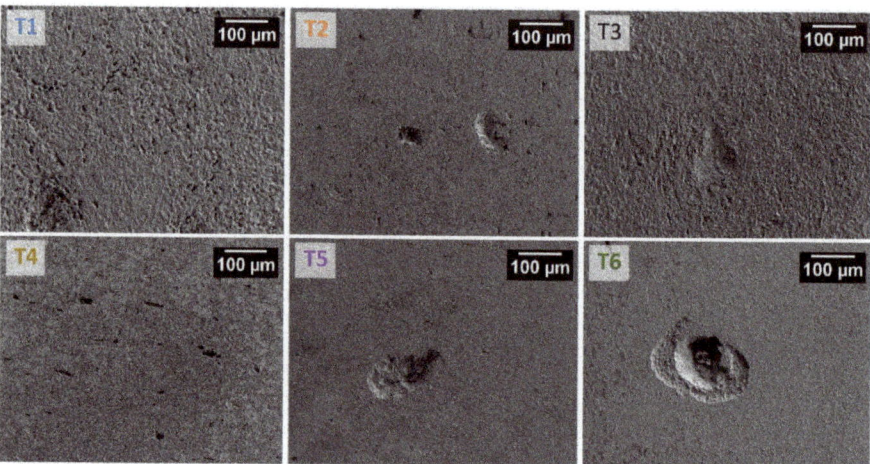

Figure 4. SEM images of surfaces of cleaned test coupons after 8-week immersion in the six different test conditions.

To further characterise the morphology of localised corrosion, 3D optical profiling was performed on one side of one of the cleaned coupons for each test condition (see examples in Figure 1). The pit density, which was calculated for all treatments based on the number of pits (>10 μm) per mm², was ~2 (pit/mm²) for T1, (uninoculated treatment), and T2, T4 and T5 (mixed consortia inoculum), and ~1 (pit/mm²) for T3 and T6 (*D. desulfuricans* alone). The coupons with the largest pits were from tests containing *D. desulfuricans* (T3, T5 and T6, see details in Figure 5). Apart from the coupon from T4 (four isolates) which had one pit deeper than 30 μm, only the coupons from tests with *D. desulfuricans* (T3 and T6) had any pits greater than 30 μm depth (five and six pits, respectively). The deepest pit found was 50.2 μm for the coupon tested in *D. desulfuricans* under anaerobic conditions (T6), compared to 27 μm for the deepest pit for the uninoculated control (T1). The average pit volumes (of five deepest pits) were nearly an order of magnitude greater for the tests containing *D. desulfuricans* (T3, T5 and T6) compared to the tests lacking *D. desulfuricans*.

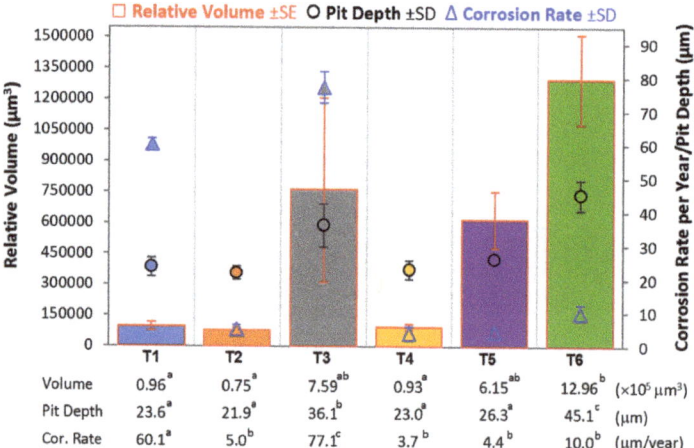

Figure 5. Summary of the corrosion attack on steel samples from the six different treatments. Averages of 5 highest values of relative volume and pit depth as well as average corrosion rate are shown with statistical analyses. Significant differences ($p < 0.05$) were shown by different series of superscript letters on each parameter.

The average corrosion rates of the DH36 coupons after 8 weeks of incubation in the different test scenarios were determined via mass loss (Figure 5). The highest corrosion rates among the six treatments obtained ($p < 0.05$) were for the T3 (*D. desulfuricans* tested aerobically) and T1 (uninoculated) test conditions. For all the other test conditions very low average corrosion rates were measured; always much less than the uninoculated control, indicating a form of corrosion inhibition. Given that the localised attack of the uninoculated control (T1) was relatively minor, the relatively high weight loss for this test condition will have been dominated by general/uniform corrosion.

3.3. Test Solutions

Over the 8-week duration of the corrosion tests, the colours of the test solutions changed (see Figure 6), which is a potential indication of the metabolic and corrosion processes that took place. Of particular interest was the general blackening of solutions for tests T2 (orange tubercle inoculum), T5 (four isolates + *D. desulfuricans* aerobic) and T6 (*D. desulfuricans* anaerobic), and the black precipitation layer observed at the bottom of test condition T3 (*D. desulfuricans* aerobic). This blackening might be an indication of sulfate reduction and FeS production.

Figure 6. Changes in appearance of test bottles of six treatments during 8 weeks of immersion.

In general, the pH values of the solutions (Table 2) were reasonably constant for all of the test conditions over the 8-week incubation period, with pH values staying within ±0.5 of the initial value of 7.54. The greatest changes observed were for T4, which went slightly more basic to 7.9, and T6 which went slightly more acidic to just below 7.0.

Table 2. Bacterial counts and pH during 8-week incubation test of six treatments (Table 1). Initial pH of sterile media before inoculation was ~7.54. Total plate count in each treatment was the sum of both aerobic and anaerobic counts on marine agar (MA) and tryptic soy agar (TSA$^+$) media, respectively.

Test (T1–T5: O$_2$; T6: AnO$_2$)	Total Bacterial Plate Count (cfu/mL)				pH Measurement	
	Initial	Week 2	Week 4	Week 8	Week 4	Week 8
T1 (uninoculated)	–	–	–	–	7.58 ± 0.02	7.55 ± 0.02
T2 (orange tubercle)	≥10^5	10^7–10^8	1 × 10^8	10^8–10^9	7.47 ± 0.03	7.32 ± 0.06
T3 (*D. desulfuricans*)	2 × 10^6	7 × 10^5	2 × 10^6	10^3–10^4	7.56 ± 0.04	7.73 ± 0.10
T4 (Four isolates)	4 × 10^6	4 × 10^7	6 × 10^7	9 × 10^7	7.74 ± 0.05	7.90 ± 0.06
T5 (Four isolates + *D. desulfuricans*)	6 × 10^6	3 × 10^7	1 × 10^8	10^6–10^7	7.32 ± 0.04	7.42 ± 0.06
T6 (*D. desulfuricans*)	2 × 10^6	5 × 10^7	5 × 10^7	10^4–10^5	6.98 ± 0.10	6.98 ± 0.09

3.4. Analysis of Microbial Populations

Total plate count results are provided in Table 2. While a reduction in counts for the *D. desulfuricans* inoculum tests (T3 and T6) was seen between the start and the end of the tests, in general the microbial populations were reasonably constant throughout incubation times. It is also worth noting that no colonies were observed in the aerobic plate tests for *D. desulfuricans* alone treatments (T3 and T6), suggesting that these solutions were free of any aerobic microbial contamination.

Figure 7 summarises the metabarcoding analysis of the mixed microbial consortium treatments (i.e., T2, T4 and T5) at week 2 (planktonic) and week 8 (planktonic and biofilm). The solution inoculated with the orange tubercle (T2) has, as expected, the most diverse community. While there were few bacterial taxa that reduced in relative abundance in the T2 solution during the test (e.g., *Exiguobacterium* spp. dropped from 13% to 3% in planktonic phase from 2 to 8 weeks), many bacteria maintained their presence in both planktonic and biofilm phases (e.g., Clostridiales and Rhizobiales). However, several bacteria were only detected after 8 weeks including *Desulfarculus baarsii*, *Desulfosporosinus* spp. and *Thioalkalivibrio* sp. Reasonable differences can also be seen between the planktonic and biofilm phases of the T2 solution at 8 weeks (e.g., Rhodobacteraceae spp. and *Clostridium thiosulfatireducens* more abundant in the biofilm phase).

The metabarcoding results for the combinations of isolates (T4 and T5) identified all of the bacteria inoculated in each test configuration in at least one of the time/phases tested. For both of these test configurations the proportion of *P. bellariivorans* tended to decrease from week 2 to week 8, and *H. korlensis* was only ever found in any reasonable level in the biofilm phase. A key difference between the two test configurations was for *B. aquimaris*, which had a strong presence in the T4 liquid (four isolates) but was only detected at low levels in the T5 liquid (four isolates + *D. desulfuricans*). *S. pontiacus* was detected at reasonably consistent levels for both test solution types throughout the tests and in both the planktonic and biofilm phases.

The metabarcoding data were further analysed for putative functions using METAGENassist to interpret microbial metabolism and other traits, the results of which are presented in Table S2. Key aspects of the sulfur cycle, including sulfate reduction (detected at high levels in all samples at week 8) and sulfur oxidation (for the orange tubercle configuration at week 8) were detected.

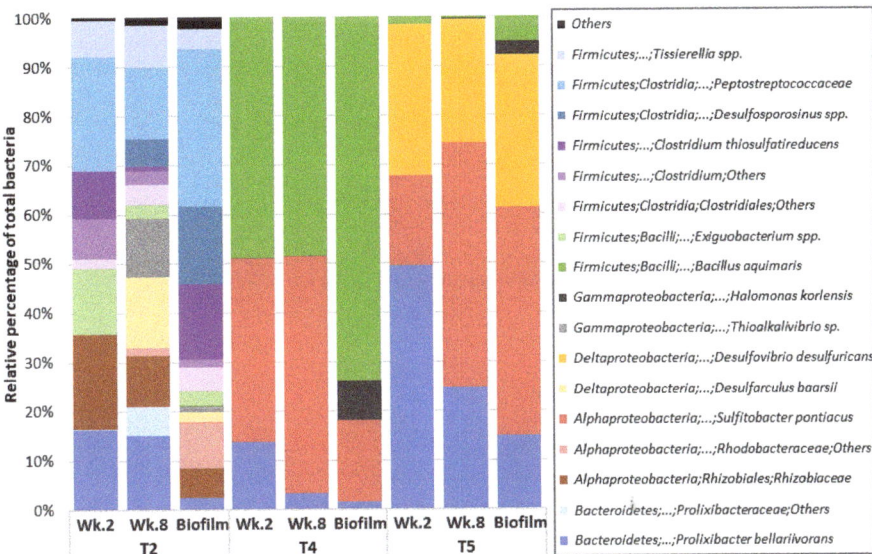

Figure 7. Microbial communities identified by 16S rRNA gene metabarcoding of mixed consortium treatments (T2, T4 and T5), for planktonic communities (at weeks 2 and 8) and for the attached biofilms obtained by sonication (at week 8).

4. Discussion

4.1. Surfaces of Coupons Before Cleaning

The black corrosion products on the surfaces and sulfur detected in the biofilms (by EDS) of coupons from several treatments (T2, T3, T5 and T6) indicates iron sulfide production, and active sulfate reduction. This matches well with the presence of *D. desulfuricans*, when specifically added to the tests (i.e., T3, T5 and T6) and the relatively high potential for sulfate reduction determined from the METAGENassist analysis of the T2 orange tubercle inoculum. The iron:sulfur ratios calculated for the biofilms indicated that the highest to lowest relative amounts of sulfur detected were for T3 > T2 > T6 > T5, which qualitatively matches the thicknesses of the biofilms observed in SEM images.

One somewhat unexpected observation was that the nominally aerobic test with the *D. desulfuricans* (T3) produced a visually thicker biofilm than that for the corresponding anaerobic test with *D. desulfuricans* (T6). While they are more commonly known as strict anaerobes, many sulfate-reducing bacteria, including the culture collection strain used in this work, can survive/grow in aerobic environments [42,43]. It is also possible that the visually thicker biofilm observed for the *D. desulfuricans* aerobic incubation (T3) is a result of a stress response by the bacteria. Although the aerobic tests (including the T3 treatment) were nominally aerobic at the start of the tests it is likely that the oxygen concentration in the solutions dropped over time due to microbial oxygen consumption, making the environment more suitable for optimal growth of *D. desulfuricans*. A suggestion for future work would be to monitor the oxygen concentration in the solutions as a function of time.

4.2. Corrosion Attack

Coupon weight loss indicates that T1 (uninoculated) and T3 (*D. desulfuricans* aerobic) treatments demonstrated greater general corrosion attack than the other treatments. However, T6 (*D. desulfuricans* anaerobic), T3 (*D. desulfuricans* aerobic) and T5 (four isolates + *D. desulfuricans* aerobic) treatments had significantly greater localised pitting than the other treatments. The pitting results for *D. desulfuricans* anaerobic incubation (T3) (average pit depth 45 μm) were greater than for previous tests for the same bacterial strain using a modified Baar's medium (average pit depth ~27 μm) with the same steel type [32]. However, the corresponding weight loss data for the previous modified Baar's medium tests with *D. desulfuricans* were much greater (~80 µm/year) than obtained in this work (10 µm/year). It is important to note that in addition to pitting, general corrosion of the steel surface was observed in the previous study. This indicates that test medium used can have an impact on steel corrosion and pitting outcomes. Previous work indicated that the addition of Fe ions/lactate to modified Baar's medium can have a significant effect on the weight loss results, which are indicative of any general corrosion taking place [40]. No addition of Fe ions to the bulk test solutions was performed in the current work.

The main hypothesis for the work was that a consortium of microbes would result in greater corrosion rates than single isolates (T3 and T6) or an uninoculated control (T1). This was because a range of phenotypes were considered important in promoting ALWC. The consortium tests (T2, T4 and T5) actually showed the three lowest weight losses (typically indicative of general corrosion) of all of the tests performed. Microbial corrosion (including ALWC), however, is more often linked to localised rather than general/uniform corrosion. In relation to localised corrosion, the average pit depths were smaller for the consortia treatments (T2, T4 and T5) than for sole *D. desulfuricans* tests (T3 and T6). The volume of pits for the T5 treatment (four isolates + *D. desulfuricans*) were similar to the sole *D. desulfuricans* tests (T3 and T6), which were much greater than for any of the other test treatments.

One possible reason for the lower levels of corrosion obtained for the mixed microbial communities compared to the uninoculated control or single *D. desulfuricans* tests was that the types of microbes present (orange tubercle, T2) or chosen (pure cultures) were inappropriate for ALWC. The orange tubercle inoculum (T2), was taken from a site which had pitting corrosion and diagnosed ALWC beneath the orange tubercle. Thus, this orange tubercle consortium should have been appropriate to generate optimal ALWC conditions in our laboratory tests. Indeed, we previously found increased corrosion when orange tubercles (taken from another site with suspected ALWC) were used as inocula in laboratory tests, compared to an uninoculated control test [25]. The metabarcoding analysis of T2 identified several microbes present in both the biofilm and planktonic phases that were capable of sulfate reduction (more detail below) and sulfur was clearly identified in the biofilm by EDS.

High corrosion rates of ~100–125 µm/year (from weight loss) were previously [25] found for tests with orange tubercle inocula compared to ~40 µm/year in uninoculated controls when using a similar test arrangement (e.g., steel type, test duration) to that used in the current work. These previous higher corrosion rates were observed for samples with uniform corrosion across sample surfaces and with little localised corrosion. Two obvious key differences between the previous [25] and current studies were the initial microbial community structure of the inocula and the types of nutrients added to the test solutions. While a Deltaproteobacteria SRB (*Desulfarculus baarsii*) was detected in the orange tubercle inoculum (T2) test, this strain has not typically been linked to corrosion. In the previous study [25], an unidentified species from the Desulfobulbaceae family of Deltaproteobacteria was found at reasonably high levels (up to ~40%) in the planktonic phase. This family includes *Desulfopila corrodens* which was directly linked to rapid corrosion in ALWC studies [23].

Glucose (a carbon and electron donor source) and yeast extract were components of the earlier test medium [25], while peptone and yeast extract were present in relatively low amounts in the current work's medium. It has been shown that the types of nutrients available for microbial growth and the physicochemical conditions can have substantial effects on the composition of microbial communities that ensue (e.g., [44,45]). These two features likely explain the microbial differences observed between the earlier [25] and the current research. Additionally, changes in the level of carbon sources present in test media can affect the extent of corrosion caused by a single SRB strain (e.g., [29]).

The majority of microbes chosen for the defined consortium tests survived in reasonable numbers throughout the test period, although there were some clear changes in the microbial species composition over the test duration (discussed in more detail below).

There was also clear evidence of sulfate reduction, as seen by the black solution and presence of sulfur in the biofilm for the defined mixed microbial treatment, which included *D. desulfuricans* (T5), indicating that *D. desulfuricans* were metabolically active. As discussed above, it is possible that relatively low levels of carbon sources/electron donors in the test media may have altered physicochemical processes (i.e., sulfate reduction) that play a role in microbial corrosion and had an effect on corrosion rates.

There have been varying reports on the effect of the presence of different species of *Bacillus* on corrosion with some indicating increased corrosion [46–48] and others showing corrosion inhibition [49–53]. Both defined mixed microbial treatments, with (T5) or without (T4) *D. desulfuricans*, had similar levels of general corrosion but qualitatively different biofilm thicknesses, where T5 was greater than T4 (Figures 1 and 2). The 16S rRNA gene metabarcoding showed very different relative amounts of *B. aquimaris*. T5 had a lower general corrosion level than either of the tests with *D. desulfuricans* alone (T3 and T6), indicating that the additional microbes reduced the corrosion mediated by *D. desulfuricans*. This could indicate corrosion inhibition, perhaps due to competition for substrates among the added pure cultures.

4.3. Analysis of Microbial Populations

Although the *D. desulfuricans* strain used is a facultative anaerobe, plate counts showed that there were still reasonable numbers (10^3–10^4 cfu/mL) of viable *D. desulfuricans* present in solutions after 8 weeks when tested in nominally aerobic test conditions. The presence of sulfur in the biofilms of the coupons from these tests (e.g., T3) indicated that *D. desulfuricans* were metabolically active despite the possible presence of oxygen in the bulk solution. Metabarcoding analysis of the planktonic and biofilm phases of the defined consortium including *D. desulfuricans* (T5) showed a relatively high abundance of *D. desulfuricans* throughout this nominally aerobic test. Given the specific inclusion of aerobic microbes in T5 it is highly likely that they used the available oxygen creating anaerobic conditions, which are optimal for the growth of *D. desulfuricans*. This is important as it shows how testing with microbial consortia can produce an environment conducive to anaerobic microbes such as *D. desulfuricans* in a more natural way rather than by exogenously producing low oxygen levels by means such as nitrogen purging. There have been some reports on the potentially important role of oxygen in MIC of SRB [22,54–57] and of other microbes [58,59]. In any case, results from the current work support further studies on this topic.

The microbial consortium composition in the orange tubercle inoculum test (T2) differed over time in the planktonic phase (between 2 weeks and 8 weeks) and also between the planktonic and biofilm phases at 8 weeks. This is expected as the environmental conditions in the laboratory test was quite different to the tubercle's native location. The nature of the community succession will be different in the planktonic and biofilm phases as there will be local differences in oxygen content and corrosion products from the carbon steel could act as a nutrient source for certain microbes [6,60].

At 8 weeks incubation both planktonic and biofilm phases in T2 contained microbes potentially capable of sulfide/sulfur oxidation (e.g., *Thioalkalivibrio* sp.) and sulfate reduction (e.g., *Desulfosporosinus* spp. and *D. baarsii*). A combination of both sulfur oxidisers and sulfate reducers could form a closed sulfur cycle, which has previously been reported to potentially lead to rapid corrosion [5,8,13–17]. However, extensive corrosion was not seen in T2. It is possible that the system may have needed more time to develop an active corrosion state. Although strains of the Deltaproteobacteria (including *D. desulfuricans*) are the most commonly regarded sulfate reducers, other microbial groups also are capable for sulfate reduction. These include *Desulfosporosinus* spp., which is a member of the Firmicutes phylum. Various microbes relevant to the nitrogen cycle (e.g., nitrogen fixation, nitrite reduction and ammonia oxidation) were identified in the liquid and biofilm of T2 at 8 weeks. Nitrifying bacteria are important as they can potentially produce fixed nitrogen that is used to support/maintain the growth of anaerobic bacteria such as SRB [14]. However, conditions would need to be aerobic to support their nitrifying phenotype.

A number of interesting spatio-temporal changes were observed in the microbial communities of the defined consortia tests (T4 without and T5 with *D. desulfuricans*). Although *H. korlensis* comprised a reasonable proportion of the initial inoculum in both T4 and T5 (7% and 5%, respectively), this bacterium was only detected in relatively small relative abundances in the biofilm phase at week 8 (<10% of total). This might suggest a preference for biofilm growth. Large differences were seen in the numbers of *B. aquimaris* present in the T4 and T5 treatments. It was the most abundant bacterium in both the planktonic phase and biofilms in T4, but it was only found in low abundances in T5. No explanation could be suggested for these differences, but since the presence of *D. desulfuricans* was the only difference, this could point to some interaction. *P. bellariivorans* made up ~1% of the initial inoculum according to viable cell counts, but it was quite abundant in T5 when in combination with the *D. desulfuricans* relative to its abundance without *D. desulfuricans* in T4. In T5, *P. bellariivorans* was more abundant than *B. aquimaris* or *H. korlensis* according to metabarcoding at weeks 2 and 8; despite that both of these latter bacteria comprised more cells in the initial inoculum relative to *P. bellariivorans*. These observed changes show how it is difficult to predict the species development in mixed consortia tests, which is likely to be affected by numerous factors including the test medium composition and physicochemical conditions. An increasing number of reports are starting to be produced investigating the development of multispecies biofilms (e.g., [61] and the references in Table S1). It will be interesting to keep a track of the development of this area of research and see what additional insights it might be able to provide microbial corrosion research.

The overall aim of this work was to investigate whether defined microbial species communities could be used to generate conditions conducive to the rapid corrosion of steel, similar to what occurs in ALWC in the marine environment. Notable differences in the biofilms and the microbial consortia were observed for the treatments tested. Lower uniform corrosion and weight loss were recorded for treatments with multispecies communities, however, the results for localised corrosion (typically more indicative of microbial corrosion) were not as straightforward, with one of the multispecies treatments (the defined consortium with *D. desulfuricans*) producing large sized pitting while another (orange tubercle inoculum) had much lower localised corrosion. Table 3 has been provided to help summarise some of the main observations of the work undertaken.

Table 3. Summary of observations from corrosion experiments using a variety of test conditions. Colours of cells have been used to highlight related results observed in different test conditions.

Test (Inoculum, Aeration Status)	Comments
T1 (uninoculated, aerobic)	Oxygen in solution led to general corrosion
	Some localised corrosion but with relatively low volumes
T2 (orange tubercle, aerobic)	Aerobic bacteria likely utilised oxygen and reduced general corrosion
	SRB present, sulfate reduction suspected
	Some localised corrosion but with relatively low volumes
T3 (*D. desulfuricans*, aerobic)	Oxygen in solution led to general corrosion
	SRB present, sulfate reduction suspected
	Localised corrosion with relatively high volumes
T4 (Four isolates, aerobic)	Aerobic bacteria likely utilised oxygen and reduced general corrosion
	Some localised corrosion but with relatively low volumes
T5 (Four isolates + *D. desulfuricans*, aerobic)	Aerobic bacteria likely utilised oxygen and reduced general corrosion
	SRB present, sulfate reduction suspected
	Localised corrosion with relatively high volumes
T6 (*D. desulfuricans*, anaerobic)	Anaerobic environment reduced general corrosion
	SRB present, sulfate reduction suspected
	Localised corrosion with relatively high volumes

5. Conclusions

The aim of this study was to investigate laboratory tests that mimic the complex communities present in real world cases of MIC/ALWC. As part of this work, a model SRB, *D. desulfuricans*, was used throughout an 8-week corrosion study. The general corrosion rates of many of the mixed microbial test configurations were lower than that of an uninoculated control. Despite that the three mixed microbial tests (T2, T4 and T5) were nominally aerobic, it is likely the test solutions became anaerobic due to oxygen use by aerobic microbes in the milieu which may have been responsible for the reduced general/uniform corrosion. Greater localised corrosion was observed for all of the tests that included *D. desulfuricans*. However, the defined mixed community with *D. desulfuricans* did not lead to increased corrosion compared to tests with *D. desulfuricans* alone. The composition of the mixed microbial communities changed over the duration of the test and key differences were observed between the planktonic and biofilm communities. More laboratory-based experimental work is required to determine optimal microbial communities and test conditions for ALWC studies. Finally, the metabolomic or transcriptomic aspects relevant to the microbe/corrosion processes should be considered for a further study. It is likely that such studies would give further useful insights into MIC and ALWC.

Supplementary Materials: The following are available online at https://www.mdpi.com/2624-5558/2/2/8/s1, Table S1: Summary of example MIC studies using defined multispecies microbial consortia (WL—weight loss test method, CR—corrosion rate), Table S2: Putative metabolic activities of microbes present in metabarcoding profiles identified using METAGENassist, Figure S1: XRD scans of products formed on DH36 coupons in six treatments at 8 weeks. The three highest peaks observed match to the Fe component of the carbon steel coupons used.

Author Contributions: Conceptualisation, all authors; methodology, all authors; validation, all authors; formal analysis, H.C.P.; investigation, H.C.P.; writing—original draft preparation, H.C.P.; writing—review and editing, all authors; visualization, H.C.P.; supervision, S.A.W. and L.L.B. All authors have read and agreed to the published version of the manuscript.

Funding: This research received no external funding.

Data Availability Statement: The data presented in this study are available on request from the corresponding author. The microbial sequencing data is planned to be deposited in the NCBI database.

Acknowledgments: We would like to thank Melanie Thomson (previously of Deakin University) for providing the *D. desulfuricans* used in this study.

Conflicts of Interest: The authors declare no conflict of interest.

References

1. Ijsseling, F. General guidelines for corrosion testing of materials for marine applications: Literature review on sea water as test environment. *Br. Corros. J.* **1989**, *24*, 53–78. [CrossRef]
2. Little, B.J.; Lee, J.S. Microbiologically influenced corrosion. In *Kirk-Othmer Encyclopedia of Chemical Technology*; Wiley-Interscience: Hoboken, NJ, USA, 2009; pp. 1–38.
3. Christie, J. The effect of MIC and other corrosion mechanisms on the global ports infrastructure. In Proceedings of the MIC—An International Perspective, Perth, Australia, 14–16 February 2007.
4. Jeffrey, R.J.; Melchers, R.E. Effect of vertical length on corrosion of steel in the tidal zone. *Corrosion* **2009**, *65*, 695–702. [CrossRef]
5. Beech, I.B.; Campbell, S.A. Accelerated low water corrosion of carbon steel in the presence of a biofilm harbouring sulphate-reducing and sulphur-oxidising bacteria recovered from a marine sediment. *Electrochim. Acta* **2008**, *54*, 14–21. [CrossRef]
6. Dang, H.; Lovell, C.R. Microbial surface colonization and biofilm development in marine environments. *Microbiol. Mol. Biol. Rev.* **2016**, *80*, 91–138. [CrossRef] [PubMed]
7. Flemming, H.-C.; Wuertz, S. Bacteria and archaea on Earth and their abundance in biofilms. *Nat. Rev. Microbiol.* **2019**, *17*, 247–260. [CrossRef] [PubMed]
8. Little, B.; Ray, R.; Pope, R. Relationship between corrosion and the biological sulfur cycle: A review. *Corrosion* **2000**, *56*, 433–443. [CrossRef]
9. Malard, E.; Gueuné, H.; Fagot, A.; Lemière, A.; Sjogren, L.; Tidblad, J.; Sanchez-Amaya, J.M.; Muijzer, G.; Marty, F.; Quillet, L.; et al. *Microbiologically Induced Corrosion of Steel Structures in Port Environment: Improving Prediction and Diagnosis of ALWC (MICSIPE)*; RFCS Publications: Luxembourg, 2013.

10. Païssé, S.; Ghiglione, J.F.; Marty, F.; Abbas, B.; Gueuné, H.; Amaya, J.M.S.; Muyzer, G.; Quillet, L. Sulfate-reducing bacteria inhabiting natural corrosion deposits from marine steel structures. *Appl. Microbiol. Biotechnol.* **2013**, *97*, 7493–7504. [CrossRef]
11. Videla, H.A.; Herrera, L.K.; Edyvean, G. An updated overview of SRB induced corrosion and protection of carbon steel. In Proceedings of the NACE Corrosion, Houston, TX, USA, 3–7 April 2005; NACE International: Houston, TX, USA, 2005; p. 05488
12. Enning, D.; Venzlaff, H.; Garrelfs, J.; Dinh, H.T.; Meyer, V.; Mayrhofer, K.; Hassel, A.W.; Stratmann, M.; Widdel, F. Marine sulfate-reducing bacteria cause serious corrosion of iron under electroconductive biogenic mineral crust. *Environ. Microbiol.* **2012** *14*, 1772–1787. [CrossRef]
13. Gehrke, T.; Sand, W. Interactions between microorganisms and physiochemical factors cause MIC of steel pilings in harbors (ALWC). In Proceedings of the NACE Corrosion, San Diego, CA, USA, 16–20 March 2003; NACE International: Houston, TX, USA, 2003; p. 03557.
14. Sand, W.; Gehrke, T. Microbially influenced corrosion of steel in aqueous environments. *Rev. Environ. Sci. Biotechnol.* **2003**, *2*, 169–176. [CrossRef]
15. Gubner, R.; Beech, I. Statistical assessment of the risk of the accelerated low-water corrosion in the marine environment. In Proceedings of the NACE Corrosion, San Antonio, TX, USA, 25–30 April 1999; NACE International: Houston, TX, USA, 1999; p. 318.
16. Javadhastri, R. *Microbiologically Influenced Corrosion—An Engineering Insight*; Springer: London, UK, 2008; pp. 125–139.
17. Smith, M.; Bardiau, M.; Brennan, R.; Burgess, H.; Caplin, J.; Ray, S.; Urios, T. Accelerated low water corrosion: The microbial sulfur cycle in microcosm. *NPJ Mater. Degrad.* **2019**, *3*, 37. [CrossRef]
18. Barco, R.A.; Hoffman, C.L.; Ramírez, G.A.; Toner, B.M.; Edwards, K.J.; Sylvan, J.B. In-situ incubation of iron-sulfur mineral reveals a diverse chemolithoautotrophic community and a new biogeochemical role for Thiomicrospira. *Environ. Microbiol.* **2017**, *19*, 1322–1337. [CrossRef]
19. Celikkol-Aydin, S.; Gaylarde, C.C.; Lee, T.; Melchers, R.E.; Witt, D.L.; Beech, I.B. 16S rRNA gene profiling of planktonic and biofilm microbial populations in the Gulf of Guinea using Illumina NGS. *Mar. Environ. Res.* **2016**, *122*, 105–112. [CrossRef] [PubMed]
20. Dang, H.; Chen, R.; Wang, L.; Shao, S.; Dai, L.; Ye, Y.; Guo, L.; Huang, G.; Klotz, M.G. Molecular characterization of putative biocorroding microbiota with a novel niche detection of Epsilon and Zetaproteobacteria in Pacific Ocean coastal seawaters. *Environ. Microbiol.* **2011**, *13*, 3059–3074. [CrossRef] [PubMed]
21. Phan, H.C.; Wade, S.A.; Blackall, L.L. Is marine sediment the source of microbes associated with accelerated low water corrosion? *Appl. Microbiol. Biotechnol.* **2019**, *103*, 449–459. [CrossRef] [PubMed]
22. Lee, J.; Ray, R.; Lemieux, E.; Falster, A.; Little, B. An evaluation of carbon steel corrosion under stagnant seawater conditions. *Biofouling* **2004**, *20*, 237–247. [CrossRef] [PubMed]
23. Marty, F.; Gueune, H.; Malard, E.; Sánchez-Amaya, J.M.; Sjögren, L.; Abbas, B.; Quillet, L.; van Loosdrecht, M.C.M.; Muyzer, G. Identification of key factors in accelerated low water corrosion through experimental simulation of tidal conditions: Influence of stimulated indigenous microbiota. *Biofouling* **2014**, *30*, 281–297. [CrossRef]
24. Pillay, C.; Lin, J. Metal corrosion by aerobic bacteria isolated from stimulated corrosion systems: Effects of additional nitrate sources. *Int. Biodeter. Biodegrad.* **2013**, *83*, 158–165. [CrossRef]
25. Wade, S.; Blackall, L. Development of a laboratory test for microbial involvement in accelerated low water corrosion. *Microbiol. Aust.* **2018**, *39*, 170–172. [CrossRef]
26. Enning, D.; Garrelfs, J. Corrosion of iron by sulfate-reducing bacteria: New views of an old problem. *Appl. Environ. Microbiol.* **2014**, *80*, 1226–1236. [CrossRef]
27. Li, Y.; Xu, D.; Chen, C.; Li, X.; Jia, R.; Zhang, D.; Sand, W.; Wang, F.; Gu, T. Anaerobic microbiologically influenced corrosion mechanisms interpreted using bioenergetics and bioelectrochemistry: A review. *J. Mater. Sci Technol.* **2018**, *34*, 1713–1718. [CrossRef]
28. Martin, A.; Auger, E.A.; Blum, P.H.; Schultz, J.E. Genetic basis of starvation survival in nondifferentiating bacteria. *Annu. Rev. Microbiol.* **1989**, *43*, 293–316. [CrossRef]
29. Xu, D.; Gu, T. Carbon source starvation triggered more aggressive corrosion against carbon steel by the *Desulfovibrio vulgaris* biofilm. *Int. Biodeter. Biodegrad.* **2014**, *91*, 74–81. [CrossRef]
30. Salgar-Chaparro, S.J.; Lepkova, K.; Pojtanabuntoeng, T.; Darwin, A.; Machuca, L.L. Nutrient level determines biofilm characteristics and subsequent impact on microbial corrosion and biocide effectiveness. *Appl. Environ. Microbiol.* **2020**, *86*, e02885–e02919. [CrossRef] [PubMed]
31. Wade, S.A.; Javed, M.A.; Palombo, E.A.; McArthur, S.L.; Stoddart, P.R. On the need for more realistic experimental conditions in laboratory-based microbiologically influenced corrosion testing. *Int. Biodeter. Biodegrad.* **2017**, *121*, 97–106. [CrossRef]
32. Javed, M.; Neil, W.; McAdam, G.; Wade, S. Effect of sulphate-reducing bacteria on the microbiologically influenced corrosion of ten different metals using constant test conditions. *Int. Biodeter. Biodegrad.* **2017**, *125*, 73–85. [CrossRef]
33. Li, H.-B.; Zhang, L.-P.; Chen, S.-F. *Halomonas korlensis* sp. nov., a moderately halophilic, denitrifying bacterium isolated from saline and alkaline soil. *Int. J. Syst. Evol. Microbiol.* **2008**, *58*, 2582–2588. [CrossRef] [PubMed]
34. Mahendran, S.; Sankaralingam, S.; Shankar, T.; Vijayabaskar, P. Alkalophilic protease enzyme production from estuarine *Bacillus aquimaris*. *World J. Fish Mar. Sci.* **2010**, *2*, 436–443.
35. Marques, J.M.; de Almeida, F.P.; Lins, U.; Seldin, L.; Korenblum, E. Nitrate treatment effects on bacterial community biofilm formed on carbon steel in produced water stirred tank bioreactor. *World J. Microbiol. Biotechnol.* **2012**, *28*, 2355–2363. [CrossRef] [PubMed]

36. Phan, H.C.; Wade, S.A.; Blackall, L.L. Microbial communities of orange tubercles in accelerated low water corrosion. *Appl. Environ. Microbiol.* **2020**, *86*, e00610–e00620. [CrossRef]
37. Cavaleiro, A.J.; Abreu, A.A.; Sousa, D.Z.; Pereira, M.A.; Alves, M.M. The role of marine anaerobic bacteria and archaea in bioenergy production. In *Management of Microbial Resources in the Environment*; Malik, A., Grohmann, E., Alves, M., Eds.; Springer: Dordrecht, The Netherlands, 2013; pp. 445–469.
38. Makita, H. Iron-oxidizing bacteria in marine environments: Recent progresses and future directions. *World J. Microbiol. Biotechnol.* **2018**, *34*, 110. [CrossRef]
39. Garrity, G.M.; Bell, J.A.; Lilburn, T. Rhodobacteraceae. In *Bergey's Manual of Systematic Bacteriology: The Proteobacteria: Part C: The Alpha-, Beta-, Delta-, and Epsilonproteobacteria*; Brenner, D.J., Krieg, N.R., Staley, J.T., Eds.; Springer Science & Business Media: New York, NY, USA, 2006; Volume 2, pp. 161–229.
40. Javed, M.A.; Stoddart, P.R.; Wade, S.A. Corrosion of carbon steel by sulphate reducing bacteria: Initial attachment and the role of ferrous ions. *Corros. Sci.* **2015**, *93*, 48–57. [CrossRef]
41. *ASTM G1 Standard Practice for Preparing, Cleaning, and Evaluating Corrosion Test Specimens*; ASTM International: West Conshohocken, PA, USA, 2017.
42. Cypionka, H. Oxygen respiration by *Desulfovibrio* species. *Ann. Rev. Microbiol.* **2000**, *54*, 827–848. [CrossRef]
43. Lobo, S.A.; Melo, A.M.; Carita, J.N.; Teixeira, M.; Saraiva, L.M. The anaerobe *Desulfovibrio desulfuricans* ATCC 27774 grows at nearly atmospheric oxygen levels. *FEBS Lett.* **2007**, *581*, 433–436. [CrossRef]
44. Kwon, M.J.; O'Loughlin, E.J.; Boyanov, M.I.; Brulc, J.M.; Johnston, E.R.; Kemner, K.M.; Dionysios, A.; Antonopoulo, D.A. Impact of organic carbon electron donors on microbial community development under iron- and sulfate-reducing conditions. *PLoS ONE* **2016**, *11*, e0146689. [CrossRef]
45. Zhan, M.; Liu, X.; Li, Y.; Wang, G.; Wang, Z.; Wen, J. Microbial community and metabolic pathway succession driven by changed nutrient inputs in tailings: Effects of different nutrients on tailing remediation. *Sci. Rep.* **2017**, *7*, 474. [CrossRef]
46. Parthipan, P.; Elumalai, P.; Ting, Y.P.; Rahman, P.K.; Rajasekar, A. Characterization of hydrocarbon degrading bacteria isolated from Indian crude oil reservoir and their influence on biocorrosion of carbon steel API 5LX. *Int. Biodeter. Biodegrad.* **2018**, *129*, 67–80. [CrossRef]
47. Rajasekar, A.; Babu, G.T.; Pandian, K.S.; Maruthamuthu, S.; Palaniswamy, N.; Rajendran, A. Biodegradation and corrosion behavior of manganese oxidizer *Bacillus cereus* ACE4 in diesel transporting pipeline. *Corros. Sci.* **2007**, *49*, 2694–2710. [CrossRef]
48. Xu, D.; Li, Y.; Song, F.; Gu, T. Laboratory investigation of microbiologically influenced corrosion of C1018 carbon steel by nitrate reducing bacterium *Bacillus licheniformis*. *Corros. Sci.* **2013**, *77*, 385–390. [CrossRef]
49. Guo, Z.; Liu, T.; Cheng, Y.F.; Guo, N.; Yin, Y. Adhesion of *Bacillus subtilis* and *Pseudoalteromonas lipolytica* to steel in a seawater environment and their effects on corrosion. *Colloids Surf. B* **2017**, *157*, 157–165. [CrossRef]
50. Jayaraman, A.; Cheng, E.T.; Earthman, J.C.; Wood, T.K. Axenic aerobic biofilms inhibit corrosion of SAE 1018 steel through oxygen depletion. *Appl. Microbiol. Biotechnol.* **1997**, *48*, 11–17. [CrossRef] [PubMed]
51. Jayaraman, A.; Hallock, P.; Carson, R.; Lee, C.-C.; Mansfeld, F.; Wood, T. Inhibiting sulfate-reducing bacteria in biofilms on steel with antimicrobial peptides generated in situ. *Appl. Microbiol. Biotechnol.* **1999**, *52*, 267–275. [CrossRef] [PubMed]
52. Karn, S.K.; Guan, F.; Duan, J. *Bacillus* sp. acting as dual role for corrosion induction and corrosion inhibition with carbon steel. *Front. Microbiol.* **2017**, *8*, 2038. [CrossRef]
53. Zuo, R.; Wood, T.K. Inhibiting mild steel corrosion from sulfate-reducing and iron-oxidizing bacteria using gramicidin-S-producing biofilms. *Appl. Microbiol. Biotechnol.* **2004**, *65*, 747–753. [CrossRef] [PubMed]
54. Beech, I.B.; Sunner, J. Sulphate-reducing bacteria and their role in corrosion of ferrous materials. In *Sulphate-Reducing Bacteria: Environmental and Engineered Systems*; Barton, L.L., Hamilton, W.A., Eds.; Cambridge University Press: Cambridge, UK, 2007; pp. 459–482.
55. Hamilton, W. Microbially influenced corrosion as a model system for the study of metal microbe interactions: A unifying electron transfer hypothesis. *Biofouling* **2003**, *19*, 65–76. [CrossRef]
56. Hardy, J.; Bown, J. The corrosion of mild steel by biogenic sulfide films exposed to air. *Corrosion* **1984**, *40*, 650–654. [CrossRef]
57. Lee, W.; Lewandowski, Z.; Morrison, M.; Characklis, W.G.; Avci, R.; Nielsen, P.H. Corrosion of mild steel underneath aerobic biofilms containing sulfate-reducing bacteria part II: At high dissolved oxygen concentration. *Biofouling* **1993**, *7*, 217–239. [CrossRef]
58. Li, X.; Xiao, H.; Zhang, W.; Li, Y.; Tang, X.; Duan, J.; Yang, Z.; Wang, J.; Guan, F.; Ding, G. Analysis of cultivable aerobic bacterial community composition and screening for facultative sulfate-reducing bacteria in marine corrosive steel. *J. Oceanol. Limnol.* **2019**, *37*, 600–614. [CrossRef]
59. Qian, H.; Ju, P.; Zhang, D.; Ma, L.; Hu, Y.; Li, Z.; Huang, L.; Lou, Y.; Du, C. Effect of dissolved oxygen concentration on the microbiologically influenced corrosion of Q235 carbon steel by halophilic archaea *Natronorubrum tibetense*. *Front. Microbiol.* **2019**, *10*, 844. [CrossRef]
60. McBeth, J.M.; Emerson, D. In situ microbial community succession on mild steel in estuarine and marine environments: Exploring the role of iron-oxidizing bacteria. *Front. Microbiol.* **2016**, *7*, 1–14. [CrossRef] [PubMed]
61. Tan, C.H.; Lee, K.W.K.; Burmølle, M.; Kjelleberg, S.; Rice, S.A. All together now: Experimental multispecies biofilm model systems. *Environ. Microbiol.* **2017**, *19*, 42–53. [CrossRef]

Influence of Organic Matter/Bacteria on the Formation and Transformation of Sulfate Green Rust

Julien Duboscq [1], Julia Vincent [2], Marc Jeannin [1], René Sabot [1], Isabelle Lanneluc [2], Sophie Sablé [2] and Philippe Refait [1,*]

1. LaSIE, UMR 7356 CNRS—La Rochelle Université, Bâtiment Marie Curie, Avenue Michel Crépeau, 17000 La Rochelle, France; julien.duboscq50@gmail.com (J.D.); mjeannin@univ-lr.fr (M.J.); rsabot@univ-lr.fr (R.S.)
2. LIENSs, UMR 7266 CNRS—La Rochelle Université, Bâtiment Marie Curie, Avenue Michel Crépeau, 17000 La Rochelle, France; julia.vincent1@univ-lr.fr (J.V.); ilannelu@univ-lr.fr (I.L.); ssable@univ-lr.fr (S.S.)
* Correspondence: prefait@univ-lr.fr; Tel.: +33-5-46-45-82-27

Abstract: The corrosion processes of carbon steel immersed in natural seawater are influenced by microorganisms due to important biological activity. An analysis of the corrosion product layers formed on carbon steel coupons in natural or artificial seawater revealed that sulfate green rust $GR(SO_4^{2-})$ was favored in natural environments. In this paper, the role of organic matter/bacteria on the formation and transformation of this compound are addressed. $GR(SO_4^{2-})$ was precipitated from Fe(II) and Fe(III) salts in the presence of various marine bacterial species not involved in the redox cycle of Fe or S. Abiotic experiments were performed for comparison, first without any organic species and then with sodium acetate added as a small organic ion. The obtained aqueous suspensions were aged at room temperature for 1 week. The number of bacteria (CFU/mL) was followed over time and the solid phases were characterized by XRD. Whatever the fate of the bacteria (no activity, or activity and growth), the formation of $GR(SO_4^{2-})$ was favored and its transformation to magnetite completely inhibited. This effect is attributed to the adsorption of organic molecules on the lateral sides of the $GR(SO_4^{2-})$ crystals. A similar effect, though less important, was observed with acetate.

Keywords: marine corrosion; steel; green rust; X-ray diffraction; microbiologically influenced corrosion; biofilm; bacterial activity

1. Introduction

Natural seawater is a biologically active medium, and it is generally acknowledged that microorganisms influence the corrosion processes of any metal or alloy immersed in a marine environment. In the case of carbon and low alloy steels, the role of sulfate-reducing bacteria (SRB) has long been recognized [1]. This has led to the hypothesis that the long-term corrosion process of carbon steel in seawater could be controlled, at least partially, by the rate of external nutrient supply that governs bacterial activity [2,3]. More generally, it has been shown that the organic molecules released by bacteria somehow influence the behavior of metals. For instance, it has been established that in some cases, extracellular polymeric substances favor the corrosion of carbon steel [4]. Conversely, it was proposed that biofilms could form along with the corrosion products, creating a protective barrier and thus decreasing the corrosion rate [5]. Enzymes, and in particular, hydrogenases that can be present in solution (i.e., out of bacterial cells), are also known to be involved [6,7].

The various molecules associated with the presence and/or activity of microorganisms can interact with the metal itself, but also with its corrosion products. One of the consequences of the activity of SRB is the formation of iron sulfide (FeS), which precipitates from Fe^{2+}_{aq} ions resulting from corrosion and the sulfide species produced by SRB. For this reason, FeS was identified as another major component of the corrosion product layer formed on carbon steel coupons after just 6–12 months of immersion in the water of a seaport [8]. It could be detected locally after only 1 month [9] of immersion, and was associated with SRB in the first de-aerated regions formed inside the corrosion product layer. Due to its electronic conductivity and low overvoltage for water reduction, FeS can act as a cathodic site and promote the formation of galvanic cells [10]. Additionally, it can facilitate the influence of electroactive SRB [11].

Most studies dealing with microbiologically influenced corrosion (MIC) are focused on the interactions between metal and bacteria or between metal and species released or produced by bacteria [1]. The interactions between these species and the corrosion products, which may modify the properties of the corrosion product layer and thus the corrosion rate, have only been rarely addressed [12,13]. In a recent work, some differences were observed between the corrosion product layers obtained in artificial and natural seawater [14]. These differences were not only related to FeS, which can only form in natural environments as it requires the presence and activity of SRB, but also to sulfate green rust, $GR(SO_4^{2-})$, which seemed to be favored in natural environments [14].

$GR(SO_4^{2-})$ is one of the main corrosion products forming on carbon steel surfaces which are permanently immersed in natural seawater [5,8,9,15–18]; it is actually the first solid obtained from dissolved Fe species [9,15,17]. It is a mixed valence Fe(II-III) double layered hydroxide with chemical formula $Fe^{II}_4Fe^{III}_2(OH)_{12}SO_4 \cdot 8H_2O$ [19,20], and contains mainly (67%) Fe(II) ions. It is oxidized by dissolved O_2, a process that leads to the formation of Fe(III)-oxyhydroxides such as goethite (α-FeOOH) and lepidocrocite (γ-FeOOH), and oxides such as magnetite Fe_3O_4 [21–23]. Consequently, any species that can affect the formation and transformation of $GR(SO_4^{2-})$ may have an indirect influence on the corrosion process.

The present study is focused on the effect of bacteria in general (i.e., bacteria not known to induce MIC processes) on the formation of $GR(SO_4^{2-})$ and the evolution of this compound in anoxic conditions. The idea was not to address the interactions between $GR(SO_4^{2-})$ and the bacteria involved in the redox cycles of Fe and S, which have already been documented [12,13,24,25], but to consider bacteria having no specific link with Fe and S, the characteristic elements of $GR(SO_4^{2-})$. This study aims to contribute to a better understanding of the interactions between complex bacterial communities and the overall corrosion system through the determination of possible interactions of selected bacteria with $GR(SO_4^{2-})$.

First, corrosion experiments were performed in artificial and natural seawater to study the composition of the corrosion product layers in each case. These layers were characterized by X-ray diffraction (XRD) and µ-Raman spectroscopy (µ-RS). Secondly, $GR(SO_4^{2-})$ was precipitated by mixing a solution of Fe(II) and Fe(III) salts with a solution of NaOH in the presence of three bacterial strains. These strains were isolated from biofilms previously formed on carbon steel coupons immersed in natural seawater [9]. The precipitates were analyzed by XRD after 1 week and 2 months of ageing at room temperature. The bacterial growth during the ageing of the precipitate was also investigated. Finally, GR was precipitated in the presence of sodium acetate to compare the effects of a small organic anion with those observed in the presence of bacteria.

2. Materials and Methods

2.1. Preliminary Corrosion Experiments

To compare the composition of the corrosion product layers formed on carbon steel in artificial and natural seawater, various S355GP steel coupons (5 cm × 5 cm × 1 cm) were exposed for 6 months in both kinds of environments. The nominal composition of this steel grade, commonly used for sea harbor sheet piles, is in wt.%: $C \leq 0.27$, $Mn \leq 1.7$, $S \leq 0.055$, $Si \leq 0.6$, $Al \leq 0.02$ and Fe for the rest. The coupons were embedded in epoxy resin so that only one side (active area of 25 cm^2) was exposed to seawater. The surface of this side was previously shot blasted (Sa 2.5, angular shot) and degreased with acetone.

First, three coupons were immersed in natural seawater for 6 months in the Minimes harbor, the marina of La Rochelle (Atlantic Ocean), using the experimental platform of the

LaSIE laboratory. The coupons were immersed vertically at a depth of ~20 cm (measured from the upper edge of the coupons). As the experimental platform floats on the sea surface, the immersion depth is constant. The temperature of the water close to the coupons was measured regularly. It varied from 9 ± 2 °C at the beginning (February) to 20 ± 2 °C at the end (July) of the experiment. Secondly, three other coupons were exposed to stagnant ASTM D1141 artificial seawater [26] in 10 L tanks. The seawater was renewed after 15 days, and monthly afterwards. The tanks were set in an unheated room so that the average temperature of the water increased from 12 °C (February) to 25 °C (July) during the experiment.

At the end of the experiment, the coupons immersed in the Minimes harbor were carried to the lab in a tank full of natural seawater sampled in situ with the coupons. Then, all coupons were removed from the seawater (artificial or natural) and rapidly transferred to a freezer and stored at −24 °C before analysis. With this procedure, already used in previous works [9,16,17], the samples can be analyzed in a frozen state so that the corrosion product layers, that contain a lot of water, can be easily handled. Due to the complexity of the corrosion product layers forming on steel in seawater [8,9,15–18], two methods were used to identify the corrosion products, namely μ-Raman spectroscopy (μ-RS) and X-ray diffraction (XRD). For each type of seawater, one coupon was analyzed by μ-RS and another by XRD.

μ-RS analysis was carried out at room temperature using a Horiba Raman spectrometer (LabRam HR evo, Horiba, Tokyo, Japan) equipped with a confocal microscope and a Peltier-based cooled charge coupled device (CCD) detector. A solid-state diode pumped green laser (wavelength = 532 nm) was used with laser power reduced to 10% (0.6 mW) of the maximum to prevent the transformation of the analyzed compounds into hematite α-Fe_2O_3. This transformation can take place due to an excessive heating [27,28]. The acquisition time depended on the nature of the analyzed phase, and thus, varied from 60 s to 2 min. At least 20 zones (diameter of 3–6 μm) were analyzed on the same sample using a 50× long working distance objective. The analysis was achieved without specific protection from air because the time required for analysis was short. Additionally, the samples remained wet during the procedure, which minimized oxidation.

For the XRD analysis, the whole corrosion product layer was scraped from the surface of the coupon and ground in a mortar (it was initially solid, as it was frozen) until a homogenous wet paste was obtained. This paste was then analyzed as described in Section 2.5.

2.2. Preparation of Green Rust Precipitates

Green rust compounds can be precipitated by mixing a solution of Fe(II) and Fe(III) salts with NaOH solution [19]. Based on this, a method was developed to prepare GR(SO_4^{2-}) under conditions simulating seawater, i.e., using Fe(II) and Fe(III) chlorides and adding sodium chloride and sodium sulfate to obtain a suspension with overall chloride and sulfate concentrations similar to those typical of seawater [23,29]. In the present study, this method was used once again. The concentrations of reactants are listed in Table 1, together with the distribution of the reactants in the two prepared solutions.

To obtain the GR precipitate, solution 1 (100 mL) was added to solution 2 (100 mL) and the overall 200 mL of suspension was vigorously stirred for 30 s at room temperature (RT = 21 ± 1 °C). After stirring, the suspension was poured into a flask, filled to the rim. The flask was hermetically sealed to avoid any oxidation by air of the precipitates during ageing periods of 1 week and 2 months. The aged precipitates were finally filtered for analysis by XRD. They were sheltered from air with a plastic membrane during filtration to avoid the oxidation of the obtained GR compounds.

Table 1. Concentrations of reactants used to prepare the initial green rust precipitate (mol L^{-1}), expressed with respect to the total volume of solution (200 mL = solution 1 + solution 2), and considered bacterial strains.

Reactants	Solution 1	Solution 2
NaOH	0.24	-
NaCl	-	0.27
Na$_2$SO$_4$·10H$_2$O	-	0.03
FeCl$_2$·4H$_2$O	-	0.04
FeCl$_3$·6H$_2$O	-	0.08
Bacteria [1]	No or *Pseudoalteromonas* IIIA004 or *Micrococcus* IVA008 or *Bacillus* IVA016	-
NaCH$_3$COO·3H$_2$O	0 or 0.06	-

[1] See text for the bacterial concentrations.

The NaOH, FeCl$_2$·4H$_2$O and FeCl$_3$·6H$_2$O concentrations used here correspond to the stoichiometry of the precipitation of the sulfate GR. This reaction can be written as:

$$4Fe^{2+} + 2Fe^{3+} + 12OH^- + SO_4^{2-} + 8H_2O \rightarrow Fe^{II}_4 Fe^{III}_2 (OH)_{12} SO_4 \cdot 8H_2O \quad (1)$$

2.3. Bacterial Strains and Culture Conditions

Considering that the influence of microorganisms/organic matter may depend significantly on the bacterial species present, three different bacterial strains (belonging to different families of bacteria) were considered: *Pseudoalteromonas* IIIA004, *Micrococcus* IVA008 and *Bacillus* IVA016. They were previously isolated from the biofilm covering carbon steel coupons immersed for 1 week (*Pseudoalteromonas*) or 2 weeks (*Micrococcus* and *Bacillus*) in natural seawater (La Rochelle marina, Atlantic Ocean) [9]. Each strain was previously identified by sequencing the 16S rRNA gene (accession numbers in the GenBank database: KJ814569 for *Pseudoalteromonas* IIIA004, KJ814564 for *Micrococcus* IVA008 and KJ814540 for *Bacillus* IVA016) [9]). These bacteria do not belong to the families of bacteria classically described as SRB, iron oxidizing bacteria (IOB) or iron reducing bacteria (IRB). The three considered strains were cultured in aerobic conditions, and consequently, were not SRB that can only grow in anaerobic conditions (or in an environment with a low oxygen concentration).

For bacterial growth, the culture medium used, called Marine Broth, was composed of ammonium nitrate 0.0016 g L^{-1}, anhydrous magnesium chloride 8.8 g L^{-1}, bacteriological peptone 5 g L^{-1}, boric acid 0.022 g L^{-1}, anhydrous calcium chloride 1.8 g L^{-1}, disodium phosphate 0.008 g L^{-1}, potassium bromide 0.08 g L^{-1}, potassium chloride 0.55 g L^{-1}, sodium bicarbonate 0.16 g L^{-1}, sodium chloride 19.4 g L^{-1}, sodium fluoride 0.0024 g L^{-1}, sodium silicate 0.004 g L^{-1}, sodium sulfate 3.24 g L^{-1}, strontium chloride 0.034 g L^{-1}, yeast extract 1 g L^{-1} and ferric citrate 0.1 g L^{-1}. The culture medium was sterilized for 20 min at 115 °C.

A concentrated suspension (5 mL) of the three bacteria was first prepared. For each strain, 200 mL of Marine Broth was inoculated with bacteria at 2%, from an overnight culture in Marine Broth, and incubated at 30 °C under constant stirring (orbital shaker, 160 rpm). After 24 h of incubation, all three bacterial suspensions were centrifuged for 20 min at 5000× g. The centrifugation pellet was finally set again in suspension in 5 mL of ASTM D1141 artificial seawater [26]. The final suspension of bacteria was then added to NaOH solution 1 (see Table 1).

2.4. Numeration of Bacteria

Quantification of bacteria was performed at four stages of the process: (1) after 24 h of growth independently for each strain, (2) once the bacteria had been concentrated, (3) right after mixing solutions 1 and 2, i.e., right after the formation of the GR precipitate, and (4) after the 1-week ageing period at RT (see Section 2.2) in a hermetically sealed flask. For aged samples, because of decantation, the solid phase(s) settled at the bottom of the flask.

Consequently, the supernatant liquid phase and the decanted precipitate (solid phase) could be sampled and analyzed separately.

In each case, a 100 µL sample was prepared by serial dilutions (10^{-1} to 10^{-6}) of the cell suspension in artificial seawater and inoculated on a solid culture medium composed of Marine Broth with 1.2% (w/v) agar. After incubation at 30 °C in aerobic conditions, bacterial growth was evaluated by counting the number of colony forming units (CFU) (three replicates). The results are expressed in CFU mL^{-1}.

2.5. XRD Analysis of the Precipitates

XRD analysis was performed with an Inel EQUINOX 6000 diffractometer (Thermo Fisher Scientific, Waltham, MA, USA) equipped with a CPS 590 detector that detects the diffracted photons simultaneously on a 2θ range of 90°. Co-Kα radiation (λ = 0.17903 nm) was used at 40 kV and 40 mA, with the XRD analysis being performed at RT with a constant angle of incidence (5 degrees) for 45 min. To prevent the oxidation of GR compounds during preparation and analysis, the wet paste obtained after filtration of the sample was mixed with a few drops of glycerol. With this procedure, the GR particles were coated with glycerol and sheltered from the oxidizing action of O_2 [30]. The angular scale was calibrated using the diffraction peaks of magnetite (if present).

The crystalline phases were identified via the ICDD-JCPDS (International Center for Diffraction Data—Joint Committee on Powder Diffraction Standards) database, and the peaks indexed according to the corresponding file. The parameters of the diffraction peaks, i.e., interplanar distance, intensity and full width at half maximum, were determined via a computer fitting of the experimental diffraction patterns. The diffraction peaks were fitted with pseudo-Voigt functions to take into account the evolution of the peak profile with increasing diffraction angle. The fitting procedure was achieved using the OriginPro 2016 software (OriginLab).

µ-Raman spectroscopy was not used for the characterization of the precipitates because (i) this method is not suitable to distinguish between the various GR compounds [31], and (ii) the bacterial cells and associated organic matter mixed with the solid phases induce an important fluorescence phenomenon that makes it difficult to acquire usable data.

3. Results

3.1. Characterization of the Corrosion Product Layers Formed in Artificial/Natural Seawater

As already reported in previous studies, the corrosion product layer forming on carbon steel in seawater is, in most cases, a bilayer, composed of an inner black stratum which is in contact with the metal surface and an outer orange stratum which is in contact with the marine medium [8,9,15–18]. The layers obtained in this study verified this general trend too.

The results given by µ-RS are described first, and two typical µ-RS spectra are displayed in Figure 1. Table 2 lists all the components identified by µ-RS in the corrosion product layers of the analyzed coupons. Because a large number of zones were analyzed in each layer, three kinds of components could be identified: the main ones, frequently occurring ones and minor ones which were rarely observed.

For the coupon left for 6 months in artificial seawater, the main identified component was magnetite Fe_3O_4. A typical spectrum is shown in Figure 1a. The three characteristic peaks of magnetite are clearly seen, with the most intense one at 671 cm^{-1} and two smaller ones at 308 and 543 cm^{-1}, as reported in literature data [27,28,32]. Magnetite was mainly identified in the inner black stratum of the corrosion product layer. Lepidocrocite γ-FeOOH and aragonite were also frequently identified, mainly in the outer orange stratum. Aragonite is not a corrosion product, as it is a form of calcium carbonate ($CaCO_3$), but it is often associated with corrosion products in the cathodic zones of the metal surface [14,17,18]. The small increase of the interfacial pH in these zones is sufficient to induce the precipitation of aragonite from the dissolved Ca^{2+} and carbonate species present in seawater. The main peak of aragonite at 1082 cm^{-1} [33] is visible on the spectrum of Figure 1a. More rarely, green rusts, i.e., ferrihydrite (FeOOH·H$_2$O) and chukanovite (Fe$_2$(OH)$_2$CO$_3$), were also identified, but only in the inner black stratum.

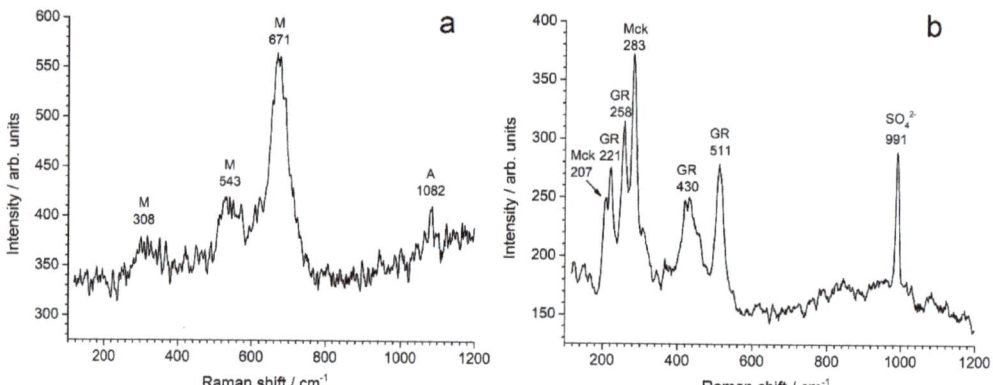

Figure 1. Raman spectra (examples) obtained during the analysis of the corrosion product layers formed after 6 months in: (**a**) artificial and (**b**) natural seawater. M = magnetite, A = aragonite, Mck = nanocrystalline mackinawite and GR = green rust, with the position (in cm^{-1}) of the corresponding Raman peak.

Table 2. µ-RS analysis of the corrosion product layers formed after 6 months in artificial/natural seawater: synthesis of the results.

Sample	Main Components	Frequently Identified Components	Minor Components
Steel coupon in artificial seawater	Magnetite	Aragonite, Lepidocrocite	Chukanovite, Ferrihydrite, Green Rust
Steel coupon in natural seawater [1]	Magnetite, Mackinawite	Green Rust	Ferrihydrite

[1] Only the black inner stratum was analyzed.

For the coupon left 6 months in natural seawater, only the inner black stratum of the corrosion product layer was analyzed. In the outer orange layer, the biofouling mixed with the corrosion products induced an important fluorescence phenomenon that made it difficult to acquire useful data. The FeOOH phases, mainly present in the outer orange layer, thus do not appear in Table 2 for this coupon (except for ferrihydrite, identified as a minor component). In this case, the main compounds identified in the black inner stratum were nanocrystalline mackinawite (FeS) and magnetite. Nanocrystalline mackinawite is the iron sulfide that forms from the dissolved Fe(II) species [34,35] produced by the corrosion of steel, and the dissolved sulfide species produced by SRB. Its Raman spectrum is characterized by two peaks, i.e., the main one at 283 cm^{-1} and the other at 207 cm^{-1} [35], as illustrated by Figure 1b. The spectrum of (well) crystallized mackinawite is slightly different, with the main peak occurring at 300 cm^{-1} [35]. GR compounds were also frequently identified. In Figure 1b, the spectral signature of nanocrystalline mackinawite is accompanied by that of a GR compound that may be GR(SO$_4^{2-}$) or GR(CO$_3^{2-}$). Both GRs have similar spectra, with two main peaks at 430–535 cm^{-1} and 510–515 cm^{-1} and two smaller ones at ~220 cm^{-1} and ~260 cm^{-1} [36,37]. The characteristic sulfate ion peak at 991 cm^{-1} does not demonstrate that this GR is GR(SO$_4^{2-}$), as it could correspond to sulfate ions adsorbed on the surface of GR(CO$_3^{2-}$) crystals.

It must be kept in mind that Figure 1 only shows one selected spectrum from each coupon. Magnetite was identified by µRS as one of the main components of the corrosion product layer in both cases, as reported in Table 2. However, in some cases, given the small size of the zone analyzed by µRS, magnetite was not observed. This means that some small regions of the corrosion product layer did not contain magnetite, as illustrated in Figure 1b.

The results given by XRD, consistent with those given by µ-RS, are presented in Figure 2. The first pattern (a) is that of the corrosion product layer formed in stagnant artificial seawater. The most intense diffraction peaks are unambiguously those of magnetite. Numerous diffraction peaks of aragonite and lepidocrocite are clearly seen. Finally, both GR(SO_4^{2-}) and GR(CO_3^{2-}) may be identified, but only owing to their main diffraction peak (GR001 or GRC003) that is very weak.

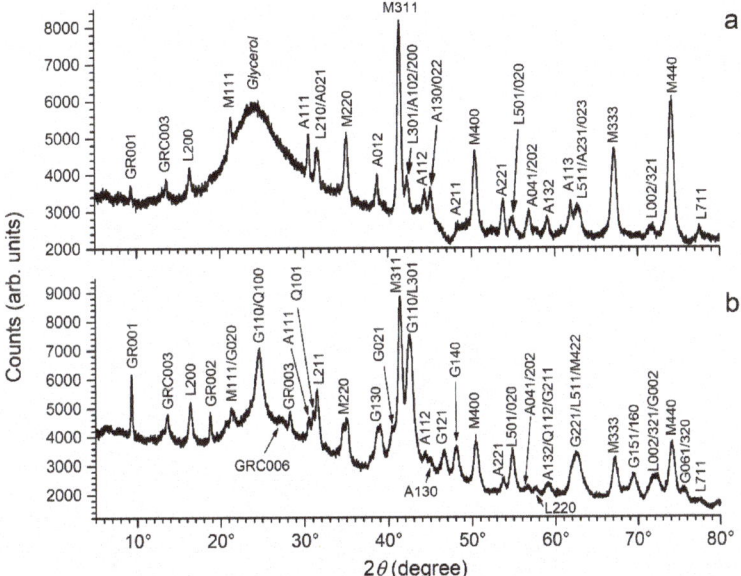

Figure 2. XRD pattern of the corrosion product layers formed after 6 months in: (**a**) artificial and (**b**) natural seawater. GR = GR(SO_4^{2-}), GRC = GR(CO_3^{2-}), A = aragonite, G = goethite, L = lepidocrocite, M = magnetite and Q = quartz, with the corresponding Miller index.

For the coupon immersed in natural seawater, corresponding to pattern (b), the main diffraction peaks are those of magnetite and goethite. If compared with pattern (a), the diffraction peaks of aragonite are slightly less intense, while those of lepidocrocite and GR(CO_3^{2-}) appear slightly more intense. Finally, the diffraction peaks of GR(SO_4^{2-}) are much more intense in pattern (b). As noted previously [8,17], the XRD analysis did not allow us to detect mackinawite FeS, because this phase remains in a nanocrystalline state.

In conclusion, the main observed difference between both kinds of coupons is the presence of FeS in the corrosion product layer formed in natural conditions a result of bacterial (SRB) activity. This FeS phase is nanocrystalline and was only identified via µ-RS analysis. However, the XRD analysis revealed other differences. In particular, it confirmed that the formation of GR(SO_4^{2-}) was indeed favored in the natural seawater of the harbor site.

3.2. Characterization of the Precipitate Obtained in Abiotic Conditions

In this section, and in Sections 3.3 and 3.4, the precipitates obtained by mixing NaOH with Fe(II) and Fe(III) salts are characterized.

Under the abiotic conditions considered here, and as previously studied [23], the initial precipitate was composed of a mixture of GR(Cl^-) and GR(SO_4^{2-}). After 1 week of ageing, the proportion of GR(Cl^-) drastically decreased. The XRD pattern presented in Figure 3 confirms this result: only the main peaks of GR(Cl^-) are seen and their intensity is very small with respect to that of the peaks of GR(SO_4^{2-}). However, the ageing induced the formation of a small amount of magnetite Fe_3O_4. This evolution was attributed to the respective stability of the three phases [23], i.e., magnetite is more stable than GR(SO_4^{2-}), which, in turn, is more stable than GR(Cl^-) in the conditions considered here.

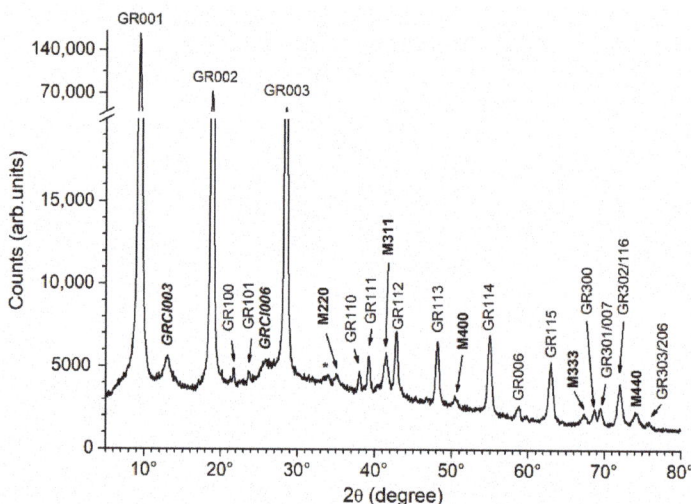

Figure 3. XRD pattern of the precipitate obtained without addition of bacteria or organic compounds, after 1 week of ageing at RT. GR = GR(SO_4^{2-}), GRCl = GR(Cl^-) and M = magnetite, with the corresponding Miller index. * = unidentified peak.

This XRD pattern will be used below as a reference. The intensities of the diffraction peaks M311, GRCl003, GR001 and GR112 were determined by computer fitting as described in Section 2.5. The results of this analysis are presented in Table 3, with the intensity of the main peak of GR(SO_4^{2-}), i.e., GR001, being arbitrarily set as 100 in each case.

Table 3. Abiotic precipitate aged for 1 week: characteristics of the diffraction peaks GR001 and GR112 of GR(SO_4^{2-}), GRCl003 of GR(Cl^-) and M311 of magnetite; 2θ = diffraction angle, in degree, d = interplanar distance (Å), FWHM = full width a half maximum, in degree, and I = peak intensity, with I = 100 for GR001.

Diffraction Peak	2θ	d	FWHM	I
GR001	9.33 ± 0.01	11.00 ± 0.01	0.27 ± 0.01	100
GR112	42.73 ± 0.01	2.455 ± 0.001	0.45 ± 0.02	3.2 ± 0.2
GRCl003	12.94 ± 0.01	7.94 ± 0.01	1.45 ± 0.02	7.0 ± 0.4
M311	41.38 ± 0.01	2.532 ± 0.001	0.65 ± 0.02	4.6 ± 0.3

The interplanar distance d_{001} obtained for GR(SO_4^{2-}) corresponded to the c parameter of the hexagonal cell. It was determined here at 11.00 Å, a value consistent with literature data [20]. The intensity of the GR112 peak was abnormally small, due to the preferential orientation of the GR particles. These particles comprised thin hexagonal platelets perpendicular to the c axis of the crystal structure [16]. For this reason, they were usually parallel to the sample holder. This preferential orientation increased the intensity of the 001 diffraction peaks. The interplanar distance d_{003} found for GR(Cl^-) was also consistent with literature data [38].

3.3. Influence of Bacteria

The results of the study of bacterial growth and quantification are presented in Table 4. Each bacterial strain grew rapidly in the culture medium, reaching between 1.7 and 3.3 × 10^9 CFU mL^{-1} after 24 h at 30 °C, and showing similar cell concentrations for the three strains. All the initial concentrated suspensions of bacteria contained more than 3 × 10^{11} CFU mL^{-1} of bacteria prior to mixing with the reagents used to prepare the GR precipitate. However, the results obtained after precipitation proved to be highly dependent on the considered bacterial strain. For *Pseudoalteromonas* IIIA004, no viable culturable bacteria could be enumerated. The same result was obtained for this strain 1 week later in both the precipitate and supernatant. In the first case, the Fe solid phases were precipitated and aged in a medium where the bacteria did not grow, likely because of cellular death due to the immersion of the bacteria in the NaOH solution 1.

Table 4. Numeration of bacteria (CFU mL^{-1}).

Bacteria	Culture of 24 h in Marine Broth	Initial Concentrated Suspensions	After Precipitation	After 1 Week of Ageing: Supernatant/Precipitate
Pseudoalteromonas IIIA004	2.2 × 10^9	>3 × 10^{11}	No growth	No growth/No growth
Micrococcus IVA008	3.3 × 10^9	>3 × 10^{11}	1.8 × 10^7	No growth/3.0 × 10^8
Bacillus IVA016	1.7 × 10^9	>3 × 10^{11}	1.3 × 10^5	No growth/3.6 × 10^5

In contrast, both *Micrococcus* IVA008 and *Bacillus* IVA016 remained viable and culturable throughout the experiments, even though the cell concentration after precipitation decreased from the initial concentrated suspension. In both cases, the results obtained in the precipitate after 1 week of ageing were similar to those obtained right after precipitation. A slight increase of the bacterial concentration was even observed, in particular for *Micrococcus* IVA008. In contrast, no viable culturable bacteria could be enumerated in the supernatant after ageing. This shows that the bacteria were mostly associated with the solid phases, more likely bound to the particles of Fe compounds. In this second case, the Fe compounds were precipitated and aged in a medium where bacteria survived and even developed.

The XRD analysis of the precipitates obtained after 1 week of ageing in the presence of bacteria provided results independent of the bacterial strain. The pattern obtained for the precipitate aged with *Micrococcus* IVA008 is displayed in Figure 4. It is mainly composed of the diffraction peaks of GR(SO$_4$$^{2-}$) that may all be clearly seen. Numerous additional, very small peaks are present, but they do not correspond to other expected Fe compounds, i.e., GR(Cl$^-$) and magnetite, that are formed in abiotic conditions (Figure 3). These peaks were likely due to the various compounds, organic and inorganic, present in the concentrated bacterial suspension introduced in the system (see Section 2.3 for instance, where the composition of the culture medium is given). Only one small peak could be tentatively identified: located at $2\theta_{hkl} = 13.756°$, i.e., $d_{hkl} = 7.47$ Å, it may correspond to the main diffraction peak of GR(CO$_3$$^{2-}$) [30,39], i.e., GRC003, as mentioned in Figure 4. NaHCO$_3$ is present in the culture medium and bacteria produce carbonate species through their metabolic activity by oxidizing organic matter.

The XRD patterns obtained for the precipitates aged with bacterial strains *Pseudoalteromonas* IIIA004 and *Bacillus* IVA016 are both displayed in Figure 5. These patterns, like the previous one, did not show any trace of the diffraction lines of GR(Cl$^-$) or magnetite. Numerous additional small peaks are also seen, located at similar positions regardless of the bacteria species. The main peak of GR(CO$_3$$^{2-}$) was not seen in the case of *Bacillus* IVA016. It was very small in the case of *Pseudoalteromonas* IIIA004, but could nonetheless be identified.

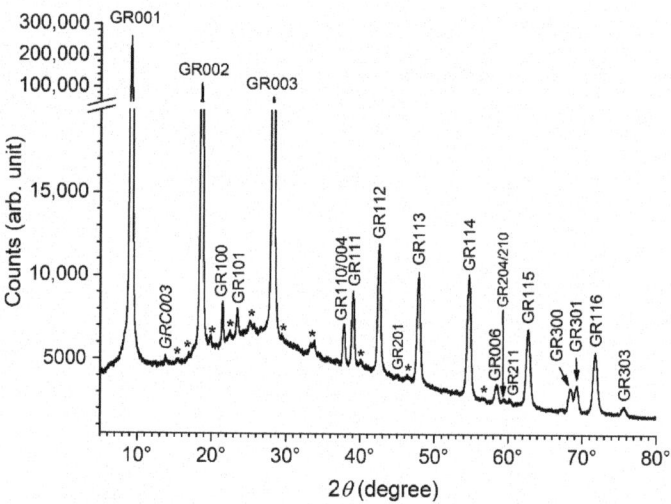

Figure 4. XRD pattern of the precipitate obtained with bacterial strain *Micrococcus* IVA008 after 1 week of ageing at RT. GR = GR(SO_4^{2-}) and GRC = GR(CO_3^{2-}), with the corresponding Miller index. * = unidentified peaks.

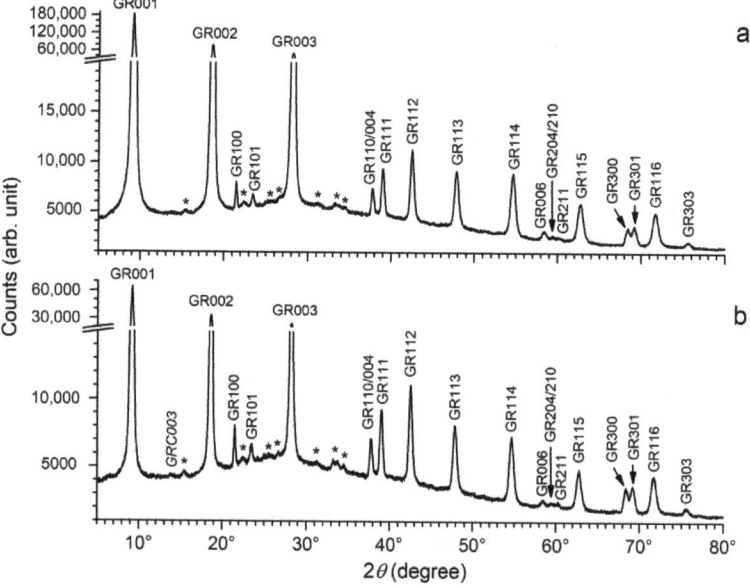

Figure 5. XRD pattern of the precipitates obtained with bacterial strain *Bacillus* IVA016 (**a**) and *Pseudoalteromonas* IIIA004 (**b**) after 1 week of ageing at RT. GR = GR(SO_4^{2-}) and GRC = GR(CO_3^{2-}), with the corresponding Miller index. * = unidentified peaks.

In conclusion, the presence of bacteria and associated organic matter prevented the formation of magnetite during ageing. The absence of GR(Cl^-) showed that the bacteria and associated organic matter either accelerated the transformation of GR(Cl^-) to GR(SO_4^{2-}) or prevented the formation of GR(Cl^-) during the precipitation reaction.

The precipitates were also aged for 2 months at RT. Once again, the results were similar for all bacterial strains. The pattern obtained for the precipitate aged 2 months with bacterial strain *Micrococcus* IVA008 is displayed in Figure 6 as an example. It was very similar to that of the precipitate aged 1 week, i.e., the main diffraction peaks were those of GR(SO_4^{2-}), and the peaks of GR(Cl^-) and magnetite were not seen. This demonstrates that the effects of the bacteria and the associated organic matter can persist for long periods. This may explain why GR(SO_4^{2-}) is favored in natural marine environments, as shown in Section 3.1.

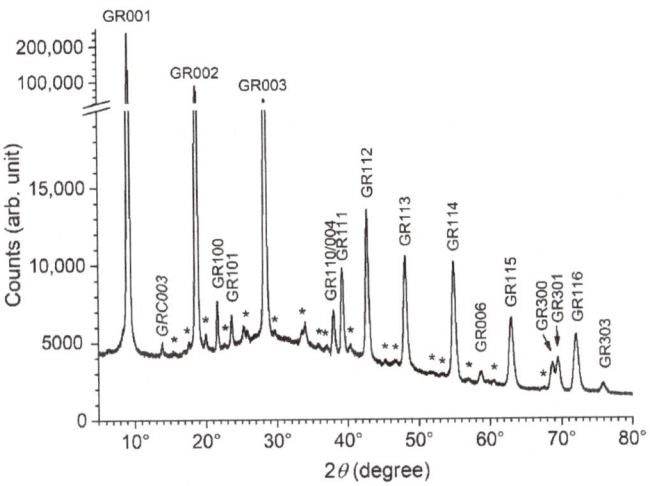

Figure 6. XRD pattern of the precipitate obtained with bacterial strain *Micrococcus* IVA008 after 2 months of ageing at RT. GR = GR(SO_4^{2-}) and GRC = GR(CO_3^{2-}), with the corresponding Miller index. * = unidentified peaks.

Finally, it can be noted that the numerous unidentified diffraction peaks, as well as the main peak of GR(CO_3^{2-}), were, compared to those of GR(SO_4^{2-}), more intense after 2 months of ageing. This can be observed visually by comparing Figures 4 and 6. To be more accurate, the intensities of the GR001 and GRC003 peaks were determined in each case by computer fitting, as described in Section 2.5. If the intensity of the GR001 peak was arbitrarily set at 100 in each case, then the intensity of the GRC003 peak slightly increased from 0.40(±0.01) to 0.45(±0.01) during ageing (from 1 week to 2 months). This may be attributed to weak bacterial activity.

3.4. Influence of Acetate Ions

The XRD pattern of the precipitate obtained with sodium acetate added as a reactant and after 1 week of ageing is displayed in Figure 7. The main diffraction peaks are once again those of GR(SO_4^{2-}), even though the acetate to sulfate concentration ratio, [CH_3COO^-]/[SO_4^{2-}], was equal to 2. Actually, it is well known that the double layered structure of GR compounds exhibits a stronger affinity for divalent anions [40,41], which explains why GR(SO_4^{2-}) forms instead of GR(Cl^-) in seawater, even though the [Cl^-]/[SO_4^{2-}] is high (about 19). The formation of GR(SO_4^{2-}) in this experiment was consistent with the findings in previous works.

As for the precipitate obtained in abiotic conditions without acetate, most of the diffraction peaks of magnetite were seen, together with the main peaks of GR(Cl^-), i.e., GRCl003 and GRCl006. From Figures 3 and 7, it can be seen that the addition of acetate decreased the intensity of the diffraction peaks of GR(Cl^-) and magnetite with respect to those of GR(SO_4^2). The data obtained via computer fitting of diffraction peaks GR001, GR112, GRCl003 and M311 confirmed this (Table 5). The intensity of the main peak of magnetite decreased from 4.6 (Table 3) to 2.0, and that of GR(Cl^-) from 7.0 to 1.8. The intensity of the GR112 lines, in contrast, increased from 3.2 to 5.9, which shows that the preferential orientation is less pronounced. If the intensity of the magnetite and GR(Cl$^-$) diffraction peaks were expressed with respect to the 112 diffraction peak of GR(SO$_4^{2-}$), the decrease due to the acetate ions would appear more significant.

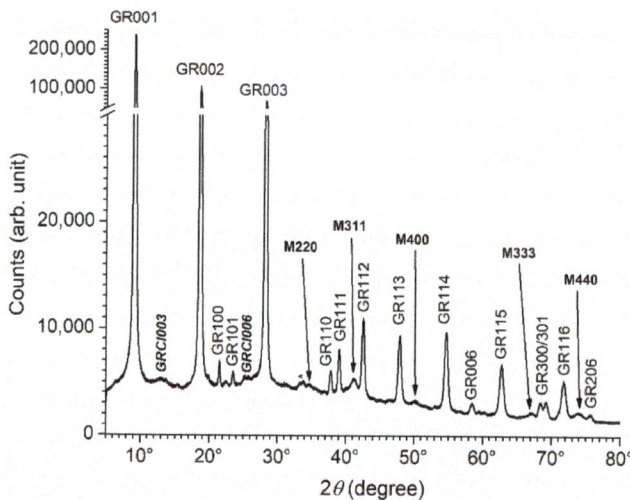

Figure 7. XRD pattern of the precipitate obtained with acetate after 1 week of ageing at RT. GR = GR(SO$_4^{2-}$), GRCl = GR(Cl$^-$), M = magnetite, with the corresponding Miller index. * = unidentified peaks.

Table 5. Precipitate obtained with acetate and aged 1 week: characteristics of the diffraction peaks GR001 and GR112 of GR(SO$_4^{2-}$), GRCl003 of GR(Cl$^-$) and M311 of magnetite; 2θ = diffraction angle, in degree, d = interplanar distance (Å), FWHM = full width a half maximum, in degree, and I = peak intensity, with I = 100 for GR001.

Diffraction Peak	2θ	d	FWHM	I
GR001	9.27 ± 0.01	11.07 ± 0.015	0.24 ± 0.01	100
GR112	42.66 ± 0.01	2.459 ± 0.001	0.47 ± 0.02	5.9 ± 0.04
GRCl003	12.91 ± 0.03	7.96 ± 0.02	1.6 ± 0.05	1.8 ± 0.01
M311	41.38 ± 0.01	2.532 ± 0.001	0.99 ± 0.02	2.0 ± 0.01

In conclusion, acetate ions induced the same effects as bacteria, but these effects were smaller and did not completely prevent the formation of magnetite and GR(Cl$^-$).

4. Discussion

GR compounds are metastable with respect to magnetite [19,42,43]. For instance, depending on pH and dissolved Fe^{2+} concentration, the precipitation of Fe^{2+} and Fe^{3+} can yield either GR(SO$_4^{2-}$) or the two-phase system Fe$_3$O$_4$ + Fe(OH)$_2$ [19]. GR compounds are also likely to be spontaneously transformed into magnetite. GR(CO$_3^{2-}$) was observed to transform spontaneously under anoxic conditions, resulting in either a mixture of magnetite and siderite (FeCO$_3$) [42] or a mixture of magnetite, chukanovite (Fe$_2$(OH)$_2$CO$_3$) and siderite [43], depending on the pH and the concentration of the carbonate species. The metastability of GR(SO$_4^{2-}$) with respect to magnetite explains why the ageing of GR(SO$_4^{2-}$) in the experimental abiotic conditions considered here induced the formation of a small proportion of Fe$_3$O$_4$.

However, it was demonstrated that the adsorption of phosphate ions on the lateral sides of the GR(CO$_3^{2-}$) particles could prevent their transformation to magnetite [44]. Similarly, lactate ions proved to have a strong effect during the oxidation of GR(SO$_4^{2-}$), which was attributed to the adsorption of the lactate ions, through their carboxyl group, on the surface of the GR crystals [37]. It can therefore be proposed that acetate ions adsorb similarly on the lateral sides of the GR(SO_4^{2-}) hexagonal platelets (Figure 8) and hinder the formation of magnetite during ageing. The organic polymeric substances associated with bacteria may therefore have similar effects. The bioreduction of lepidocrocite γ-FeOOH by *Shewanella putrefaciens* was studied, and it was observed that the formation of GR compounds was favored, with respect to the formation of magnetite, in the presence of polyacrylic acid or polyacrylamide [45]. Polyacrylic acid, which can model extracellular polymeric substances found in biofilms, was also observed to inhibit, albeit moderately, the reactivity of GR compounds towards methyl red [46]. This reactivity was assumed to mainly involve the Fe(II) reactive sites present on the lateral sides of the GR crystals. Polyacrylic acid, like acetate ions, carries a negatively charged carboxyl group. This confirms that GR compounds can be stabilized by carboxylates.

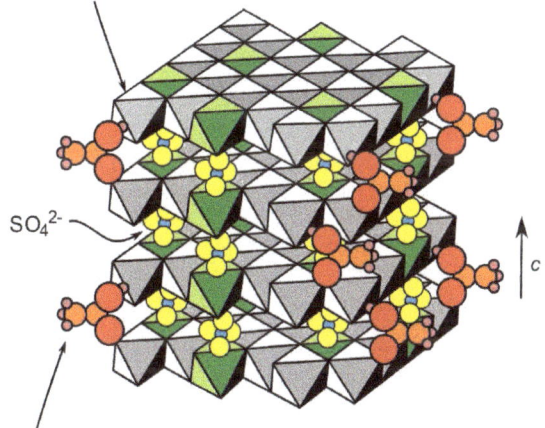

Figure 8. Schematic representation of a GR(SO_4^{2-}) crystal with adsorbed acetate ions. The interlayers are actually composed of two layers of SO_4^{2-} ions and include water molecules that were omitted for clarity. The octahedra in green are those built on Fe(III) cations.

Our results showed that the presence of bacteria cells led to stronger effects, because it prevented the formation of magnetite even after 2 months of ageing in anoxic conditions. This effect was the same whether the bacteria were dead (*Pseudoalteromonas* IIIA004) or alive and active (*Micrococcus* IVA008 and *Bacillus* IVA016). The only difference observed was the formation of a very small amount of GR(CO_3^{2-}), which resulted from the presence of NaHCO$_3$ in the culture medium and/or the oxidation of organic substances to carbonate by the microorganisms. In the study of the reactivity of GR compounds toward methyl red, it was demonstrated that bacterial cells had a stronger effect on the reactivity of GR(SO_4^{2-}) than polyacrylic acid [46]. This is fully consistent with what was observed here by comparing the results obtained in the presence of bacteria and those obtained with acetate. The stronger effect of bacterial cells and/or associated organic species, in particular with respect to the much smaller CH_3COO^- ions, may have been due to a higher steric effect that would more efficiently hinder the interaction between the GR crystal surfaces (in particular lateral sides) and solution, as illustrated schematically in Figure 8. The transformation of GR(SO_4^{2-}) to magnetite requires the release in solution of the SO_4^{2-} ions present in the interlayers of the GR structure, which was assumed to be hindered by the adsorption of anionic species on the lateral sides of the crystals [44]. Voluminous adsorbed species obviously induce a stronger barrier effect than small ones.

The presence of the bacteria, and to a lesser extent, of acetate ions, also led to the absence of GR(Cl$^-$) after 1 week of ageing. Two mechanisms can be proposed to explain this. First, it can be proposed that the bacterial cells and/or the associated organic species accelerate the transformation of GR(Cl$^-$) to GR(SO$_4^{2-}$). However, since bacteria cells tend to stabilize GR crystals, preventing their transformation, this first assumption seems unlikely. The second hypothesis is that bacterial cells and/or associated organic species inhibit the formation of GR(Cl$^-$). During the precipitation reaction, all of the anionic species present in solution may initially adsorb on the Fe-hydroxide sheets that constitute the basic elements of the GR structure [18] (see also Figure 8). In the reference abiotic experiment, only Cl$^-$ ions could compete with SO$_4^{2-}$ to adsorb on these hydroxides sheets, which led to the formation of a small amount of GR(Cl$^-$), together with GR(SO$_4^{2-}$). Over time, GR(Cl$^-$) transformed spontaneously to the more stable GR(SO$_4^{2-}$) [23]. When bacteria or acetate were added to the reactants, other species could compete with Cl$^-$ and SO$_4^{2-}$. It can therefore be postulated, in particular with bacteria and the numerous associated organic species, that the preferential adsorption of monovalent organic anionic species on the hydroxide sheets prevented the formation of GR(Cl$^-$) and favored the formation of GR(SO$_4^{2-}$), with SO$_4^{2-}$ being the main divalent available anionic species.

5. Conclusions

Aqueous suspensions of GR(SO$_4^{2-}$) were obtained by mixing a solution of Fe(II) and Fe(III) salts with a NaOH solution and ageing for 1 week in anoxic conditions. Bacterial cells (dead or still alive and active) and/or associated organic species, for instance, extracellular polymeric substances, prevented the transformation of GR(SO$_4^{2-}$) to magnetite in anoxic conditions. Similar effects, albeit less important, were observed with acetate ions added to the system. It is proposed that the adsorption of organic species on the lateral sides of the GR(SO$_4^{2-}$) crystals hinders the release of SO$_4^{2-}$ ions into solution, a process required for the transformation of GR(SO$_4^{2-}$) to magnetite. After further ageing (up to two months) in the presence of bacterial cells, magnetite was not observed either.

GR(Cl$^-$) was observed as a minor component together with GR(SO$_4^{2-}$) in the absence of bacteria or acetate. Acetate ions decreased the amount of obtained GR(Cl$^-$) whereas bacteria prevented completely its formation. This suggests that monovalent organic anionic species compete with Cl$^-$ during the precipitation reaction and prevent the formation of GR(Cl$^-$), consequently favoring that of the more stable GR(SO$_4^{2-}$).

The corrosion product layer formed on carbon steel in a natural marine environment appeared to be enriched in GR(SO$_4^{2-}$) with respect to the layer formed in artificial seawater. This may be a consequence of the interaction between GR(SO$_4^{2-}$), and, more generally, GR compounds, with bacterial cells and associated organic matter.

Author Contributions: Conceptualization, P.R., M.J., R.S., I.L. and S.S.; methodology, P.R., M.J., R.S., I.L. and S.S.; validation, P.R., M.J., R.S., I.L. and S.S.; formal analysis, J.D., J.V., P.R., M.J., I.L. and S.S.; investigation, J.D., J.V., P.R. and M.J.; data curation, P.R., M.J., I.L. and S.S.; writing—original draft preparation, P.R.; writing—review and editing, P.R., M.J., I.L., R.S. and S.S.; visualization, P.R.; supervision, P.R., M.J., I.L., R.S. and S.S.; project administration, P.R., M.J., R.S., I.L. and S.S.; funding acquisition, P.R. All authors have read and agreed to the published version of the manuscript.

Funding: This research received no external funding.

Institutional Review Board Statement: Not applicable.

Informed Consent Statement: Not applicable.

Data Availability Statement: The data presented in this study are available on request from the corresponding author.

Conflicts of Interest: The authors declare no conflict of interest.

References

1. Lee, W.; Lewandowski, Z.; Nielsen, P.H.; Hamilton, W.A. Role of sulfate-reducing bacteria in corrosion of mild steel: A review. *Biofouling* **1995**, *8*, 165–194. [CrossRef]
2. Melchers, R.E.; Wells, T. Models for the anaerobic phases of marine immersion corrosion. *Corros. Sci.* **2006**, *48*, 1791–1811. [CrossRef]
3. Melchers, R.E.; Jeffrey, R. Corrosion of long vertical steel strips in the marine tidal zone and implications for ALWC. *Corros. Sci.* **2012**, *65*, 26–36. [CrossRef]
4. Beech, I.B.; Zinkevich, V.; Tapper, R.; Gubner, R. The direct involvement of extracellular compounds from a marine sulphate-reducing bacterium in deterioration of steel. *Geomicrobiol. J.* **1998**, *15*, 119–132. [CrossRef]
5. Malard, E.; Kervadec, D.; Gil, O.; Lefevre, Y.; Malard, S. Interactions between steels and sulphide-producing bacteria-Corrosion of carbon steels and low-alloy steels in natural seawater. *Electrochim. Acta* **2008**, *54*, 8–13. [CrossRef]
6. Kumar, A.V.R.; Singh, R.; Nigam, R.K. Mössbauer spectroscopy of corrosion products of mild steel due to microbiologically influenced corrosion. *J. Radioanal. Nucl. Chem.* **1999**, *242*, 131–137. [CrossRef]
7. Mahanna, M.; Basseguy, R.; Delia, M.-L.; Girbal, L.; Demuez, M.; Bergel, A. New hypotheses for hydrogenase implication in the corrosion of mild steel. *Electrochim. Acta* **2008**, *54*, 140–147. [CrossRef]
8. Pineau, S.; Sabot, R.; Quillet, L.; Jeannin, M.; Caplat, C.; Dupont-Morral, I.; Refait, P. Formation of the Fe(II-III) hydroxysulphate green rust during marine corrosion of steel associated to molecular detection of dissimilatory sulphite-reductase. *Corros. Sci.* **2008**, *50*, 1099–1111. [CrossRef]
9. Lanneluc, I.; Langumier, M.; Sabot, R.; Jeannin, M.; Refait, P.; Sablé, S. On the bacterial communities associated with the corrosion product layer during the early stages of marine corrosion of carbon steel. *Int. Biodeterior. Biodegrad.* **2015**, *99*, 55–65. [CrossRef]
10. Smith, J.S.; Miller, J.D.A. Nature of sulfides and their corrosive effect on ferrous metals: A review. *Br. Corros. J.* **1975**, *10*, 136–143. [CrossRef]
11. Enning, D.; Venzlaff, H.; Garrelfs, J.; Dinh, H.T.; Meyer, V.; Mayrhofer, K.J.J.; Hassel, A.W.; Stratmann, M.; Widdel, F. Marine sulfate-reducing bacteria cause serious corrosion of iron under electroconductive biogenic mineral crust. *Environ. Microbiol.* **2012**, *14*, 1772–1787. [CrossRef]
12. Zegeye, A.; Huguet, L.; Abdelmoula, A.; Carteret, C.; Mullet, M.; Jorand, F. Biogenic hydroxysulfate green rust, a potential electron acceptor for SRB activity. *Geochim. Cosmochim. Acta* **2007**, *71*, 5450–5462. [CrossRef]
13. Langumier, M.; Sabot, R.; Obame-Ndong, R.; Jeannin, M.; Sablé, S.; Refait, P. Formation of Fe(III)-containing mackinawite from hydroxysulphate green rust by sulphate reducing bacteria. *Corros. Sci.* **2009**, *51*, 2694–2702. [CrossRef]
14. Duboscq, J.; Sabot, R.; Jeannin, M.; Refait, P. Localized corrosion of carbon steel in seawater: Processes occurring in cathodic zones. *Mater. Corros.* **2019**, *70*, 973–984. [CrossRef]
15. Duan, J.; Wu, S.; Zhang, X.; Huang, G.; Du, M.; Hou, B. Corrosion of carbon steel influenced by anaerobic biofilm in natural seawater. *Electrochim. Acta* **2008**, *54*, 22–28. [CrossRef]
16. Refait, P.; Nguyen, D.D.; Jeannin, M.; Sablé, S.; Langumier, M.; Sabot, R. Electrochemical formation of green rusts in deaerated seawater-like solutions. *Electrochim. Acta* **2011**, *56*, 6481–6488. [CrossRef]
17. Refait, P.; Grolleau, A.-M.; Jeannin, M.; François, E.; Sabot, R. Localized corrosion of carbon steel in marine media: Galvanic coupling and heterogeneity of the corrosion product layer. *Corros. Sci.* **2016**, *111*, 583–595. [CrossRef]
18. Refait, P.; Grolleau, A.M.; Jeannin, M.; Rémazeilles, C.; Sabot, R. Corrosion of carbon steel in marine environments: Role of the corrosion product layer. *Corros. Mater. Degrad.* **2020**, *1*, 198–218. [CrossRef]
19. Refait, P.; Géhin, A.; Abdelmoula, M.; Génin, J.-M.R. Coprecipitation thermodynamics of iron(II–III) hydroxysulphate green rust from Fe(II) and Fe(III) salts. *Corros. Sci.* **2003**, *45*, 656–676. [CrossRef]
20. Simon, L.; François, M.; Refait, P.; Renaudin, G.; Lelaurain, L.; Génin, J.M. Structure of the Fe(II-III) layered double hydroxysulphate green rust two from Rietveld analysis. *Solid State Sci.* **2003**, *5*, 327–334. [CrossRef]
21. Detournay, J.; De Miranda, L.; Dérie, R.; Ghodsi, M. The region of stability of green rust II in the electrochemical potential-pH equilibrium diagram of iron in sulphate medium. *Corros. Sci.* **1975**, *15*, 295–306. [CrossRef]
22. Olowe, A.A.; Génin, J.-M.R. The mechanism of oxidation of ferrous hydroxide in sulphated aqueous media: Importance of the initial ratio of reactants. *Corros. Sci.* **1991**, *32*, 965–984. [CrossRef]
23. Refait, P.; Sabot, R.; Jeannin, M. Role of Al(III) and Cr(III) on the formation and oxidation of the Fe(II–III) hydroxysulfate green rust. *Colloids Surf. A Phys. Eng. Asp.* **2017**, *531*, 203–212. [CrossRef]
24. Ona-Nguema, G.; Carteret, C.; Benali, O.; Abdelmoula, M.; Génin, J.-M.R.; Jorand, F. Competitive formation of hydroxycarbonate green rust 1 vs hydroxysulphate green rust 2 in *Shewanella putrefaciens* cultures. *Geomicrobiol. J.* **2004**, *21*, 79–90. [CrossRef]
25. Zegeye, A.; Ona-Nguema, G.; Carteret, C.; Huguet, L.; Abdelmoula, M.; Jorand, F. Formation of hydroxysulphate green rust 2 as a single iron(II-III) mineral in microbial culture. *Geomicrobiol. J.* **2005**, *22*, 389–399. [CrossRef]
26. *ASTM D1141-98(2013)*; Standard Practice for the Preparation of Substitute Ocean Water. ASTM International: West Conshohocken, PA, USA, 2013.
27. De Faria, D.L.A.; Silva, S.V.; Oliveira, M.T.D. Raman micro spectroscopy study of some iron oxides and oxyhydroxides. *J. Raman Spectrosc.* **1997**, *28*, 873–878. [CrossRef]
28. Shebanova, O.N.; Lazor, P. Raman study of magnetite (Fe_3O_4): Laser-induced thermal effects and oxidation. *J. Raman Spectrosc.* **2003**, *34*, 845–852. [CrossRef]

29. Refait, P.; Duboscq, J.; Aggoun, K.; Sabot, R.; Jeannin, M. Influence of Mg^{2+} ions on the formation of green rust compounds in simulated marine environments. *Corros. Mater. Degrad.* **2021**, *2*, 46–61. [CrossRef]
30. Hansen, H.C.B. Composition, stabilisation, and light absorption of Fe(II)-Fe(III) hydroxycarbonate (green rust). *Clay Miner.* **1989**, *24*, 663–669. [CrossRef]
31. Refait, P.; Grolleau, A.-M.; Jeannin, M.; François, E.; Sabot, R. Corrosion of carbon steel at the mud zone/seawater interface: Mechanisms and kinetics. *Corros. Sci.* **2018**, *130*, 76–84. [CrossRef]
32. Chicot, D.; Mendoza, J.; Zaoui, A.; Louis, G.; Lepingle, V.; Roudet, F.; Lesage, J. Mechanical properties of magnetite (Fe_3O_4), hematite (α-Fe_2O_3) and goethite (α-FeO·OH) by instrumented indentation and molecular dynamics analysis. *Mater. Chem. Phys.* **2011**, *129*, 862–870. [CrossRef]
33. Tomic, Z.; Makreski, P.; Gajic, B. Identification and spectra–structure determination of soil minerals: Raman study supported by IR spectroscopy and X-ray powder diffraction. *J. Raman Spectrosc.* **2010**, *41*, 582–586. [CrossRef]
34. Ohfuji, H.; Rickard, D. High resolution transmission electron microscopic study of synthetic nanocrystalline mackinawite. *Earth Planet. Sci. Lett.* **2006**, *241*, 227–233. [CrossRef]
35. Bourdoiseau, J.A.; Jeannin, M.; Sabot, R.; Rémazeilles, C.; Refait, P. Characterisation of mackinawite by Raman spectroscopy: Effects of crystallisation, drying and oxidation. *Corros. Sci.* **2008**, *50*, 3247–3255. [CrossRef]
36. Legrand, L.; Sagon, G.; Lecomte, S.; Chausse, A.; Messina, R. A Raman and infrared study of a new carbonate green rust obtained by electrochemical way. *Corros. Sci.* **2001**, *43*, 1739–1749. [CrossRef]
37. Sabot, R.; Jeannin, M.; Gadouleau, M.; Guo, Q.; Sicre, E.; Refait, P. Influence of lactate ions on the formation of rust. *Corros. Sci.* **2007**, *49*, 1610–1624. [CrossRef]
38. Refait, P.; Abdelmoula, M.; Génin, J.-M.R. Mechanisms of formation and structure of green rust one in aqueous corrosion of iron in the presence of chloride ions. *Corros. Sci.* **1998**, *40*, 1547–1560. [CrossRef]
39. McGill, J.R.; McEnaney, B.; Smith, D.C. Crystal structure of green rust formed by corrosion of cast iron. *Nature* **1976**, *259*, 200–201. [CrossRef]
40. Miyata, S. Anion-exchange properties of hydrotalcite-like compounds. *Clays Clay Miner.* **1983**, *31*, 305–311. [CrossRef]
41. Mendiboure, A.; Schöllhorn, R. Formation and anion exchange reactions of layered transition metal hydroxides $[Ni_{1-x}M_x](OH)_2(CO_3)_{x/2}(H_2O)_z$ (M=Fe,Co). *Rev. Chim. Miner.* **1986**, *23*, 819–827. [CrossRef]
42. Benali, O.; Abdelmoula, M.; Refait, P.; Génin, J.-M.R. Effect of orthophosphate on the oxidation products of Fe(II)-Fe(III) hydroxycarbonate; the transformation of green rust to ferrihydrite. *Geochim. Cosmochim. Acta* **2001**, *65*, 1715–1726. [CrossRef]
43. Refait, P.; Reffass, M.; Landoulsi, J.; Sabot, R.; Jeannin, M. Role of nitrite species during the formation and transformation of the Fe(II-III) hydroxycarbonate Green Rust. *Colloids Surf. A Phys. Eng. Asp.* **2014**, *459*, 225–232. [CrossRef]
44. Bocher, F.; Géhin, A.; Ruby, C.; Ghanbaja, J.; Abdelmoula, M.; Génin, J.-M.R. Coprecipitation of Fe(II-III) hydroxycarbonate green rust stabilised by phosphate adsorption. *Solid State Sci.* **2004**, *6*, 117–124. [CrossRef]
45. Jorand, F.P.A.; Sergent, A.-S.; Rémy, P.-P.; Bihannic, I.; Ghanbaja, J.; Lartiges, B.; Hanna, K.; Zegeye, A. Contribution of anionic vs. neutral polymers to the formation of green rust 1 from γ-FeOOH bioreduction. *Geomicrobiol. J.* **2013**, *30*, 600–615. [CrossRef]
46. Rémy, P.-P.; Etique, M.; Hazotte, A.A.; Sergent, A.-S.; Estrade, N.; Cloquet, C.; Hanna, K.; Jorand, F.P.A. Pseudo-first-order reaction of chemically and biologically formed green rusts with Hg^{II} and $C_{15}H_{15}N_3O_2$: Effects of pH and stabilizing agents (phosphate, silicate, polyacrylic acid, and bacterial cells). *Water Res.* **2015**, *70*, 266–278. [CrossRef] [PubMed]

Localized Corrosion of Mooring Chain Steel in Seawater

Xiaolong Zhang [1,*], Nanni Noël-Hermes [1], Gabriele Ferrari [1] and Martijn Hoogeland [2]

1 Endures BV, Bevesierweg 1 DC002, 1781 CA Den Helder, The Netherlands; nanni.noel@endures.nl (N.N.-H.); g.ferrari@quicknet.nl (G.F.)
2 TNO, Leeghwaterstraat 44, 2628 CA Delft, The Netherlands; martijn.hoogeland@tno.nl
* Correspondence: xiaolong.zhang@endures.nl

Abstract: Corrosion of mooring chains is regarded as one of main threats to the offshore mooring systems. Localized corrosion is even more dangerous than uniform corrosion because it may not show significant mass loss but it can cause stress concentration and initiate cracks under force, leading to accelerated degradation of mooring chains. Localized corrosion of steel in seawater is influenced by many factors such as the local heterogeneities of the steel, and the local electrochemical and microbiological environments. It is difficult to predict and the mechanism is not fully understood. The aim of this work was to study the mechanism of localized corrosion on mooring chain steel in seawater which is helpful in the search for corresponding monitoring tools and mitigation methods. The corrosion behavior of chain steel grade R4 was studied in artificial seawater and artificial seawater containing microorganisms collected from a practice field. The corrosion behavior of the steel was studied using different techniques such as potentiodynamic polarization, linear polarization resistance measurements and electrochemical impedance spectroscopy. The microstructures such as inclusions and compositions of the chain steel were studied using SEM: Scanning Electron Microscope and EDS: Energy Dispersive Spectroscopy. The microbial cells were observed using epi-fluorescence microscopy. The corrosion morphology and pit geometry were investigated using photo-microscopy. The localized corrosion rate has been found to be much higher than the uniform corrosion rate of the steel in the seawater in the presence of bacteria. In the case of localized corrosion, applying uniform corrosion measurement techniques and formulas is not considered representative. The representative areas have to be introduced to match physical results with the measurements. Inclusions, such as MnS and TiVCr found in the steel have a critical influence on localized corrosion. The corrosion mechanism of the steel in seawater is discussed.

Keywords: localized corrosion; mooring chain; MIC; SEM; steel

1. Introduction

Mooring chains are widely used to fix a floating production, storage and offloading (FPSO) system. Mooring chain steel has to withstand seawater corrosion and cyclic force loading during service. Marine corrosion, in particular localized corrosion, combined with mechanical loading is the main reason for mooring chain failures [1–4]. Ma et al. compiled a historical review of integrity issues of mooring systems [1]. They found that the chain, connector and wire rope are the top three components causing more incidents. Fontaine et al. undertook an industry survey of past failures and degradations for mooring systems of floating production units. They found that almost half of all failures were associated with chains and two out of three chain failures were related to corrosion and fatigue [2].

Many factors affect the corrosion of metals in marine environments. Marine environments include a number of zones, such as atmospheric, splash, tidal, submerged and bottom sediment areas. Some recovered chain links showed uniform corrosion in the splash zone and pitting corrosion in the submerged near-surface zone [5]. Microbial activity in concentrations around and below the low water zone leads to an aggressive form of concentrated corrosion, known as accelerated low water corrosion, which has been identified as microbiologically influenced corrosion (MIC) [6,7].

The steels for making mooring chains are classified so far by specified minimum ultimate tensile strength into five grades (R3, R3s, R4, R4s and R5) [8]. To achieve the required strength the chemical compositions and manufacture processes have to comply with the approved specifications [8,9]. Even so, the steels showed different corrosion rates in different seawater areas. It was reported that 120 mm diameter R4 chain serviced in the North Sea located in the Pierce field for 13 years has a corrosion rate of 0.53 mm/y of diameter reduction in the pitted area (for the worst case) [10]. Fontaine al. inspected 76 mm-diameter chain links of type R3 and ORQ grades in West African waters [11]. The long-term corrosion rate of these links was approximately 1.5 mm/y of dimeter reduction in the pitted areas, which is significantly higher than the corrosion wear allowance of 1 mm/y required for tropic waters [12,13].

Local corrosion attack may initiate at inclusions or grain boundaries due to a local electrochemical potential difference [14–16]. Avci et al. investigated MnS-mediated pit initiation and propagation in carbon steel in an anaerobic sulfidogenic media. They found that pitting on 1018 carbon steel was initiated within a 20–30 nm zone at the MnS inclusion boundary [17].

Jeffery et al. investigated the effect of microbiological involvement on the topography of corroding mild steel in coastal seawater and found that microbiological factors are responsible for the more severe pitting observed on the natural seawater coupons [18].

Melchers et al. investigated the corrosion of a working chain continuously immersed in seawater and developed a model to predict short- and long-term corrosion rates based on general corrosion loss [19]. For pitting corrosion, field data are necessary for calibration of the pit depth model [20].

A long-term field exposure test is a simple and valid method to verify the long-term corrosion performance of the steel but it is a time-consuming process. Mass loss gives an average corrosion rate which cannot reflect the localized corrosion rate. Sample surface analysis is necessary. Corrosion in seawater is an electrochemical process. Is there any electrochemical method that can be used to monitor the corrosion of offshore steel structures?

Potentiodynamic polarization (PDP) curve measurements under sliding were employed by López et al. to investigate the tribocorrosion of mooring high–strength, low-alloy steels (grade R4 and R5) in synthetic seawater [21]. Based on the mass loss it was found that both the R4 and R5 steels have the same triboccorosoion behaviour in seawater. The PDP method is a destructive method that does not fit long-term monitoring.

Linear polarization resistance (LPR) and electrochemical impedance spectroscopy (EIS) are less destructive (compared to PDP) to the system to be studied since only a small dc or ac potential amplitude is applied. These techniques only give corrosion resistance of the steel. To convert the corrosion resistance to corrosion rate, Tafel slopes are necessary and are obtained from the PDP curves. Using these electrochemical techniques combined with microbiological and surface analytical techniques may be helpful in studying MIC [22].

So far, localized corrosion has been difficult to predict, and the mechanism is not fully understood. The aim of this work was to study the mechanism of localized corrosion on mooring chain steel in seawater which is helpful for the search of corresponding monitoring tools and mitigation methods.

Investigation into the mechanism of local corrosion was carried out by exposing samples of mooring chain steel grade R4 (named R4 according to the International Classification Society of Offshore Systems) in artificial seawater (SW) in the laboratory. During and after exposure, electrochemical and microstructure analyses were performed. Since MIC is expected to be one of the main causes of local corrosion, tests were also run with the addition of microorganisms cultured in the laboratory. These microorganisms, collected at Makassar Strait (Indonesia), contained different types of corrosive organisms.

The corrosion behavior of the steel in seawater was investigated using PDP, LPR measurements and EIS. Microbial attachment and biofilm formation were studied using fluorescent dye and epi-fluorescence microscopy [23]. The surface microstructures and compositions were analyzed using SEM and EDS, the corrosion morphology using photo-microscopy.

2. Experimental

The experiments were designed to investigate local corrosion and how that is influenced by micro-organisms. This means that electrochemical measurements and microbial growth needed to be combined. For practical relevance, chain steel material as served in the North Sea was used and a bacterial culture was enriched from a representative offshore site. Various methods were applied to measure and analyze the results which are described below.

2.1. Materials

Steel samples were cut from a chain link (R4, Φ120 mm) used in the North Sea at a depth of 85 m for 13 years, provided by a project partner. The nominal composition of the steel is presented in Table 1. The dimensions of steel samples were $25 \times 20 \times 10$ mm^3. A copper wire was connected with the steel sample for an electrical connection. Steel samples and connections were embedded in epoxy resin and polished using sandpaper up to 1200 grit. Then the samples were cleaned ultrasonically in ethanol for 2 min and blow dried in air.

Table 1. Nominal composition of the grade R4 steel in wt.% (the rest is Fe).

QR4	C	Mn	Si	P	S	Ni	Mo	Cr	Al	Cu	Sn	V	Ti	As
min.	0.18	0.85	0.15			0.50	0.20	0.90	0.015			0.04		
max.	0.24	1.20	0.35	0.020	0.015	0.80	0.40	1.25	0.040	0.25	0.030	0.10	0.015	0.025

2.2. Experimental Set-Up

In Figure 1 a typical vessel for test exposure is shown. Each vessel contains two identical steel samples (duplicates, as working electrodes (WEs)) and a platinum counter electrode (CE). A reference electrode (KCl saturated Ag/AgCl) was inserted just before electrochemical measurements and taken out after the measurements. The reference electrode was always cleaned in alcohol before it was inserted in a glass bottle through the hole in the rubber cover to prevent the interference of microorganisms from outside of the vessels. The vessels were closed during the tests. In Table 2 an overview of the test methods, including samples, electrolytes (artificial seawater (SW) and artificial seawater with addition of bacteria (SW + bacteria (MIC))), and test period is given. The steel samples P1 and P2 were for PDP measurements, samples B1–B4 for combined LPR and EIS measurements, and samples with initial code S for only LPR measurements.

Table 2. Experimental methods and conditions.

Samples	SW	SW + Bacteria (MIC)	Test Duration (Day)
P1, P2	PDP		
B1, B2		LPR + EIS	28
B3, B4	LPR + EIS		28
S5, S6	LPR		21
S8, S9		LPR	7
S10, S11		LPR	21
S12, S13		LPR	28
S14, S15	LPR		7

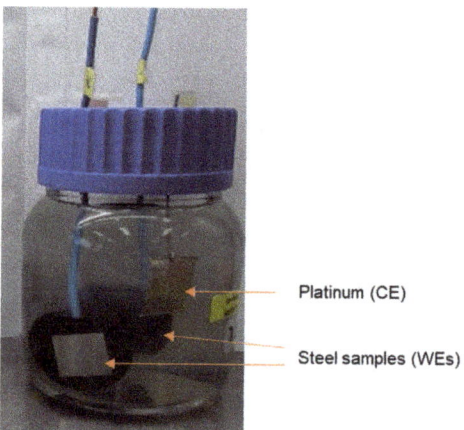

Figure 1. A typical sample exposure vessel for linear polarization resistance (LPR) and electrochemical impedance spectroscopy (EIS) measurements.

Samples were exposed to stagnant electrolytes (600 mL) at room temperature. The pH of the solutions at the start was 8.0. The electrolytes were described in Section 2.3. Using stagnant electrolytes aimed to decrease disturbance to the growth of biofilms.

Different test durations were designed to study how biofilm and corrosion develop in different time durations, e.g., tests for samples S8, S9, S14 and S15 were stopped after exposure for 7 days and they were taken out for surface analysis. This information may help us to understand how biofilms build up and about the initiation and progress of localized corrosion is in different media.

2.3. Electrolytes

Two types of electrolytes were used in this study. One electrolyte was low-nutrient loaded artificial seawater (SW) that was used for exposure of samples without MIC contribution. The SW was freshly prepared using chemicals presented in Table 3 [24].

Table 3. Chemical concentration (g/L) in the artificial seawater.

NaCl	$MgCl_2$	$CaCl_2$	Na_2SO_4	KCl	$NaHCO_3$	KBr	H_3BO_3	NaF	$SrCl_2$	Yeast	Lactate
23.93	5.07	1.15	4.01	0.68	0.197	0.099	0.03	0.01	0.14	0.01	4.2 mL

The other electrolyte was SW+bacteria (MIC), which was for investigating the susceptibility to MIC. In this electrolyte, microorganisms collected at Makassar Strait and cultured in the laboratory were added to the artificial seawater.

To support the bacteria to grow, an additional nutrient supply of 0.004 g $FeSO_4 \cdot 7H_2O$, 0.30g $Na_3C_6H_5O_7 \cdot 2H_2O$ and 0.10 g $C_6H_8O_6$ dissolved in 10 mL deionised water was added through a 0.2 µm pore size filter.

Moreover, once a week 250 mL of the electrolyte was exchanged with fresh solution to supply enough nutrients for a continuous microbial growth.

2.4. Inoculation of Microorganisms

Different groups of corrosion relevant microorganisms have been detected from mooring chain environment in Makassar Strait and were enriched under laboratory conditions. Bacteria included sulfate-reducing bacteria (SRB), iron-reducing bacteria, sulfur-oxidizing bacteria, acid-producing bacteria, slime-formers and manganese-oxidizing bacteria. The bacteria were grown in specific media to keep them active until the start of the experiment. A total amount of 6×10^6 cells/mL were added to the vessel (counted using a Thoma counting chamber).

2.5. Electrochemical Measurements

2.5.1. Potentiodynamic Polarization (PDP) Curve Measurements

The PDP curve measurements were performed to obtain the Tafel slopes of polarization curves to calculate the corrosion current density and corrosion rate. These slopes were used to calculate the corrosion rate from the corrosion resistance measured by the LPR and EIS measurements. The PDP measurements were carried out after holding the cells at open circuit for 1 h and the open circuit potential (OCP) was measured in SW, open to air. The polarization curves were measured by scanning the potential, started at −0.25 V vs. OCP and ended at 0.35 V vs. OCP. The scan rate was 0.167 mV/s.

2.5.2. Linear Polarization Resistance (LPR) and Electrochemical Impedance Spectroscopy (EIS) Measurements

The LPR measurements were performed in closed vessels (see Figure 1) after holding open the circuit for 0.5 h, and OCP was measured. A linear polarization line was scanned from −0.01 V vs OCP to +0.01 V vs OCP. The EIS were measured using ac, amplitude 0.01 V, in frequency range 0.01–100,000 Hz. The LPR and EIS measurements were carried out after 7, 14, 21 and/or 28 days of exposure.

2.6. Surface Analysis

2.6.1. Epi-Fluorescence Microscopy

After the exposure test, samples were taken out and stained by a fluorescent dye to discriminate cells in active (green) or inactive (red) cells. Stained microbial cells were made visible by exciting the DNA/stain with ~490 nm blue light and observing the emitted green or red fluorescence under the microscope. In the case of no bacteria, no fluorescence will be detected.

2.6.2. Photo-Microscopy

The exposed sample surfaces were cleaned first in 15% HCl solution with the addition of 0.5% hexamethylenetetramine for 10 min, then rinsed in tap water, ultrasonically in alcohol for 2 min. and finally dried in blowing air. The topography of the exposed samples was analysed using optical microscopy to see if corrosion took place uniformly or locally at sample surfaces. The optical microscopes used were Olympus (DP200, for low magnification) and Leica (Reichert MEF 4 M, for high magnification) with the Infinity X camera and DeltaPix software. In the case of localized corrosion, pit depth was first measured manually using microscope Leica by turning the "Fine adjustment" from focusing on the sample surface to focusing on the bottom of the pits. The depth was calculated from the turning scales on the "Fine adjustment" which was calibrated. The deepest pits were chosen to gain a cross-sectional view in order to measure the real depth to make sure no deposit at the pit bottom might hinder the light reaching the real bottom. Four deep pits on each sample were measured by cross-sectional view.

2.6.3. Scanning Electron Microscopy (SEM)

The steel sample surface was first ground using grinding paper (SiC) till 2500 grit and then polished up to 1 μm. The microstructure such as inclusions and the compositions at the steel surface were analysed using SEM in combination with energy-dispersive spectroscopy (EDS). SEM was undertaken using a Jeol JSM 5800LV instrument equipped with a Noran instrument EDS system.

3. Results
3.1. Electrochemical Measurements
3.1.1. PDP Curve Measurements

Figure 2 shows polarization curves in semilogarithmic plots for both polished samples P1 and P2. Small (Tafel) slopes in the anodic polarization parts and large slopes in the cathodic polarization parts are observed, approximately 50 mV more active than the corrosion potentials. This means that the corrosion is controlled by the cathodic reactions for the steel in the seawater.

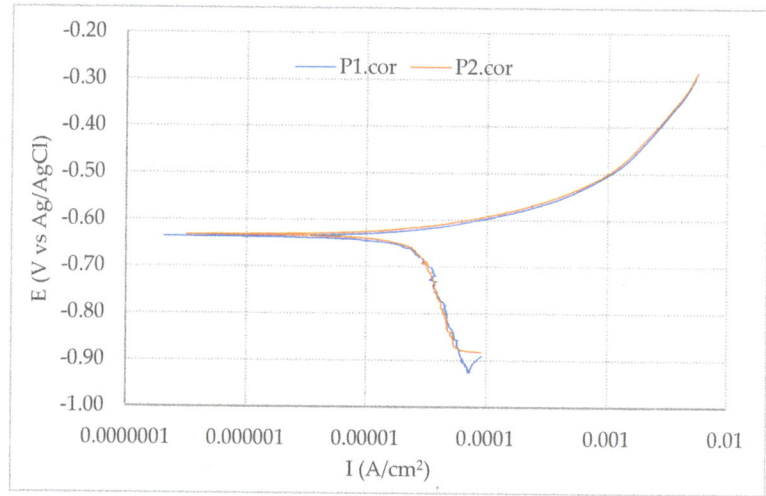

Figure 2. Polarization curves for the polished samples in artificial seawater (open to air) at 20 °C.

The anodic reaction is iron dissolution [25],

$$Fe \Rightarrow Fe^{2+} + 2e^- \qquad (1)$$

The cathodic reactions include here mainly the reduction of dissolved oxygen,

$$O_2 + 2H_2O + 4e^- \Rightarrow 4OH^- \qquad (2)$$

and, possibly, hydrogen evolution in anaerobic environments, e.g., under depsoits and biofilm

$$2H^+ + 2e^- \Rightarrow H_2 \qquad (3)$$

or,

$$2H_2O + 2e^- \Rightarrow H_2 + 2OH^- \qquad (4)$$

Corrosion current density i_{corr} can be calculated from the Tafel slopes in the polarization curves using the Stern–Geary equation [26],

$$i_{corr} = b_a \times b_c / (2.3 \cdot R_p \cdot (b_a + b_c)) \qquad (5)$$

where b_a and b_c are anodic and cathodic slopes, respectively, in the polarization curves; R_p is the polarization resistance.

The general corrosion rates (CR) can be calculated from the corrosion current densities using the following equation,

$$CR = 3267 \cdot (i_{corr} \cdot M_{eq})/\rho \ (mm/y) \qquad (6)$$

where i_{corr} is current density (A/cm^2), M_{eq} equivalent mass (g), and ρ density of the materials (g/cm^3).

The corrosion potentials and corrosion current densities are presented in Table 4. The average corrosion rate is about 0.3 mm/y. The corresponding corrosion resistance is about 1030 $\Omega \cdot$cm^2 (as a reference for further comparison with LPR results).

Table 4. Corrosion potentials and corrosion rates for the polished samples in seawater.

Steel	E_c (V)	i_{corr} (μA/cm^2)	b_a (mV/dec)	b_c (mV/dec)	CR (mm/y)
P1	−0.64	26	68	618	0.30
P2	−0.63	25	65	676	0.29

3.1.2. LPR Measurements

The OCP values as a function of time for the steel in the different electrolytes are shown in Figure 3. The OCP values for the samples in the seawater without bacteria are around −0.6 V. With bacteria, the OCP values moved from −0.67 V to −0.6 V within 7 days. This suggests that the corrosion systems are unstable in the initial period (within 1 week).

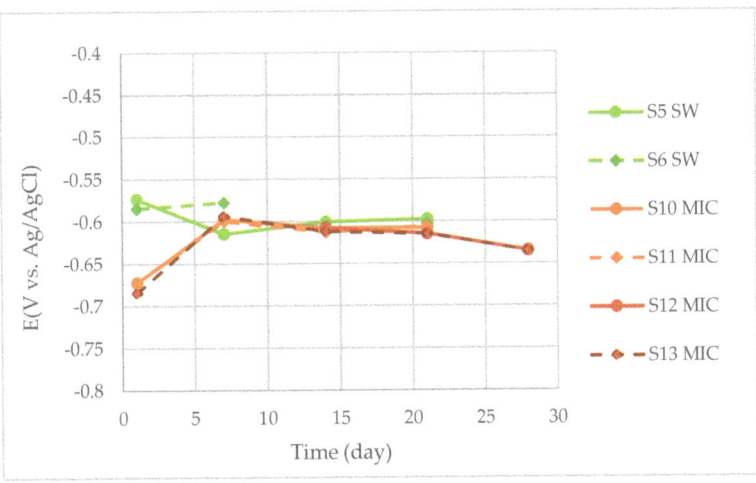

Figure 3. Open circuit potential (OCP) values as a function of time measured for R4 steel samples in seawater with and without bacteria.

A typical linear polarization (LP) curve for a sample (S5) in SW exposed for 14 days is shown in Figure 4. The LPR ($R_p = \Delta E/\Delta I$) was calculated from the line in ±5 mV near the corrosion potential (zero current).

A typical LP curve for a sample (S10) in the SW + bacteria is shown in Figure 5. The linear range of the LP curve near the corrosion potential (−0.61 V) is much narrower compared to that of the S5 in SW near the corrosion potential (zero current). In the presence of bacteria, biofilms formed at the sample surfaces, which exhibit capacitive behaviour. In this case, the corrosion resistance was estimated in the anodic part.

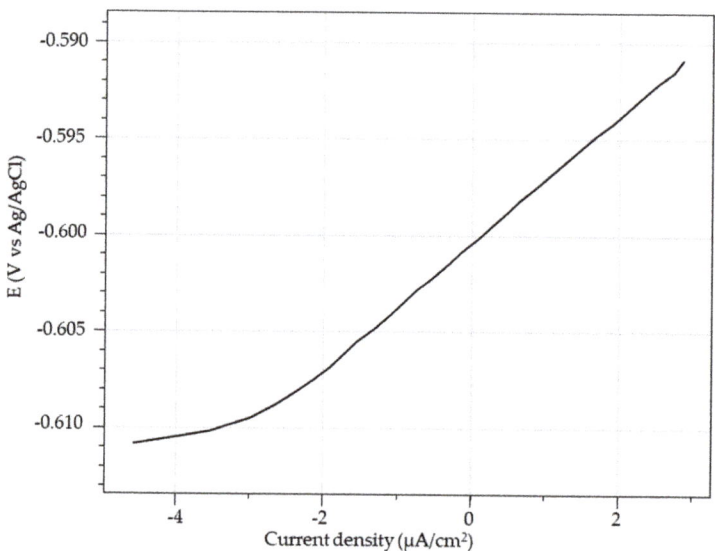

Figure 4. A typical linear polarization curve for a sample (S5) in seawater (SW), exposed for 14 days.

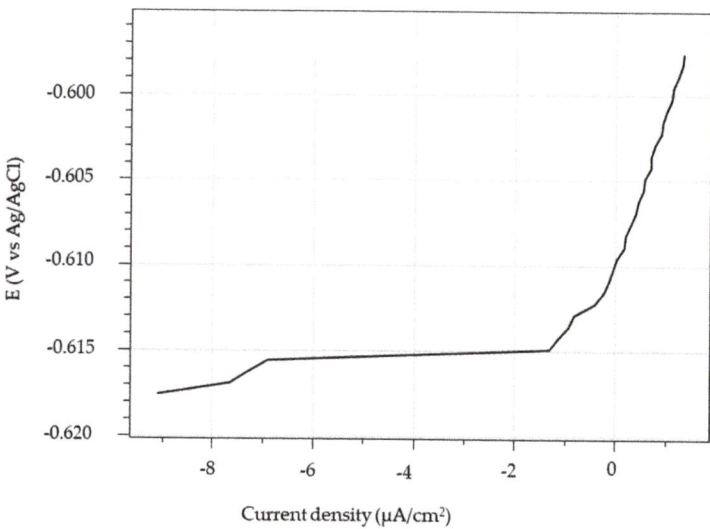

Figure 5. A typical polarization curve for a sample (S10) in SW+bacteria (MIC), exposed for 14 days.

Figure 6 shows the polarization resistance after various exposure times measured for R4 steel samples in SW and in SW+bacteria (MIC). The polarization resistance for the samples in the seawater without bacteria (S5-6 SW) increased from 2.4 kΩ·cm^2 to 3.6 kΩ·cm^2, while for the samples in the seawater with bacteria (MIC) it decreased from 10 kΩ·cm^2

to 1 kΩ·cm² in 21 days. This indicates that the sample surfaces exposed to seawater with bacteria became more active in 3 weeks, compared to those exposed to seawater only. Originally the designed LPR test duration was for 21 days. Samples S12 and S13 were added and extended the test duration to 28 days to see if their LPR would further decrease in the SW + bacteria. (No sample was added to SW for 28 days, since the LPR in SW did not change much within 21 days).

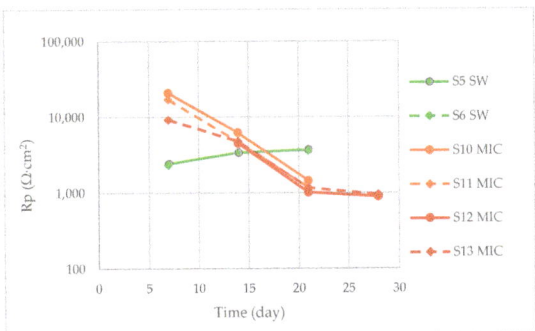

Figure 6. Polarization resistance measured by LPR as a function of time measured for R4 steel samples in seawater with and without bacteria. S5 and S6 in SW showed the same LPR values.

3.1.3. EIS Measurements

To better study the corrosion resistance, EIS measurements were performed for samples B1–B4. The OCP values for the samples varied around −0.62 V (±0.05 V).

Figure 7 shows the Nyquist plots (a) and Bode plots (b) for the samples exposed to SW and SW+bacteria (MIC) for 14 days. The semicircles of B3 and B4 (in SW) in the Nyquist plots are larger than those of B1 and B2 in SW+bacteria (MIC). The amplitude of the impedance for the samples (B3–B4) in SW is higher than in the SW+bacteria (B1–B2 MIC) at low frequency side (0.01 Hz). The phase peaks shifted to the low-frequency side for the samples in the SW with bacteria, which suggests that the capacitive behaviour is significant.

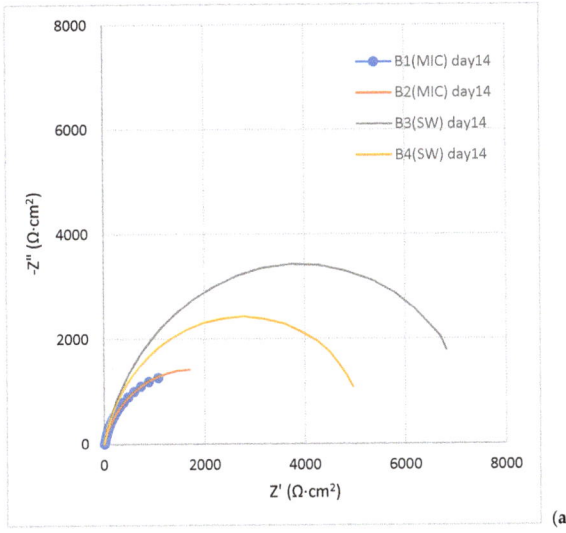

(a)

Figure 7. Cont.

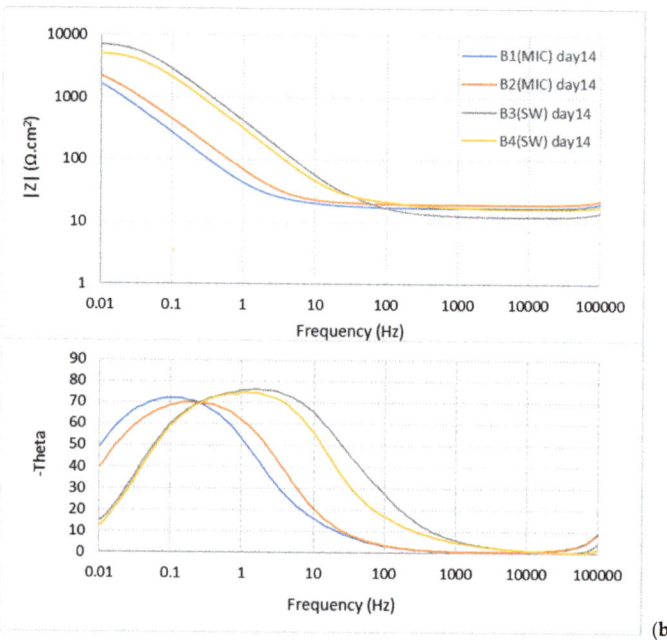

Figure 7. Nyquist (**a**) and Bode (**b**) impedance plots for samples exposed to SW (B3, B4) and to SW + bacteria (B1, B2 MIC) for 14 days.

Figure 8 shows a comparison of Nyquist (a) and Bode (b) impedance plots of samples exposed to SW+bacteria (B1 MIC) and exposed to SW (B3) for different time durations. The impedance module at 0.01 Hz decreased over time in both conditions. The phase angle peak of B1 (MIC) shifted in the low-frequency direction with the increase of duration.

The impedance of the samples at a frequency of 0.01 Hz measured with different exposure durations is presented in Table 5. The electrochemical impedance value at the low-frequency side is related to the corrosion resistance of the steel. The samples exposed to the seawater with microorganisms have smaller impedance values and larger phase angle than without microorganisms.

Table 5. Impedance values and phase angles at 0.01 Hz.

Sample	Solution	Time (Day)	$|Z|$ ($\Omega \cdot cm^2$)	Theta
B1	SW + bacteria	14	1629	−49
		21	980	−66
		28	392	−73.2
B2	SW + bacteria	14	2268	−40
		21	478	−74.6
		28	460	−73
B3	SW	14	7036	−14.7
		21	5038	−12.5
		28	3227	−14.9
B4	SW	14	5093	−12.3
		21	3880	−13
		28	3275	−16

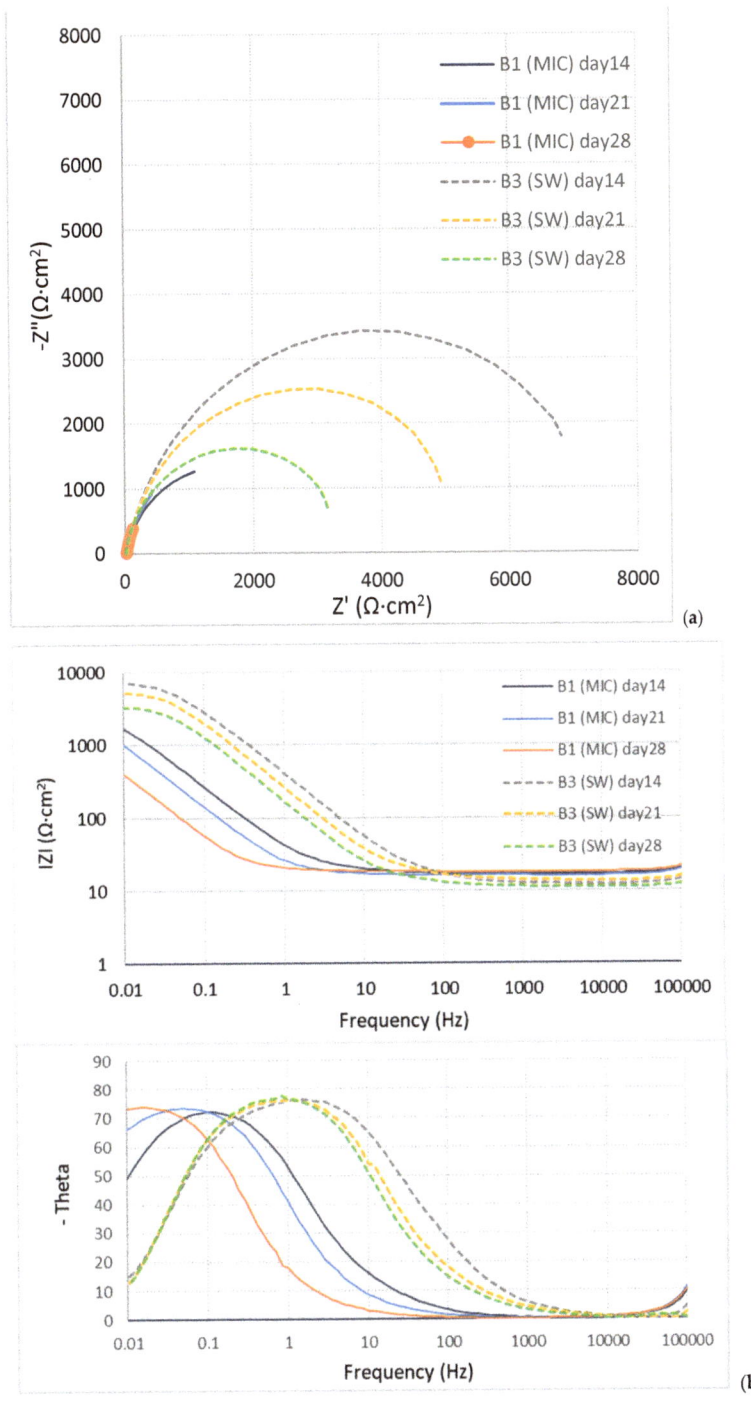

Figure 8. A comparison of Nyquist (**a**) Bode (**b**) impedance plots of samples exposed to SW + bacteria (B1 (MIC)) and exposed to SW (B3) for different time durations.

The impedance module can be influenced by resistance, capacitance and even inductance in a corrosion cell. To further analyse the capacitive and resistive behaviour of the corrosion cells, the impedance data were fitted with an equivalent circuit presented in Figure 9. The capacitive elements are submitted by constant phase elements (CPE) Qc and Qdl. The impedance of a CPE can be calculated by the equation:

$$Z_{CPE} = Y_0^{-1}(j\omega)^{-n} \tag{7}$$

where Y_0 is the admittance constant of the CPE (in s^n/Ω); ω is the angular frequency (rad/s); n is the CPE exponent, and $n = \alpha/(\pi/2)$ (α is the constant phase angle of the CPE). When n = 1, the CPE becomes a pure capacitor [27].

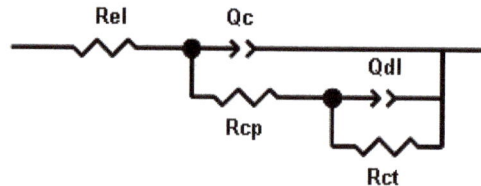

Rel: electrolyte resistance,
Qc: constant phase element for the oxide layer,
Rcp: pore resistance,
Qdl: constant phase element for the double layer,
Rct: charge transfer resistance.

Figure 9. Equivalent circuit used for fitting the impedance data.

The fitting results are presented in Table 6. The fitting results show that the resistance attributed to the surface layer (Rcp) is very small, compared to the charge transfer resistance Rct. Thus, the polarization resistance is in the same order of the Rct. After exposure for 28 days the corrosion resistance is approaching the same level (3.5 kΩ·cm^2) for the samples in SW and in SW+bacteria (MIC). The corrosion resistance was calculated using apparent surface area of the samples, since the real active corrosion area was unknown.

Table 6. Parameters and fitting results of the impedance data using an equivalent circuit, R in Ω·cm^2 and C in (F·cm^{-2}).

Sample	Time (Day)	C_c	n_1	R_{cp}	C_{dl}	n_2	R_{ct}	χ^2 ($\times 10^{-4}$)
B1 (MIC)	14	0.00227	0.893	15	0.00125	0.899	3210.8	0.6
	21	0.00673	0.869	17	0.00211	0.9998	4328.6	1.2
	28	0.01239	1	3	0.01007	0.868	3624.8	3
B2 (MIC)	14	0.00104	1	3	0.00052	0.808	3654	0.9
	21	0.00754	0.931	4	0.01378	0.932	4216.5	8
	28	0.01590	0.997	7	0.00410	0.814	4477.5	1.8
B3 (SW)	14	0.00020	0.88	39	0.00007	0.965	7825.5	4.5
	21	0.00023	0.932	14	0.00018	0.922	5494.3	8
	28	0.00038	0.94	10	0.00030	0.93	3487.4	5.9
B4 (SW)	14	0.00015	0.872	12	0.00019	0.943	5512.6	5.2
	21	0.00021	0.922	12	0.00031	0.931	4222.6	4.4
	28	0.00043	0.931	11	0.00038	0.93	3569	3.6

The capacitance was also calculated ($C = Y_0^{1/n} \cdot R^{(1-n)/n}$, [28]), using apparent surface area. The steel samples in SW+bacteria have a larger capacitance of the double layer, than in SW. A larger capacitance results in a lower impedance module at the low-frequency side (Table 5).

3.2. Surface Analysis

3.2.1. Epifluorescence Microscopy

After 7 days of incubation, microorganisms were regularly found on the metal coupon surface. Biofilms covered the damaged area. Active cells (green) were mostly found in and around the pits. The outer part of the biofilm is inactive because the surface was often covered by red (inactive) cells (Figure 10). This means that microorganisms initially attached to the entire surface but could only grow in limited areas where they could form a biofilm. These preferred spots for microbial attachment may contain the right (metallic) nutrients to encourage the growth of bacteria or deliver attractive sites for attachment. The electrons given out by iron at a corrosion spot can be harvested by an SRB film via extracellular electron transfer, which accelerates the cathodic depolarization [29]. In such locations they play a role in the local corrosion process.

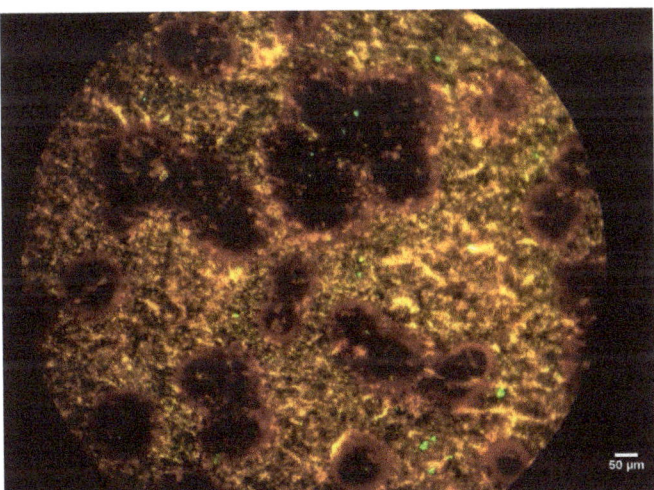

Figure 10. Micrograph of metal coupon (S9) after 7 days of exposure (before cleaning). A biofilm was detected on the metal coupon surface. Active cells (green) were mostly found in the pits whereas red (inactive) cells were located around the pits.

3.2.2. Corrosion Morphology

Steel in Seawater without Bacteria

After exposure for 7 days samples S8, S9, S14 and S15 were taken out for surface analysis. After cleaning, the surface of sample S15 is shown in Figure 11a. The upper part of the sample corroded less than the lower part, e.g., little corrosion started on the upper-left corner in the image, where the original polishing pattern can be recognized. This is attributable to the fact that the oxygen is more easily accessible near the water surface than in the lower part. Localized corrosion is visible on the bottom-left corner (Figure 11a), although a large area of the sample S15 showed general corrosion. Small pits are visible in the magnified image Figure 11b. These small pits were in the initial stage. Inside the pits corrosion involved metal hydrolysis and pH decrease. The surrounding area of pits acted as cathodes. The halos surrounding the small pits in the Figure 11b are evidence. The small pits may grow in depth and laterally, becoming big pits or connecting in surface area as general corrosion.

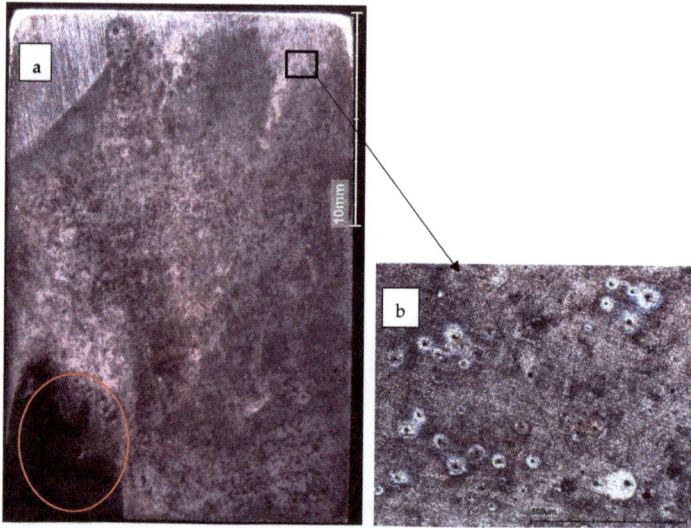

Figure 11. An image for the sample S15 exposed to seawater for 7 days (after cleaning the surface). Local corrosion attack was found near the bottom-left corner (**a**) and small pits near the upper-right corner zoomed in the image (**b**).

Figure 12 shows an image for the sample B3 exposed to seawater for 28 days after cleaning. The whole surface was corroded. Most areas of the steel surface showed general corrosion. The different colour on the surface is due to the rough surface after corrosion attack and possibly remaining deposits which were not completely removed in the standard cleaning procedure. Localized corrosion spots are visible at the sample surface. The corrosion products rolling down from the upper part affected the corrosion at the lower part of the sample (e.g., lower right corner in Figure 12).

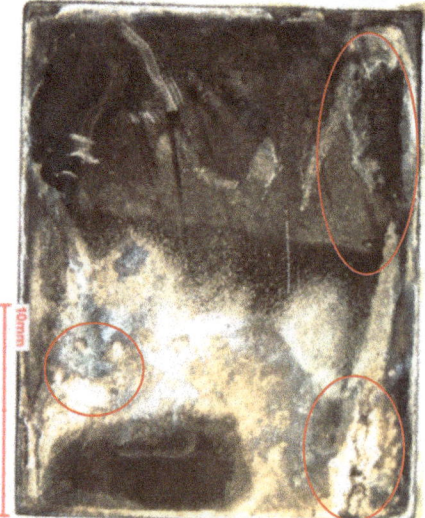

Figure 12. An image for sample B3 exposed to seawater for 28 days (after cleaning). Localized corrosion attacks were found in the steel (red circle). The lower part of the steel sample was affected by corrosion products rolling down.

The maximum depth of the pits was measured as 46 μm in sample B3 (Figure 13). It corresponds to 0.6 mm/y, assuming that the local corrosion attack keeps occurring at the same speed.

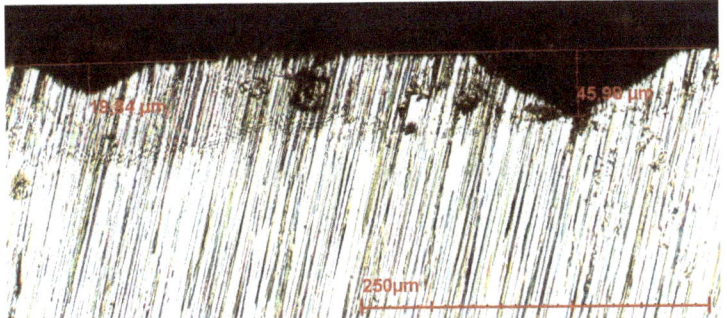

Figure 13. Cross-sectional view at pits for samples B3 in SW for 28 days.

Steel in Seawater with Bacteria

For sample S9 (Figure 14) exposed to SW + bacteria for 7 days, many small pits and a few big pits (about ϕ 0.6 mm) were found. The maximum depth was about 13 μm. More than 80% of the sample surface area showed micro pits; only about 5% surface area showed uniform corrosion; about 10% of the area did not show corrosion.

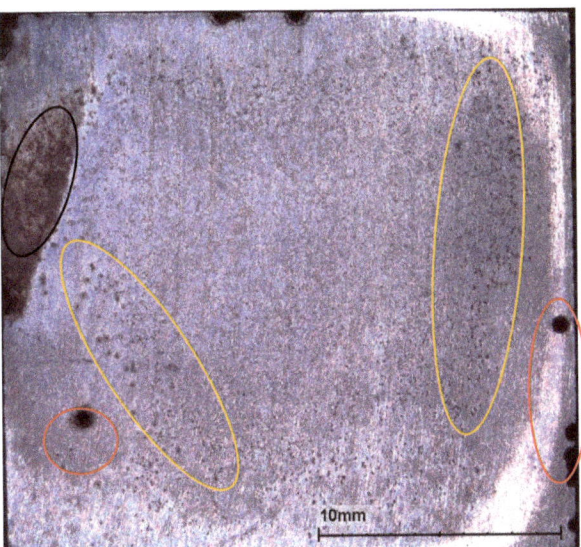

Figure 14. An image for the sample S9 exposed to SW + bacteria for 7 days (after cleaning the surface). Many small pits (yellow, diameter < 0.1 mm, >80% area) and a few big pits (red, diameter > 0.5 mm, ~3% area) were in the steel. Uniform corrosion took place in a small area (black, ~5% area).

With the addition of bacteria in SW, typical localized corrosion due to MIC was observed (Figure 15). The corrosion boundary has a round shape. Approximately 40% surface area was corroded. Assuming one third of surface area was corroding, the real corrosion current density would be three times of the average corrosion current density, which means the corrosion rate would be three times of the measured average corrosion rate.

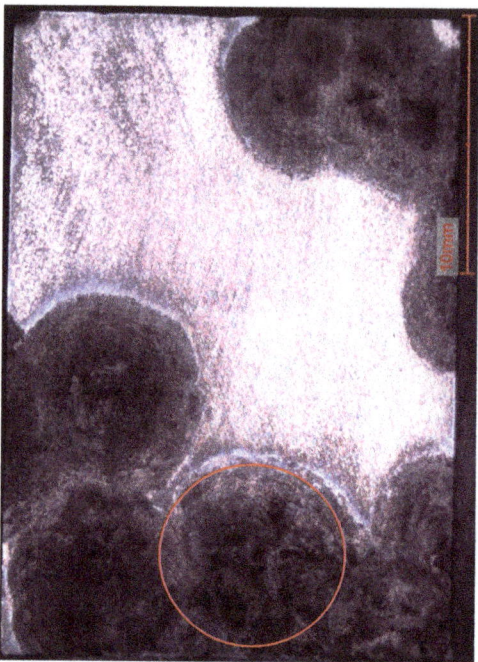

Figure 15. An image of sample S13 exposed to seawater with bacteria for 28 days (after cleaning). Half of the surface area was attacked by local corrosion.

Similar corrosion morphology was found for the samples B1 and B2 exposed to SW+bacteria (MIC) for 28 days. The maximum pit depth is 63 μm (Figure 16), which corresponds to 0.82 mm/y, assuming that the corrosion keeps occurring at the same speed. This corrosion rate is higher than that measured in field assessment (0.53 mm/y in diameter reduction in the worst case), which is attributable to those corrosive bacteria added in the seawater and the temperature in the lab test being higher than in North Sea waters.

Figure 16. Cross-sectional view at two pits for sample B2 after exposure to the SW + bacteria for 28 days. The maximum pit depth is 63 μm.

The relative percentage of the corroded area and maximum pit depth in samples B1–B4 were evaluated using photo microscopy and are given in Table 7. The areas of micro-pits and macro-pits were estimated by taking a few images and adjusting the black/white contrast. The samples (B1–B2) exposed to SW+bacteria showed a greater percentage of localized corrosion area, while the samples (B3–B4) exposed to SW showed a higher percentage of uniform corrosion and micro-pit area.

Table 7. Percentage of corrosion area (%) and maximum pit depth of samples B1–B4.

Sample	Condition	Time (Day)	Uniform Corrosion	Macro-Pits	Micro-Pits	Corrosion Products Affected Area	Intact Area	Maximum Pit Depth (µm)
B1	MIC	28	-	30	70	-	-	42
B2	MIC	28	-	20	70	-	10	63
B3	SW	28	30	10	50	10	-	46
B4	SW	28	40	10	50	-	-	39

3.2.3. Microstructure and Inclusions in the Steel

Figure 17 shows the microstructure of the steel. It shows a typical fine grain microstructure, composed of tempered martensite and bainite.

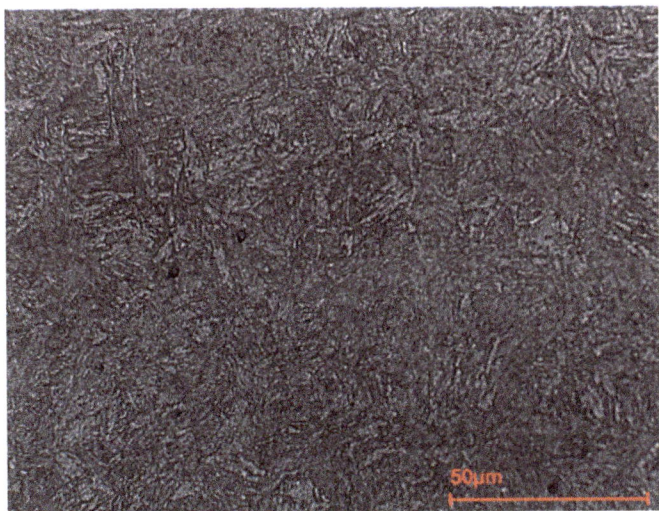

Figure 17. Microstructure of the chain grade R4 steel, composed of tempered martensite and bainite.

Inclusions have been found in the steel sample. Figure 18 shows a backscatter image at a cross section (a) and EDS plots at a particle (b) and nearby area (c) for an exposed steel sample (S8). The composition analysed using EDS (see Table 8) indicates that the dark-grey particle in Figure 18a is an MnS inclusion. This inclusion is about 20 µm in length and 5 µm in width. The elements detected at position (3) are related to the composition of the steel matrix.

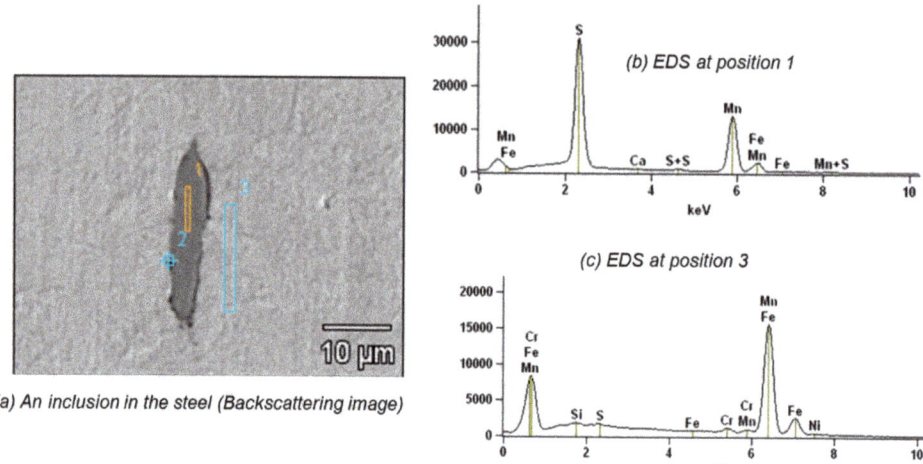

Figure 18. An inclusion (backscattering image (a), EDS plots in the position 1 (b) and at position 3 (c) for an exposed sample (S8) after exposure to seawater without bacteria for 7 days.

Table 8. Element percentage (wt. %) measured at positions in Figure 18a.

Element	Si	S	Ca	Cr	Mn	Fe	Ni
Position 1	6.2	38.0	0.2		53.2	2.4	
Position 3	0.5	0.4		2.0	1.2	95.3	0.6

TiVCr-enriched particles were also found in this steel sample (see Figure 19 and Table 9). These particles are known for initiating local corrosion due to their potential difference with regard to that of the surrounding steel matrix.

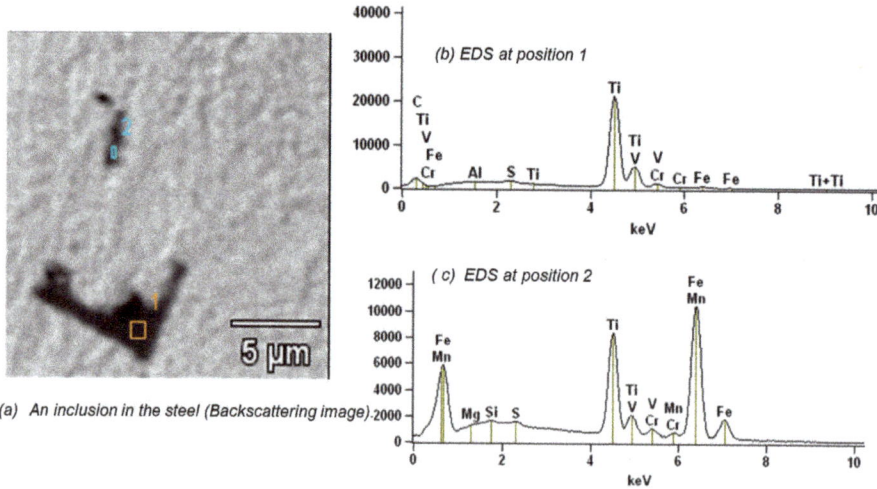

Figure 19. An inclusion (backscattering image (a), EDS plots in the position 1 (b) and at position 3 (c) for an exposed sample (S8) after exposure to seawater without bacteria for 7 days.

Table 9. Element percentage (wt.%) measured at positions in Figure 19a.

Element	C	Mg	Al	Si	S	Ti	V	Cr	Mn	Fe
Point 1	9.1		0.1		1.0	75.0	8.6	3.5		2.7
Point 2		0.2		0.3	0.5	24.6	1.7	1.8	1.2	69.7

4. Discussion

Mooring chain steel samples were exposed to artificial seawater and artificial seawater with bacteria for different durations. Different techniques were used to study the localized corrosion phenomena associated with these conditions.

PDP curve measurements show that the fresh steel surface has an average corrosion rate of 0.3 mm/y in seawater (open to air). However, the corrosion rate changes with time because of the surface condition changes over time.

In the presence of micro-organisms, OCP showed a large scatter during the first period (a few days) of exposure. This has to be attributed to irregular attachment of the organisms at the surface and formation of biofilm which disturbs the balance of electrochemical reactions. After 2 to 3 weeks all samples reached a stable value of circa -0.6 $V_{Ag/AgCl}$ which is usually measured for this steel in seawater.

The LPR (R_p) measured (in closed vessels) was relatively stable during exposure to SW in the absence of microorganisms. In the closed system the corrosion rate measured using LPR was one third of that measured using PDP curves in an open system. The Rp value 3.6 kΩ·cm^2 corresponds with circa 0.1 mm/y corrosion rate for this steel in seawater, in case general corrosion is assumed. On the other hand, in the presence of microorganisms the Rp decreased over time. In all cases good reproducibility was found, but questions were raised about the meaning of the Rp in the case of local corrosion, the subject of this study. This parameter and the way to measure it is completely based on the theory of uniform corrosion. In fact, pits were found in all samples, and that means localized corrosion. The local corrosion rate cannot be established via Rp since the representative area is not clear. If only one third of the surface area is corroding, the real corrosion rate will be three times the measured average corrosion rate.

The reason for applying OCP and LPR in this investigation was that these techniques are relatively easy to perform and therefore suitable to be used for in situ monitoring. However, LPR creates errors due to the non-linearity when biofilms are present at steel surface. EIS gives not only information of polarization resistance, but also information of capacitive behaviour of the surface layers. The polarization resistance measured by EIS was approaching 3.5 kΩ·cm^2 in 28 days. The capacitance for the steel in the SW with bacteria was larger than in SW without bacteria, which could be attributable to a larger charged surface area in the presence of biofilm. Thus, EIS gives more information about the surface conditions, although more data fitting is needed. Next to EIS measurements, a detailed pit analysis is also required.

In all cases pits were found in the exposed steel surfaces. A distinction can be made between two different types of pits:

(a) Relatively small pits which occur in large areas of the steel. Local corrosion attacks initiate at defects such as the grain boundaries or inclusions. According to literature these "micro" pits are formed very quickly after immersion. Most of these pits reach the depth of 100–200 µm and then stop propagating [20,30]. Pits can continue their growth only under a layer of corrosion products or biofilms. The observed pits are the locations where anodic reactions occur, the rest of the surface being the cathodic part. Sometimes after the start of exposure, corrosion products are formed in the pits increasing the electrical resistance and inhibiting the access of oxidizing agents. Then the reactions stop and start elsewhere, but with lower driving forces.

(b) A limited number of clearly larger pits is found on the surface of all samples. In contrast to the small pits described above, these pits are found in limited locations. These relatively large corrosion spots initiate from small pits and grow in depth and laterally due to high local driving forces such as local metallic inclusions. Relatively large inclusions are found in the steel (MnS and TiVCr, 5–20 µm, see Figures 18 and 19); inclusions are known to have different potentials with regard to the matrix and cause local galvanic corrosion. Therefore, it is obvious that a link exists between the inclusions found and the large pits.

The exposure of coupons without bacteria indicated the aforementioned formation of pits. The steel was fine-grain treated. It contains Al, Ti, V, Cr, Ni and Mo alloy elements, apart from Mn. The microstructure was composed of martensite and bainite. A tiny difference in local chemical difference can initiate small pits as demonstrated in Figure 11. The large pits and the underlying corrosion mechanism is attributed to the exists of inclusions. The pit size depends on the geometry and orientation of the inclusions. The inclusions were not uniformly distributed. The number of inclusions per unit area was not determined in this work. The correlation between the locations of the localized corrosion and inclusions deserves further investigation.

The important question is if the "large" pits will propagate because of MIC or other local causes such as, for example, oxygen depletion ("crevice corrosion"). In the presence of microorganisms, biofilms are formed on the surface. Biofilms can include elements which contribute to the corrosion mechanism but can also function as a barrier to oxygen. One mechanism of MIC is the oxygen differential cell formed under the biofilm which accelerates the local corrosion. Results of epi-fluorescence microscopy showed local concentrations of active organisms (near pits), which implies also local activity. Thus, it is evidence that active organisms preferentially settle in the neighbourhood of pits indicating their possible role in the corrosion process.

Concerning the mechanisms of MIC, a number of theories and models are reported, such as cathodic depolarization theory (CDT), iron sulphide mechanism, anodic depolarization, biomineralization, Romero's mechanism etc. [31]. However, Blackwood examined the CDT theory and reported that both the CDT and direct electron transfer from the metal into the cell for the role of SRB in the corrosion of carbon steel were incorrect [32]. The MIC process is so complicated that to understand the mechanism needs more effort by materials scientists, electrochemists and biologists working together [32,33].

Results of this study, in particular those of the surface investigation after exposure, prove that surface properties of the steel have an essential role in starting a local corrosion attack. The microstructure and composition heterogeneities at the matrix such as grain boundary (which will be investigated in future) and inclusions generate local corrosion cells because a small difference in composition or microstructures generates an electrochemical difference (e.g., potential difference). These local corrosion cells are likely onsets of local corrosion.

5. Conclusions

From the results described above conclusions can be drawn on the corrosion behaviour of R4 steel in seawater:

(a) Localized corrosion has been found in the absence as well as in the presence of microorganisms, and occurs from the start of the exposure.

(b) Inclusions of MnS and TiVCr have been detected in the R4 steel. These inclusions formed during the manufacture of the chain steel have a critical influence on the local corrosion attack.

(c) With the addition of bacteria, already after 7 days of incubation an active biofilm was detected on the surface of the coupons with favoured locations in and around the pits.

(d) The localized corrosion rate was as high as 0.82 mm/y in the SW in the presence of bacteria. In the case of local corrosion, applying uniform corrosion measurement techniques and formulas are not considered representative. This paper shows that representative areas have to be introduced to match physical results with the measurements.

Author Contributions: Conceptualization, X.Z. and G.F.; methodology, X.Z., N.N.-H. and G.F.; validation, X.Z. and N.N.-H.; formal analysis, X.Z. and N.N.-H.; investigation, X.Z. and N.N.-H.; resources, M.H.; data curation, X.Z.; writing—original draft preparation, X.Z. and G.F.; writing—review and editing, X.Z., N.N.-H. and M.H.; project administration, M.H.; funding acquisition, G.F. and M.H. All authors have read and agreed to the published version of the manuscript.

Funding: This research was co-funded by TKI Maritiem grant number TKIM-2014-47, Localized Mooring Chain Corrosion (LMCC). The partners are Bluewater, SBM, Bumi Armada, SOFEC, Asian Star, Franklin, Boskalis, Corrosion, TNO, ABS and DNV.

Institutional Review Board Statement: Not applicable.

Informed Consent Statement: Not applicable.

Data Availability Statement: Not applicable.

Acknowledgments: The authors acknowledge the project partners for the financial support, materials supply and fruitful discussions.

Conflicts of Interest: The authors declare no conflict of interest.

References

1. Ma, K.T.; Shu, H.; Smedley, P.; L'Hostis, D.; Duggal, A. *A Historical Review on Integrity Issues of Permanent Mooring Systems*; OnePetro: Houston, TX, USA, 2013.
2. Fontaine, E.; Kilner, A.; Carra, C.; Washington, D.; Ma, K.T.; Phadke, A.; Laskowski, D.; Kusinski, G. *Industry Survey of Past Failure, Pre-Emptive Replacements and Reported Degradations for Mooring Systems of Floating Production Units*; OnePetro: Houston, TX, USA, 2014; p. 14.
3. Zhang, X.; Hoogeland, M. Influence of deformation on corrosion of mooring chain steel in seawater. *Mater. Corros.* **2019**, *70*, 962–972. [CrossRef]
4. Melchers, R.E.; Jeffrey, R.; Fontaine, E. Corrosion and the structural safety of FPSO mooring systems in Tropical waters. In *Australasian Structural Engineering Conference 2012: The Past, Present and Future of Structural Engineering*; Engineers Australia: Barton, Australian, 2012; p. 748.
5. Fontaine, E.; Rosen, J.; Potts, A.; Ma, K.T.; Melchers, R. SCORCH JIP-Feedback on MIC and Pitting Corrosion from Field Recovered Mooring Chain Links. In Proceedings of the Offshore Technology Conference, Houston, TX, USA, 8 May 2014.
6. Little, B.J.; Ray, R.I.; Lee, J.S. Microbiologically influenced corrosion of pilings. *VTS Navig. Mooring Berthing* **2013**, *60*, 69–71.
7. GL Noble Denton. *Microbiologically Influenced Corrosion of Mooring Systems for Floating Offshore Installations*; HSE: London, UK, 2017.
8. Veritas, D.N. *Offshore Mooring Chain*; Offshore Standard DNV-OS-E302; DNV: Bærum, Norway, 2013.
9. ABS. Guide for the Certification of Offshore Mooring Chain. 2009. Available online: https://ww2.eagle.org/content/dam/eagle/rules-and-guides/current/survey_and_inspection/39_certificationoffshoremooringchain_2017/Mooring_Chain_Guide_e-May17.pdf (accessed on 11 October 2021).
10. Vicinay. Haewene Brim Chain Accessment, Report. 2013. Available online: https://vicinayinnovacion.com/case-studies/ (accessed on 11 October 2021).
11. Fontaine, E.; Potts, A.; Ma, K.T.; Arredondo, A.; Melchers, R.E. SCORCH JIP: Examination and Testing of Severely-Corroded Mooring Chains from West Africa. In Proceedings of the Offshore Technology Conference, Houston, TX, USA, 30 April 2012.
12. DNVGL. Position Mooring, Offshore Standards. 2018. Available online: https://rules.dnv.com/docs/pdf/DNV/os/2018-07/dnvgl-os-e301.pdf (accessed on 11 October 2021).
13. Shu, H.; Yao, A.; Ma, K.T.; Ma, W.; Miller, J. API RP 2SK 4th Edition-An Updated Stationkeeping Standard for the Global Offshore Environment. In Proceedings of the Offshore Technology Conference, Houston, TX, USA, 30 April 2018.
14. Melchers, R.E.; Chaves, I.A.; Jeffrey, R. A Conceptual Model for the Interaction between Carbon Content and Manganese Sulphide Inclusions in the Short-Term Seawater Corrosion of Low Carbon Steel. *Metals* **2016**, *6*, 132. [CrossRef]
15. Man, C.; Dong, C.; Xiao, K.; Yu, Q.; Li, X. The Combined Effect of Chemical and Structural Factors on Pitting Corrosion Induced by MnS-(Cr, Mn, Al)O Duplex Inclusions. *Corrosion* **2017**, *74*, 312–325. [CrossRef]
16. Li, T.; Wu, J.; Frankel, G. Localized corrosion: Passive film breakdown vs. Pit growth stability, Part VI: Pit dissolution kinetics of different alloys and a model for pitting and repassivation potentials. *Corros. Sci.* **2021**, *182*, 109277. [CrossRef]
17. Avci, R.; Davis, B.; Wolfenden, M.; Beech, I.; Lucas, K.; Paul, D. Mechanism of MnS-mediated pit initiation and propagation in carbon steel in an anaerobic sulfidogenic media. *Corros. Sci.* **2013**, *76*, 267–274. [CrossRef]
18. Jeffrey, R.J.; Melchers, R.E. The effect of microbiological involvement on the topography of corroding mild steel in coastal seawater. In Proceedings of the CORROSION 2010, San Antorio, TX, USA, 14 March 2010.
19. Melchers, R.E.; Moan, T.; Gao, Z. Corrosion of working chains continuously immersed in seawater. *J. Mar. Sci. Technol.* **2007**, *12*, 102–110. [CrossRef]
20. Melchers, R.E. Pitting corrosion of mild steel in marine immersion environment-Part 1: Maximum pit depth. *Corrosion* **2014**, *60*, 824–836. [CrossRef]

21. López-Ortega, A.; Bayón, R.; Pagano, F.; Igartua, A.; Arredondo, A.; Arana, J.; Gonzalez, J.J. Tribocorrosion behaviour of mooring high strength low alloy steels in synthetic seawater. *Wear* **2015**, *338-339*, 1–10. [CrossRef]
22. Mansfeld, F.; Little, B. A technical review of electrochemical techniques applied to microbiologically influenced corrosion. *Corros. Sci.* **1991**, *32*, 247–272. [CrossRef]
23. Webb, D.J.; Brown, C.M. Epi-Fluorescence Microscopy. In *Cell Imaging Techniques*; Taatjes, D.J., Roth, J., Eds.; Humana Press Totowa, NJ, USA, 2012; pp. 29–59.
24. Heyer, A. *Microbiological Influenced Corrosion in Ship Ballast Tanks*; Haveka Holding, B.V.: Alblasserdam, Zuid-Holland, The Netherlands, 2013; ISBN 978-90-820590-0-7.
25. Uhlig, H.H. *Corrosion and Corrosion Control*, 2nd ed.; John Wiley & Sons Inc.: Hoboken, NJ, USA, 1971.
26. Stern, M.; Geary, A.L. Electrochemical Polarization 1. Theoretical analysis of the shape and polarization curves. *J. Electrochem. Soc.* **1957**, *104*, 56–63. [CrossRef]
27. van Westing, E.P.M.; Ferrari, G.M.; De Wit, J.H.W. The determination of coating performance with impedance measurements -I. Coating polymer properties. *Corros. Sci.* **1993**, *34*, 1511–1530. [CrossRef]
28. Buchheit, R.G.; Cunningham, M.; Jensen, H.; Kendig, M.W. A Correlation between Salt Spray and Electrochemical Impedance Spectroscopy Test Results for Conversion-Coated Aluminum Alloys. *Corrosion* **1998**, *54*, 61–69. [CrossRef]
29. Gu, T.; Jia, R.; Unsal, T.; Xu, D. Toward a better understanding of microbiologically influenced corrosion caused by sulfate reducing bacteria. *J. Mater. Sci. Technol.* **2018**, *35*, 631–636. [CrossRef]
30. Butler, G.; Stretton, P.; Beynon, J.G. Initiation and growth of pits on high-purity iron and its alloys with chromium and copper in neutral chloride solutions. *Br. Corros. J.* **1972**, *7*, 168–173. [CrossRef]
31. Khan, M.A.A.; Hussain, M.; Djavanroodi, F. Microbiologically influenced corrosion in oil and gas industries: A review. *Int. J. Corros. Scale Inhib.* **2021**, *10*, 80–106. [CrossRef]
32. Blackwood, D.J. An Electrochemist Perspective of Microbiologically Influenced Corrosion. *Corros. Mater. Degrad.* **2018**, *1*, 5. [CrossRef]
33. Little, B.; Blackwood, D.; Hinks, J.; Lauro, F.; Marsili, E.; Okamoto, A.; Rice, S.; Wade, S.; Flemming, H.-C. Microbially influenced corrosion—Any progress? *Corros. Sci.* **2020**, *170*, 108641. [CrossRef]

Article

Influence of Mg²⁺ Ions on the Formation of Green Rust Compounds in Simulated Marine Environments

Philippe Refait *, Julien Duboscq, Kahina Aggoun, René Sabot and Marc Jeannin

Laboratoire des Sciences de l'Ingénieur pour l'Environnement (LaSIE), UMR 7356 CNRS-La Rochelle Université, Av. Michel Crépeau, CEDEX 01, F-17042 La Rochelle, France; julien.duboscq50@gmail.com (J.D.); kahina.aggoun@univ-lr.fr (K.A.); rsabot@univ-lr.fr (R.S.); mjeannin@univ-lr.fr (M.J.)
* Correspondence: prefait@univ-lr.fr; Tel.: +33-5-46-45-82-27

Abstract: Green rust compounds (GR), i.e., Fe(II-III) layered double hydroxides, are important transient compounds resulting from the corrosion of steel in seawater. The sulfated variety, GR(SO$_4^{2-}$), was reported as one of the main components of the corrosion product layer, while the chloride variety, GR(Cl$^-$), was more rarely observed. The carbonate variety, GR(CO$_3^{2-}$), is favored by an increase in pH and forms preferentially in the cathodic areas of the metal surface. Since Mg(II) is abundant in seawater, it may have a strong influence on the formation of GR compounds, in particular as it can be incorporated in the hydroxide sheets of the GR crystal structure. In the present work, the influence of Mg^{2+} on the precipitation reaction of GR(SO$_4^{2-}$) was investigated. For that purpose, Mg^{2+} was substituted, partially or entirely, for Fe^{2+}. The GR was then prepared by mixing a solution of FeCl$_3$·6H$_2$O, Na$_2$SO$_4$·10H$_2$O, NaCl, FeCl$_2$·4H$_2$O and/or MgCl$_2$·4H$_2$O with a solution of NaOH. The precipitation of the GR was followed or not by a 1-week aging period. The obtained precipitate was characterized by X-ray diffraction. It was observed that Mg(II) favored the formation of chloride green rust GR(Cl$^-$) and magnetite Fe$_3$O$_4$ at the detriment of GR(SO$_4^{2-}$). The proportion of GR(Cl$^-$) and Fe$_3$O$_4$ increased with the Mg(II):Fe(II) substitution ratio. Without Fe(II), the precipitation reaction led to iowaite, i.e., the Mg(II)-Fe(III) compound structurally similar to GR(Cl$^-$). It is forwarded that the presence of Mg^{2+} cations in the hydroxide sheets of the GR crystal structure is detrimental for the stability of the crystal structure of GR(SO$_4^{2-}$) and favors the formation of other mixed valence Fe(II,III) compounds.

Keywords: carbon steel; marine corrosion; seawater; green rust; magnesium; magnetite; X-ray diffraction

---

1. Introduction

Green rust compounds (GR) are common and important corrosion products of steel exposed to marine environments [1]. They are mixed valence Fe(II,III) hydroxysalts and a particular case of layered double hydroxide (LDH). LDH compounds can be based on various divalent and trivalent cations, for instance, Mg(II), Ni(II), Zn(II), Al(III), Cr(III), etc., and can incorporate various monovalent and divalent anions, e.g., Cl$^-$, SO$_4^{2-}$, and CO$_3^{2-}$. Actually, when carbon steel is immersed in seawater, the sulfated green rust GR(SO$_4^{2-}$) with composition Fe$^{II}_4$Fe$^{III}_2$(OH)$_{12}$SO$_4$·8H$_2$O is the first corrosion product that forms [2]. As it contains mainly Fe(II) cations, GR(SO$_4^{2-}$) is readily oxidized by dissolved O$_2$, a process that leads to Fe(III)-oxyhydroxides and/or magnetite (Fe$_3$O$_4$) [3,4]. This process explains why the corrosion product layer formed on carbon steel permanently immersed in seawater is mainly composed of (at least) two strata. First, an inner dark stratum is present at the metal surface. It contains the Fe(II)-based corrosion products (e.g., the sulfated green rust) forming from the dissolution of the metal. Second, an orange-brown outer stratum is present on top of the dark inner stratum. It contains mainly Fe(III)-oxyhydroxides resulting from the oxidation of Fe(II)-based corrosion products [2,5–7].

The formation of the carbonated green rust GR(CO_3^{2-}), i.e., $Fe^{II}_4Fe^{III}_2(OH)_{12}CO_3 \cdot 2H_2O$ is favored when a cathodic polarization is applied to steel [8]. As a result, pyroaurite was observed at the surface of steel structures under cathodic protection [9]. This compound is similar to GR(CO_3^{2-}), with Mg^{2+} cations substituted for Fe^{2+} cations. The formation of pyroaurite is the consequence of the presence of Mg^{2+} ions in seawater ([Mg^{2+}] ~ 0.053 mol/kg, the second most abundant cation after Na^+ [10]). This finding suggests that, even at the open circuit potential (OCP), some Mg^{2+} cations could be incorporated in the crystal structure of green rust compounds, thus influencing more or less importantly the nature and properties of various components of the corrosion product layer. The main aim of the present study was to determine whether Mg^{2+} ions could indeed have an important role on the formation of GRs, and in particular GR(SO_4^{2-}), a question that has not yet been addressed.

In the present study, GR(SO_4^{2-}) was formed by precipitation from dissolved Fe(II) and Fe(III) species, i.e., no metal (Fe^0) was used. The GR was then prepared by mixing a solution of Fe(III), Fe(II), and/or Mg(II) salts (chlorides and/or sulfates) with a solution of NaOH. This precipitation reaction is assumed to mimic the process leading from the dissolved species produced by the corrosion of steel to the GR compound. It corresponds to the first step of the formation of the corrosion product layer that covers steel surfaces immersed in seawater [1,2]. The aim of the study was then to determine the effects of dissolved Mg(II) species on the precipitation reaction. For that purpose, Mg^{2+} cations were partially or totally substituted for Fe^{2+}. The solid phases obtained for various Mg(II):Fe(II) substitution ratios were characterized by X-ray diffraction (XRD), immediately after precipitation or after one week of ageing. To simulate a marine environment, the overall chloride and sulfate concentrations were adjusted at values typical of seawater.

2. Materials and Methods

2.1. Synthesis of (Fe,Mg)^{II}-Fe^{III} LDH

Five precipitates, called M0–M4 were precipitated by mixing a solution (100 mL) of $FeCl_3 \cdot 6H_2O$, $FeCl_2 \cdot 4H_2O$ and/or $MgCl_2 \cdot 4H_2O$, NaCl and $Na_2SO_4 \cdot 10H_2O$ with a solution (100 mL) of NaOH. All the chemicals had a purity higher or equal than 99%. The experiments were performed at room temperature (RT = 22 ± 1 °C).

The considered concentrations are given in Table 1. They are expressed with respect to the overall amount of the solution, i.e., 200 mL, and are based on previous work [11]. The overall chloride concentration is 0.55 mol/L, whereas the sulfate concentration is 0.03 mol L^{-1}. They are both similar to the Cl^- and SO_4^{2-} concentrations characteristic of seawater [10]. M0 is the reference experiment performed without Mg(II). M1–M3 are experiments performed with increasing Mg(II):Fe(II) concentration ratios, i.e., 1:3 for M1, 1:1 for M2, and 3:1 for M3. M4 is the experiment performed without Fe(II).

Table 1. Concentrations of reactants (mol L^{-1}) used for the various experiments M0–M4.

Reactants	Concentrations (mol L^{-1})					
	M0	M1	M2	M3	M4	M4s [1]
NaOH	0.24	0.24	0.24	0.24	0.24	0.24
NaCl	0.19	0.19	0.19	0.19	0.19	0
$Na_2SO_4 \cdot 10H_2O$	0.03	0.03	0.03	0.03	0.03	0
$FeCl_2 \cdot 4H_2O$	0.12	0.09	0.06	0.03	0	0
$MgCl_2 \cdot 4H_2O$	0	0.03	0.06	0.09	0.12	0
$FeCl_3 \cdot 6H_2O$	0.04	0.04	0.04	0.04	0.04	0
$MgSO_4 \cdot 7H_2O$	0	0	0	0	0	0.12
$Fe_2(SO_4)_3 \cdot 5H_2O$	0	0	0	0	0	0.12

[1] M4s: Specific experiment without Fe(II) and Cl^-.

M4s is an additional experiment performed without Fe(II) and Cl^- ions, i.e., using Mg(II) and Fe(III) sulfates and omitting NaCl.

The precipitation reaction of GR(SO$_4^{2-}$) can be written as follows:

$$4Fe^{2+} + 2Fe^{3+} + 12OH^- + SO_4^{2-} + 8H_2O \rightarrow Fe^{II}_4 Fe^{III}_2 (OH)_{12} SO_4 \cdot 8H_2O \quad (1)$$

According to this reaction, stoichiometric conditions correspond to [Fe^{II}]/[OH$^-$] = 1/3, [Fe^{II}]/[Fe^{III}] = 2, and [Fe^{II}]/[SO$_4^{2-}$] = 4. The experimental conditions considered to precipitate M0 correspond to [Fe^{II}]/[OH$^-$] = 1/2, [Fe^{II}]/[Fe^{III}] = 3, and [Fe^{II}]/[SO$_4^{2-}$] = 4, i.e., to an excess of Fe(II) with respect to Fe(III) and OH$^-$. As observed in [11], this situation leads to an excess of dissolved Fe(II) (and SO$_4^{2-}$) species in the solution and hinders the formation of magnetite Fe$_3$O$_4$. The precipitation reaction is then, for the experimental conditions considered in the present study (omitting Cl$^-$ and Na$^+$ ions that do not participate in the reaction though present in the solution):

$$6Fe^{2+} + 2Fe^{3+} + 12OH^- + \frac{3}{2}SO_4^{2-} + 8H_2O \rightarrow Fe^{II}_4Fe^{III}_2(OH)_{12}SO_4 \cdot 8H_2O + 2Fe^{2+} + \frac{1}{2}SO_4^{2-} \quad (2)$$

The suspensions were stirred for 1 min and aged 1 week at RT in a flask filled to the rim. The flask was then hermetically sealed to avoid any oxidation by air of the precipitates. The aged precipitates were finally filtered for analysis by XRD. They were sheltered from air with a plastic membrane during filtration to avoid the oxidation of the obtained GR compounds. The pH of the suspensions after ageing was measured close to neutrality (6.5 to 7.3). The pH of the solution has an influence on the evolution of the precipitate during the ageing procedure. The experimental conditions of the present study were chosen to avoid the transformation of GR to magnetite [11].

Additional experiments were performed similarly to analyze the unaged precipitate. In this case, the suspension was filtered immediately after the 1 min-stirring.

2.2. XRD Analysis

The solid phases obtained with various Mg(II):Fe(II) ratios were analyzed by X-ray diffraction (XRD), a method suitable for distinguishing between the various types of green rusts [7–9]. The other method usually used to characterize the corrosion products of steel, often coupled to XRD for that purpose, is µ-Raman spectroscopy [2,5,7–9]. However, the Raman spectra of the various GR compounds are similar and this method is not adequate to identify unambiguously a given type of GR [12]. Fourier transform infrared (FTIR) spectroscopy was also considered but the few tests we performed revealed that the small amounts of magnetite identified via XRD in some samples were difficult to detect.

X-ray diffraction (XRD) analysis was achieved with an Inel EQUINOX 6000 diffractometer (Thermo Fisher Scientific, Waltham, MA, USA) using the Co-Kα radiation (λ = 0.17903 nm) at 40 kV and 40 mA. The diffractometer is equipped with a CPS 590 detector that detects the diffracted photons simultaneously on a 2θ range of 90°. To prevent the oxidation of Fe(II)-based compounds during preparation and analysis, the samples were mixed with a few drops of glycerol in a mortar before being crushed until a homogenous oily paste was obtained. With this procedure, the various particles that constitute the sample are coated with glycerol and thus, sheltered from the oxidizing action of O$_2$ [13]. Glycerol may only give rise to a very broad "hump" visible on the XRD pattern between $2\theta \sim 25°$ and $2\theta \sim 35°$.

Mg(II)-Fe(III) compounds (M4 and M4s experiments), that cannot be further oxidized by O$_2$, were analyzed whether as a wet paste immediately after filtration or as a dry powder after drying in air. In this last case, sodium salts such as NaCl are present together with the Mg-Fe compounds.

The analysis was performed in any case at RT with a constant angle of incidence (5°) during 45 min.

The various obtained solid phases were identified via the ICDD-JCPDS (International Center for Diffraction Data—Joint Committee on Powder Diffraction Standards) database, and the peaks indexed according to the corresponding file. Moreover, the parameters, i.e., interplanar distance, intensity and full width at half maximum, of the diffraction peaks, were determined via a computer fitting of the experimental diffraction patterns. The diffraction peaks were fitted in any case with pseudo-Voigt functions to take into account the evolution of the peak profile with increasing diffraction angle. The fitting procedure was achieved using the OriginPro 2016 software (OriginLab).

3. Results

3.1. XRD Analysis of Aged Precipitates

Figure 1 displays the XRD pattern of precipitate M0 after 1 week of ageing. In this first case, Mg(II) cations were not present and the obtained compound is then a Fe(II)-Fe(III) LDH.

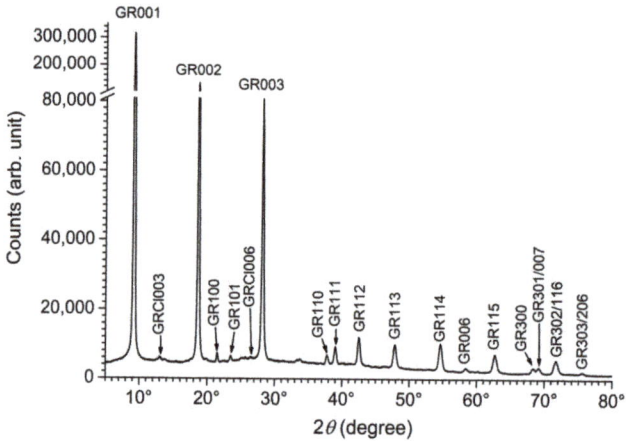

Figure 1. XRD pattern of reference precipitate M0 ([Mg^{II}] = 0) after 1 week of ageing at room temperature (RT). GR = GR(SO$_4^{2-}$), GRCl = GR(Cl$^-$), with the corresponding Miller index.

In agreement with the previous work [11], the XRD pattern reveals that the solid phase is mainly composed of GR(SO$_4^{2-}$), i.e., the Fe(II)-Fe(III) SO$_4$-LDH. The two main peaks of the chloride green rust are seen together with those of GR(SO$_4^{2-}$), but their intensity is very low. Using the fitting procedure described in Section 2.2, the intensity ratio between the main peak of GR(SO$_4^{2-}$) (GR001, at 2θ = 9.2°) and the main peak of GR(Cl$^-$) (GRCl003, at 2θ = 12.9°) is determined at 93:1.

Figure 2 displays the XRD pattern of precipitate M4 after 1 week of ageing. This second case corresponds to the situation where Fe(II) cations are not present. The obtained compound is consequently a Mg(II)-Fe(III) LDH. Strikingly, its XRD pattern drastically differs from that of GR(SO$_4^{2-}$). The main diffraction peak, which corresponds to the distance between two consecutive Fe planes in the LDH structure, is located at about 2θ = 13°. This leads to an interplanar distance of 8 Å, rather typical of GR(Cl$^-$). By comparison, the main diffraction peak of GR(SO$_4^{2-}$) is found at 9.2° (Figure 1), which corresponds to an interplanar distance of 11.15 Å. The diffraction peaks of the obtained Mg(II)-Fe(III) LDH actually correspond to the mineral iowaite, that is the Mg(II)-Fe(III) Cl-LDH similar to GR(Cl$^-$) [14,15] with the chemical formula Mg$_6$Fe$_2$(OH)$_{16}$Cl$_2$·4H$_2$O [15]. In the experimental conditions considered here, when Mg(II) is substituted for Fe(II), a Cl-LDH is formed rather than a SO$_4$-LDH. Note that the solid phase was analyzed as a dry powder so that the diffraction lines of NaCl are also seen.

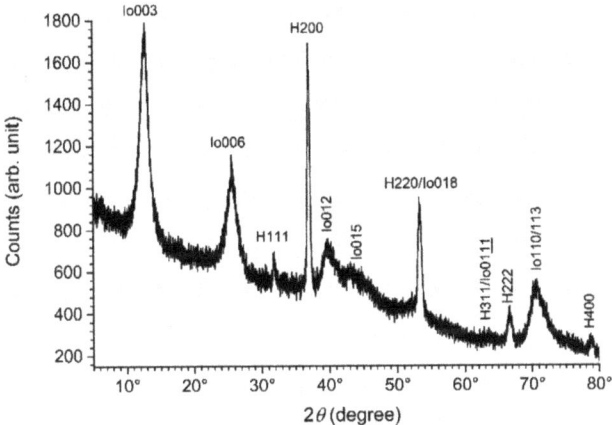

Figure 2. XRD pattern of precipitate M4 ([Fe^{II}] = 0) after 1 week of ageing at RT. The precipitate was analyzed as a dry powder. Io: Iowaite, H: Halite NaCl, with the corresponding Miller index.

It can finally be observed that the diffraction peaks of the obtained iowaite are much broader than those of the sulfated GR obtained in the absence of Mg(II) (Figure 1). This shows that the average crystal size, or more exactly the mean coherent domain size, of the Mg(II)-Fe(III) Cl-LDH is much smaller than that of GR(SO$_4^{2-}$), i.e., the Fe(II)-Fe(III) SO$_4$-LDH.

The XRD pattern of the precipitate obtained with equal amounts of Fe(II) and Mg(II), i.e., precipitate M2, is displayed in Figure 3. Both GR(SO$_4^{2-}$) and GR(Cl$^-$) are identified, and found in similar proportions according to the respective intensity of their main peaks. Note that both compounds are likely to comprise not only Fe(II) cations, but Mg(II) cations too. Consequently, they may not be green rust compounds sensu stricto. However, for clarity, this terminology will be used in the following to designate the Fe^{II}-(Mg^{II})-Fe^{III} SO$_4$-LDH and Cl-LDH.

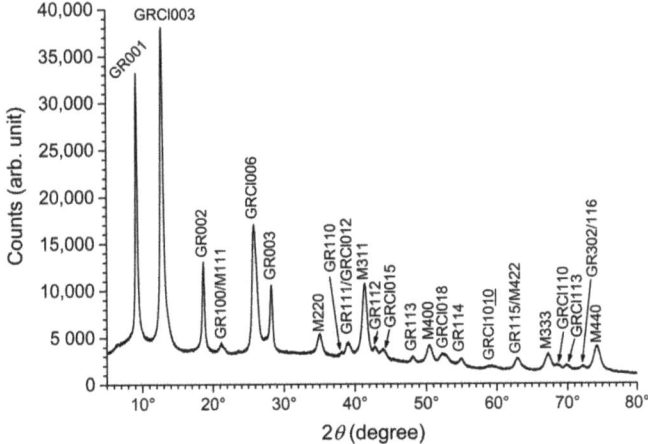

Figure 3. XRD pattern of precipitate M2 ([Mg^{II}]/[Fe^{II}] = 1) after 1 week of ageing at RT. GR = GR(SO$_4^{2-}$), GRCl = GR(Cl$^-$), M = Fe$_3$O$_4$, with the corresponding Miller index.

Magnetite, the Fe(II-III) mixed valence oxide with chemical formula Fe$_3$O$_4$, is also identified. This shows that the presence of Mg(II) cations has induced in this case the formation of both GR(Cl$^-$) and Fe$_3$O$_4$.

Figure 4 displays the XRD patterns of precipitates M1 and M3 after 1 week of ageing. These data confirm that Mg(II) favors the formation of GR(Cl$^-$) and magnetite. Actually, for the high substitution ratio [Mg^{II}]/[Fe^{II}] = 3 (precipitate M3), the main obtained LDH is GR(Cl$^-$). The intensity of the diffraction peaks of GR(SO$_4^{2-}$) is very weak, even with respect to that of the main peak of magnetite (M311, at 2θ = 41.3°). The intensity ratio between the main peak of GR(SO$_4^{2-}$) and the main peak of GR(Cl$^-$) is now determined at 1:32. Conversely, for the low substitution ratio [Mg^{II}]/[Fe^{II}] = 1/3 (precipitate M1), the diffraction peaks of both GR(Cl$^-$) and magnetite remain very small. However, the intensity ratio between the main peak of GR(SO$_4^{2-}$) and the main peak of GR(Cl$^-$) is equal to 22:1 in this case, while it was 93:1 in the absence of Mg(II) cations. The influence of Mg(II) is small but nonetheless detectable.

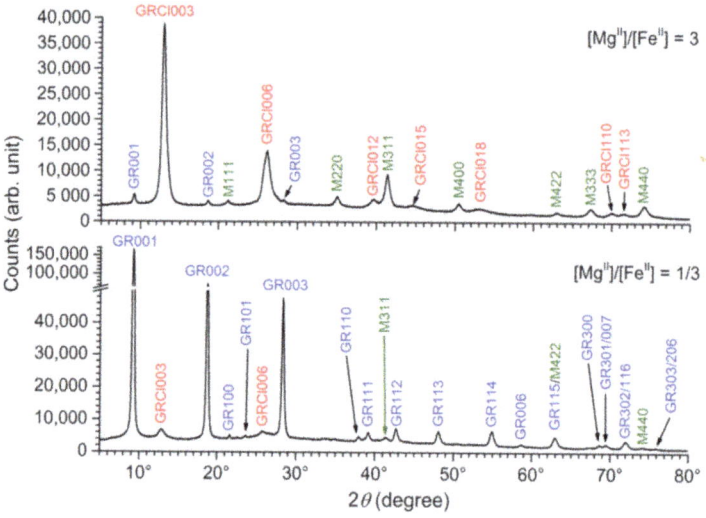

Figure 4. XRD pattern of precipitates M1 ([Mg^{II}]/[Fe^{II}] = 1/3) and M3 ([Mg^{II}]/[Fe^{II}] = 3) after 1 week of ageing at RT. GR = GR(SO$_4^{2-}$), GRCl = GR(Cl$^-$), M = Fe$_3$O$_4$, with the corresponding Miller index.

A detailed analysis of the XRD data was achieved to obtain further information, in particular about a possible variation of the GR lattice parameters with the Mg(II):Fe(II) concentration ratio. For that purpose, the angular regions where the two main peaks of GR(SO$_4^{2-}$) and GR(Cl$^-$) are present were computer fitted (see Section 2.2). The result obtained for precipitate M2 in the 24–30° 2θ region of the GRCl006 peak is displayed in Figure 5 as an example.

Since the GRCl006 peak overlaps slightly with the GR003 peak, both peaks were taken into account. However, the experimental curve could not be adequately fitted and an additional broad peak had to be added. The position of this peak was determined through the fitting procedure at 2θ = 27.52°, a diffraction angle associated with an interplanar distance of 3.76 Å. It corresponds exactly to the 006 diffraction peak of the carbonated green rust GR(CO$_3^{2-}$) [13,16]. This finding actually shows that a very small amount of GR(CO$_3^{2-}$) has formed together with GR(SO$_4^{2-}$), GR(Cl$^-$), and magnetite, although carbonate species were not added specifically to the system. These carbonate species could originate in (i) the dissolution of CO$_2$ in the solution and (ii) some impurities present in the chemicals used. It happened that the NaOH pellets used for this study contained a small proportion of Na$_2$CO$_3$.

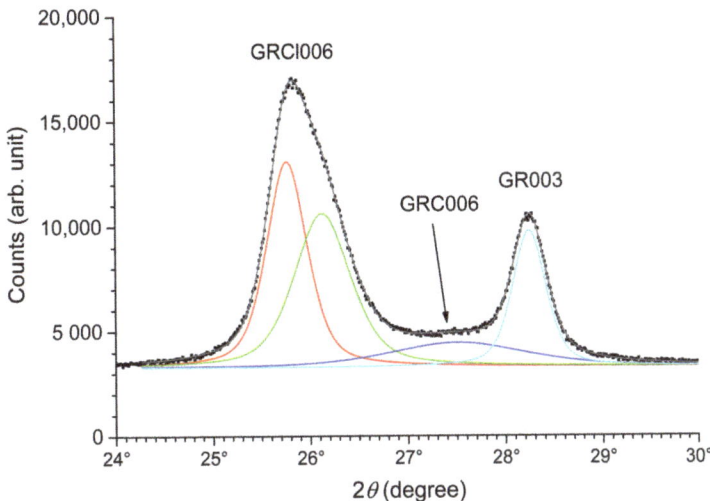

Figure 5. Fitting of the XRD pattern of precipitate M2 ($[Mg^{II}]/[Fe^{II}] = 1$) after 1 week of ageing at RT: Detail of the 24–30° angular region. GR = GR(SO_4^{2-}), GRCl = GR(Cl^-), GRC = GR(CO_3^{2-}), with the corresponding Miller index.

However, the presence of the weak GRC006 peak cannot explain the important asymmetry of the 006 diffraction peak of GR(Cl^-). As it can be seen in Figure 5, the computer fitting procedure had to be achieved with two pseudo-Voigt functions in the case of the GRCl006 diffraction peak. Such an asymmetry was not observed for the diffraction peaks of GR(SO_4^{2-}), as illustrated by the GR003 peak in Figure 5.

All the results obtained with the fitting of the XRD patterns are listed in Table 2. The data corresponding to the traces of carbonate GR, identified in each case, are omitted as they are only the consequence of the presence of carbonate traces (CO_2 and impurities) in the system.

Table 2. Characteristics of the two main diffraction peaks of GR(SO_4^{2-}) and GR(Cl^-)/iowaite for the aged M0-M4 precipitates; d: Interplanar distance (Å), I: Peak intensity, with $I = 100$ for the most intense peak of the considered compound, and FWHM: Full width at half maximum, in degrees. GR = GR(SO_4^{2-}) and GRCl = GR(Cl^-)/iowaite.

Diffraction Peak	Parameter	M0	M1	M2	M3	M4
GR001	d	11.18 Å	11.14 Å	11.13 Å	11.16 Å	-
	I	100	100	100	100	-
	FWHM	0.21°	0.24°	0.34°	0.35°	-
GR002	d	5.53 Å	5.51 Å	5.52 Å	5.52 Å	-
	I	51	51	40	52	-
	FWHM	0.25°	0.28°	0.36°	0.41°	-
GRCl003	d	-	8.01 Å	8.04 Å	7.96 Å	8.14 Å
	I	-	100	100	100	100
	FWHM	-	0.80°	0.46°	0.64°	1.53°
GRCl006	d_1	-	4.01 Å	4.01 Å	4.02 Å	4.04 Å
	I_1	-	82	27	7	57
	$FWHM_1$	-	0.80°	0.49°	0.53°	1.79°
	d_2	-	-	3.96 Å	3.95 Å	-
	I_2	-	-	28	35	-
	$FWHM_2$	-	-	0.68°	0.81°	-

First, these results show that the 001 and 002 interplanar distances of GR(SO$_4^{2-}$) linked to the c parameter of the hexagonal cell, are not influenced by the [Mg^{II}]/[Fe^{II}] substitution ratio. They vary slightly around an average of 11.16 ± 0.02 Å for d_{001} and 5.52 ± 0.01 Å for d_{002} with no apparent link with [Mg^{II}]/[Fe^{II}]. However, a clear trend is observed for the width of those peaks. FWHM increases significantly with the proportion of Mg(II), which shows that the growth of the GR(SO$_4^{2-}$) crystals, and/or the increase of crystallinity of GR(SO$_4^{2-}$), is hindered by the presence of the Mg(II) cations.

In contrast, more important changes are observed for the diffraction peaks of GR(Cl$^-$). The data obtained for precipitate M4, that is for the Mg(II)-Fe(III) Cl-LDH, are indeed characteristic of iowaite [14]. It can then be noted that the lattice parameters of iowaite differ from those of GR(Cl$^-$). The d_{003} and d_{006} interplanar distances are linked to the c parameter of the conventional hexagonal cell. They lead to an average c/3 value of 8.11 ± 0.03 Å (average of d_{003} and 2 × d_{006}) comparable to the values reported in previous works for iowaite, which are between 8.04 [14] and 8.11 Å [15]. The c/3 parameter of GR(Cl$^-$) is smaller, about 7.95 Å [17].

The main peak GRCl003 of the chloride GR, though slightly asymmetric, could be fitted in any case with only one pseudo-Voigt function. However, the corresponding interplanar distance was observed between 7.96 Å for M3 and 8.04 Å for M2, and up to 8.14 Å for M4. The two extreme values are typical of GR(Cl$^-$) and iowaite [14,15,17]. The important asymmetry of the GRCl006 diffraction peak implied the use of two pseudo-Voigt functions. Actually, variations of d_{hkl} are associated with larger variations of $2\theta_{hkl}$ in the angular region corresponding to the GRCl006 peak, which may explain that the asymmetry of the GRCl003 peak was smaller. The phenomenon was more pronounced in the case of precipitate M2 (Figure 5) and led to two peaks with a similar intensity (Table 2). The corresponding d_{006} distances were determined at 4.01–4.02 and 3.95–3.96 Å. They lead to values of 8.03 ± 0.01 and 7.91 ± 0.01 Å, respectively. Though the asymmetry of the GRCl006 peak may have various origins, a heterogeneous Mg(II) content could lead to a variation of the c lattice parameter of the conventional hexagonal cell, this parameter increasing with the Mg(II) content, as illustrated by the difference between the c lattice parameter of GR(Cl$^-$) and that of iowaite.

The width of the GRCl peaks also varies with the [Mg^{II}]/[Fe^{II}] substitution ratio. As already noted, FWHM is very high in the absence of Fe(II), that is for iowaite. The influence of Mg(II) is also illustrated by the increase of FWHM from M2 to M3. However, the width of the GR(Cl$^-$) peaks is larger for M1 even though the [Mg^{II}]/[Fe^{II}] ratio is smaller.

3.2. XRD Analysis of Unaged Precipitates

Some solid phases may result from the precipitation reactions, while other phases may form during ageing via the transformation of initially precipitated compounds. The evolution with time of precipitate M0, previously studied [11], showed, for instance, that the amount of GR(Cl$^-$) decreased upon ageing, which implied that part of the initially formed GR(Cl$^-$) transformed to GR(SO$_4^{2-}$). Consequently, only traces of GR(Cl$^-$) remained after 1 week (as seen in Figure 1). Similarly, it was observed that, in the absence of excess dissolved Fe(II) species, part of the initially precipitated GR(SO$_4^{2-}$) could transform into magnetite [11].

Figure 6 displays the XRD patterns of unaged precipitates M1 and M2. In both cases, the diffraction peaks are clearly broader than those of the aged compounds (Figures 3 and 4). This illustrates a well-known effect of ageing, i.e., the increase of crystallinity and crystal size with time. In the case of M1, only two phases are detected, namely GR(SO$_4^{2-}$) and GR(Cl$^-$). After 1 week of ageing, magnetite was present. This result shows that magnetite results in this case from the ageing procedure. The intensity ratio between the main peak of GR(SO$_4^{2-}$) and the main peak of GR(Cl$^-$) is determined before ageing at 5.5:1. It was determined (see previous Section 3.1) at 22:1 after 1 week of ageing. This shows that the proportion of GR(Cl$^-$) decreased significantly during ageing, as observed for M0 [11], i.e., in the absence of Mg(II) cations. For the lowest [Mg^{II}]/[Fe^{II}] ratio of 1/3, GR(Cl$^-$) may then have also transformed to GR(SO$_4^{2-}$).

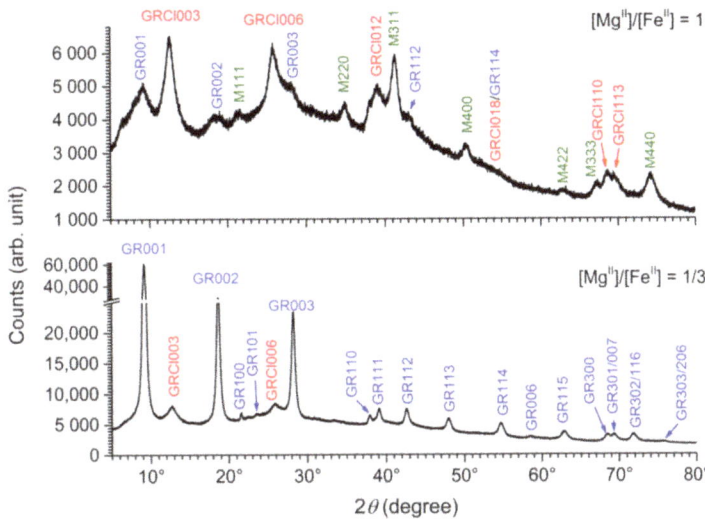

Figure 6. XRD pattern of unaged precipitates M1 ([Mg^{II}]/[Fe^{II}] = 1/3) and M2 (Mg^{II}]/[Fe^{II}] = 1). GR = GR(SO$_4^{2-}$), GRCl = GR(Cl$^-$), M = Fe$_3$O$_4$, with the corresponding Miller index.

In the case of M2, magnetite is already present among the solid phases that compose the unaged precipitate. Consequently, the three phases observed after ageing, i.e., GR(SO$_4^{2-}$), GR(Cl$^-$), and Fe$_3$O$_4$, result from the precipitation process. The intensity ratio between the main peak of GR(SO$_4^{2-}$) and the main peak of GR(Cl$^-$) is determined at 1:1.5 for the unaged precipitate and 1:1.8 for the aged precipitate (Figure 3). The variation is slight and may not be significant. In any case, it shows that the proportion of GR(Cl$^-$) remained constant or increased slightly upon ageing, in contrast with what was observed without Mg(II) (precipitate M0, [11]) or with the lowest [Mg^{II}]/[Fe^{II}] substitution ratio (precipitate M1). This shows that the presence of Mg(II) not only favors the precipitation of the Cl-LDH, but also increases its stability with respect to the SO$_4$-LDH.

3.3. Analysis of the Mg(II)-Fe(III) Solid Phases Obtained in the Absence of Chloride

The first XRD pattern, shown in Figure 7, is that of precipitate M4s aged 1 week and analyzed immediately after filtration as a wet paste. The obtained Mg(II)-Fe(III) compound is poorly crystallized and its pattern is similar to that of GR(SO$_4^{2-}$), i.e., the main diffraction peak is located at 9.0°. This pattern was indexed according to the ICCD-JCPDS file of wermlandite Mg$_7$Al$_{1.14}$Fe$_{0.86}$(OH)$_{18}$Ca$_{0.6}$Mg$_{0.4}$(SO$_4$)$_2$(H$_2$O)$_{12}$, a mineral structurally similar to GR(SO$_4^{2-}$) [18]. Wermlandite includes Al^{3+} and Ca^{2+} ions and not only Mg^{2+} and Fe^{3+} cations. In our experiment, Al^{3+} and Ca^{2+} ions were not present and the obtained compound is then a Mg(II)-Fe(III) SO$_4$-LDH. Other SO$_4$-LDH are also characterized by this type of structure, where two consecutive metal cations planes are separated by ~11 Å. An example is hydrohonnessite, where the cations present in the hydroxide layers are Ni^{2+} and Fe^{3+} [19].

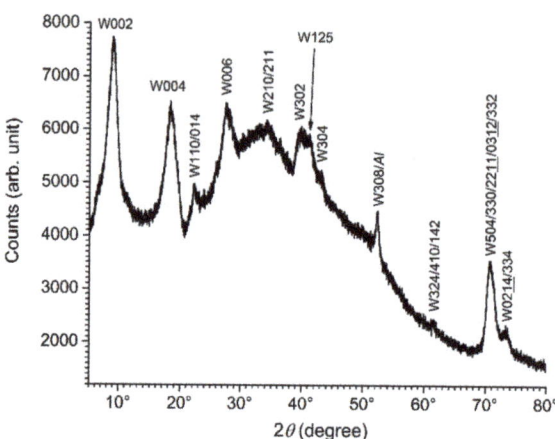

Figure 7. XRD pattern of precipitate M4s ([Fe^{II}] = 0 and [Cl$^-$] = 0) after 1 week of ageing at RT. W = Mg(II)-Fe(III) hydroxysulfate similar to wermlandite, with the corresponding Miller index.

This result clearly shows that a Mg(II)-Fe(III) SO$_4$-LDH similar to GR(SO$_4{}^{2-}$) can be obtained if Cl$^-$ ions are not available for the formation of a Cl-LDH. The distance between two consecutive planes of metal cations is determined at 11.56 Å, which shows that, as for the Cl-LDH, the substitution of Fe(II) by Mg(II) cations leads to an increase of the c lattice parameter of the hexagonal cell. Actually, the ionic radius of Mg(II) is smaller than that of Fe(II) [20], which induces a decrease of the a lattice parameter of Mg(II)-Fe(III) LDHs with respect to Fe(II)-Fe(III) LDHs [21]. However, the c lattice parameter is nonetheless higher with Mg(II) [21]. This illustrates how the cationic composition of the hydroxide layer influences the electrostatic interactions that bind together the hydroxide sheets and the interlayers and ensures the stability of the crystal structure [21]. This crucial point is further discussed in Section 4.

The second XRD pattern, shown in Figure 8, was obtained with the same aged M4s precipitate. However, the wet paste obtained after filtration was dried in air and the solid phase was analyzed as a dry powder 10 days later. The result of the drying is a change in the structure of the Mg(II)-Fe(III) SO$_4$-LDH. This new compound can be considered as a second form of SO$_4$-LDH and will be called in the following the Mg(II)-Fe(III) hydroxysulfate-b. Its new structure seems similar to that of GR(Cl$^-$) and iowaite and was then indexed similarly. The main peak of the Mg(II)-Fe(III) hydroxysulfate-b is then the 003 peak. Actually, this second type of SO$_4$-LDH was already reported [19]. Honessite, a Ni(II)-Fe(III) SO$_4$-LDH, for example, is characterized by a distance between two consecutive planes of metal cations of 8.7 Å [19].

The main diffraction peak of the Mg(II)-Fe(III) hydroxysulfate-b obtained here is located at a position $2\theta = 11.74°$. This corresponds to a $c/3$ distance of 8.75 Å, very similar to that of honessite. The transformation from one type of structure to the other, associated with the drying of the solid phase, is due to the release of water molecules initially present in the interlayers [22].

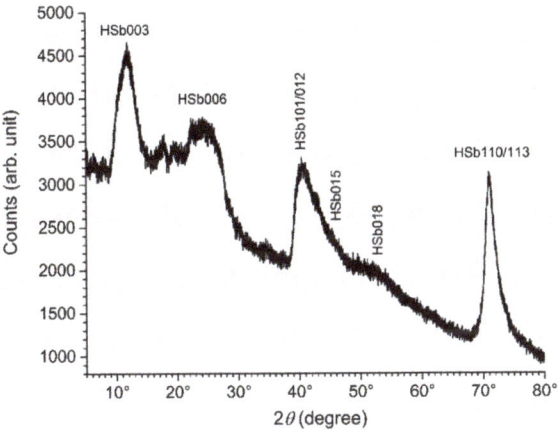

Figure 8. XRD pattern of precipitate M4s ([Fe^{II}] = 0 and [Cl$^-$] = 0) after 1 week of ageing at RT, filtration, and drying in air (10 days). HSb = Mg(II)-Fe(III) hydroxysulfate-b (see text) with the corresponding Miller index.

4. Discussion

In the considered experimental conditions, SO_4^{2-} and Cl^- were the only anions available for the formation of LDH compounds. Consequently, the only Fe(II)-Fe(III) mixed valence compounds that could possibly form were GR(SO_4^{2-}), GR(Cl^-), and Fe_3O_4. The traces of GR(CO_3^{2-}) detected in each case are due to CO_2 and/or chemical impurities and the formation of this phase will not be further discussed. These experimental conditions were chosen so that in the absence of Mg(II) cations, the Fe(II)-Fe(III) SO_4-LDH, i.e., GR(SO_4^{2-}), was obtained, only accompanied by traces of the Fe(II)-Fe(III) Cl-LDH, i.e., GR(Cl^-). The aim was to reproduce the first stage of the corrosion process of carbon steel in seawater, which leads to GR(SO_4^{2-}) [1,2], via a precipitation reaction involving dissolved Fe(II) and Fe(III) species, OH- ions, and the main anionic species of seawater, i.e., Cl^- and SO_4^{2-}.

The first and more important effect of Mg(II) cations is to favor the formation of a Cl-LDH at the detriment of the SO_4-LDH obtained with Fe(II) and Fe(III). This is clearly illustrated by the increase of the proportion of GR(Cl^-) with the increase of the [Mg^{II}]/[Fe^{II}] substitution ratio and the formation of iowaite, the Mg(II)-Fe(III) Cl-LDH, when [Fe^{II}] = 0.

A Mg(II)-Fe(III) SO_4-LDH could be obtained when Cl^- ions were removed from the system. However, the solid phase identified in an aqueous suspension, structurally similar to GR(SO_4^{2-}), underwent a transformation upon drying, which led to a SO_4-LDH structurally closer to GR(Cl^-).

The main difference between the two GR structures is the organization of the interlayers, that involve two planes of anions and water molecules in GR(SO_4^{2-}) [23] and only one plane in GR(Cl^-) [17] (and in GR(CO_3^{2-}), as well). Figure 9 displays a schematic representation of these structures. GR(Cl^-) and GR(SO_4^{2-}) were initially called GR-1 and GR-2 [24] and a similar terminology can be retained to distinguish the structure of GR(Cl^-) and GR(CO_3^{2-}) from that of GR(SO_4^{2-}). It must be noted that for the GR-1 rhombohedral $R\bar{3}m$ structure of GR(Cl^-) [17], the stacking sequence is AcB i BaC i CbA i, where A, B, C are planes of OH$^-$ ions, a, b, c planes of Fe atoms, and i corresponds to the interlayers. In the case of the $P\bar{3}m1$ trigonal structure of GR(SO_4^{2-}) [23], i.e., GR-2, the stacking sequence is AcB i AcB.

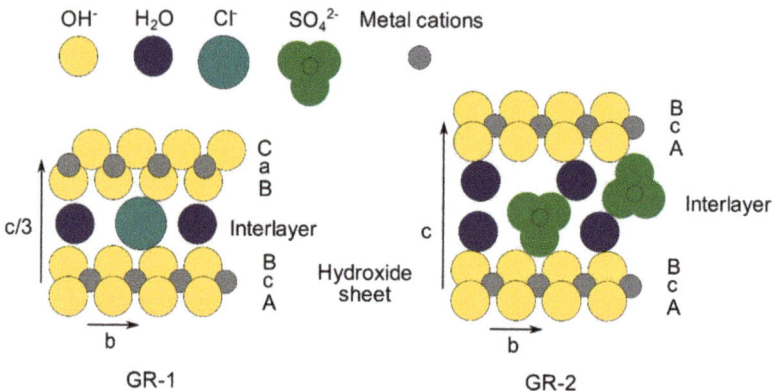

Figure 9. Schematic representations of the GR-1 and GR-2 structures, drawn according to the crystal structures of GR(Cl$^-$) given in [17] and GR(SO$_4^{2-}$) given in [23].

The results obtained here show that Mg(II) cations favor the GR-1 structure. From a fundamental point of view, the cohesion of a LDH structure is due to (i) the water molecules of the interlayers that interact with the adjacent hydroxide layers and the intercalated anions via hydrogen bonds and (ii) the intercalated anions that interact with the hydroxide layers via electrostatic interactions and hydrogen bonds [21,25]. Changes in the hydroxide layers necessarily have an influence on the bonds linking these layers and the species (anions and water molecules) present in the interlayers. They have thus an influence on the cohesion of the LDH structure. The dependence between the cationic composition of the hydroxide layer and the structural stability has been studied and modelled in [21]. This study demonstrated how important the nature of cations for the stability of the crystal structure was.

The thorough analysis of the diffraction data showed that the c lattice parameter of GR(SO$_4^{2-}$) did not vary with the [Mg^{II}]/[Fe^{II}] substitution ratio. However, the distance between two planes of metal cations is higher for the Mg(II)-Fe(III) SO$_4$-LDH, with 11.56 Å vs. 11.16 Å for GR(SO$_4^{2-}$) (Table 2). In contrast, the diffraction peaks of GR(Cl$^-$) proved to be influenced by the [Mg^{II}]/[Fe^{II}] ratio. This suggests that the Mg(II) cations are not present, or only in a small amount, in the hydroxide layers of GR(SO$_4^{2-}$). Consequently, they would be preferentially incorporated in the GR(Cl$^-$) structure or left in the solution. The small amount of Mg(II) possibly present in the hydroxide layers of GR(SO$_4^{2-}$) would explain the decrease of crystal/mean coherent domain size observed with the increasing Mg(II)/Fe(II) concentration ratio (Table 2).

An interesting first case is the [Mg^{II}]/[Fe^{II}] ratio of 1/3. With this Mg(II) amount, only a minor proportion of GR(Cl$^-$) is present after 1 week of ageing, while 25% of Fe(II) is substituted by Mg(II). According to the initial amounts of reactants, the precipitation reaction could be written as:

$$\frac{9}{2}Fe^{2+} + \frac{3}{2}Mg^{2+} + 2Fe^{3+} + 12OH^- + SO_4^{2-} + 8H_2O \rightarrow Fe^{II}_4Fe^{III}_2(OH)_{12}SO_4 \cdot 8H_2O + \frac{1}{2}Fe^{2+} + \frac{3}{2}Mg^{2+} \quad (3)$$

This writing shows that for this [Mg^{II}]/[Fe^{II}] ratio, all the Mg^{2+} ions could be released into the solution, more likely during the ageing procedure where GR(Cl$^-$) transforms to GR(SO$_4^{2-}$). It can then be forwarded that in this first case, the SO$_4$-LDH is close to GR(SO$_4^{2-}$) and contains a very small proportion of Mg(II). The GR-2 structure is obtained since Mg(II) cations are preferentially found in the solution and in the small amount of the remaining GR(Cl$^-$) (or more exactly Cl-LDH).

In contrast, for the higher [Mg^{II}]/[Fe^{II}] ratios of 1 and 3, an important amount of Mg(II) is necessarily incorporated in the solid phase, which implies that the GR-1 structure is favored leading to the predominance of the Cl-LDH similar to GR(Cl$^-$). Both GR(Cl$^-$) and iowaite are characterized by a (Fe,Mg)(II) to Fe(III) cation ratio of 3:1 [14,16], which implies that all divalent cations should be incorporated into the solid phase in the considered experimental conditions. For instance, for the highest [Mg^{II}]/[Fe^{II}] ratio considered here, the precipitation reaction of the Cl-LDH can be written, neglecting the small amount of GR(SO$_4^{2-}$) that forms, as:

$$\frac{3}{2}Fe^{2+} + \frac{9}{2}Mg^{2+} + 2Fe^{3+} + 16OH^- + 2Cl^- + 4H_2O \rightarrow Fe^{II}_{1.5}Mg^{II}_{4.5}Fe^{III}_2(OH)_{16}Cl_2 \cdot 4H_2O \quad (4)$$

However, Mg(II) cations also promoted the formation of magnetite. Looking to reaction (4), it is seen that in the considered experimental conditions, which correspond to a (Fe,Mg)(II) to Fe(III) cation ratio of 3:1, the precipitation of a Cl-LDH having the same (Fe,Mg)(II) to Fe(III) cation ratio of 3:1 does not leave any divalent cations in the solution. In a previous study [11], it was demonstrated that in this case, the ageing of the suspension led to the formation of magnetite. Moreover, it must be noted that the experimental conditions considered here correspond to an [OH] to [Fe^{II}+Mg^{II}+Fe^{III}] ratio of 3 to 2. Reaction (4) requires an [OH] to [Fe^{II}+Mg^{II}+Fe^{III}] ratio of 4 to 2. Consequently, both divalent and trivalent cations are in excess with respect to the OH$^-$ ions available. It can then be forwarded that the excess Fe(II) and Fe(III) cations react with water molecules to form a small proportion of magnetite, according to the following reaction:

$$Fe^{2+} + 2Fe^{3+} + 4H_2O \rightarrow Fe^{II}Fe^{III}_2O_4 + 8H^+ \quad (5)$$

The present findings can be connected with more applied aspects of marine corrosion and cathodic protection of steel structures. Actually, the concentration of Mg^{2+} in seawater is important, about 0.053 mol/kg [10]. Therefore, the smallest [Mg^{II}]/[Fe^{II}] ratio of 1/3 considered here would correspond to a Fe^{2+} concentration of 0.16 mol/kg in the bulk seawater, which is rather high. However, GR(SO$_4^{2-}$) is the main GR compound identified in the corrosion product layers formed on steel immersed in seawater [1,2,5–7]. At the vicinity of the steel/seawater interface, where the Fe^{2+} cations are produced, the [Mg^{II}]/[Fe^{II}] ratio is necessarily lower than in the bulk seawater and it can be forwarded that the formation of GR(SO$_4^{2-}$) only takes place close to the steel surface.

Our results also explain more clearly why an anodic polarization favors the formation of GR(SO$_4^{2-}$) with respect to any other Fe(II,III) mixed valence compounds [1,7]. An anodic polarization decreases the interfacial [Mg^{II}]/[Fe^{II}] ratio and thus prevents the influence of Mg^{2+} cations.

In contrast, pyroaurite, the Mg(II)-Fe(III) CO$_3$-LDH was observed on a steel surface under cathodic protection [9]. In this case, due to the low dissolution rate of iron, the [Mg^{II}]/[Fe^{II}] ratio is necessarily higher, even at the steel/seawater interface. The increase of the interfacial pH associated with the cathodic polarization tends to favor GR(CO$_3^{2-}$) with respect to GR(SO$_4^{2-}$) [8] even if Mg^{2+} cations are not present. However, the formation of the Mg(II)-Fe(III) LDH rather than the Fe(II,III) LDH confirms that Mg(II) cations can favor the formation of LDH phases characterized by the GR1-structure, i.e., Cl-LDH and CO$_3$-LDH.

5. Conclusions

- For [Mg^{II}]/[Fe^{II}] ratios higher than 1, the influence of Mg(II) is strong and induces the formation of GR(Cl$^-$) and magnetite. In the absence of Fe(II), the Mg(II)-Fe(III) Cl-LDH, i.e., iowaite, is the only solid phase obtained.
- The influence of Mg^{2+} cations on the formation of the sulfated GR is not significant up to a [Mg^{II}]/[Fe^{II}] ratio of 1/3, where only a slight increase of the proportion of GR(Cl$^-$) is observed. In the absence of Mg(II), only GR(SO$_4^{2-}$) is obtained, with only traces of GR(Cl$^-$), in agreement with the previous work [11].
- It is forwarded that the presence of Mg^{2+} cations in the hydroxide layers of the LDH structure of GR compounds favors the Cl- and CO$_3$-GR-1 structure, thus hindering the formation of the SO$_4$-GR-2 structure.

Author Contributions: Conceptualization, P.R., M.J., and R.S.; methodology, P.R., M.J., and R.S.; validation, P.R, J.D., K.A., R.S., and M.J.; formal analysis, P.R, J.D., K.A., R.S., and M.J.; investigation, P.R, J.D., K.A., R.S., and M.J.; writing—original draft preparation, P.R.; writing—review and editing, P.R., M.J., and R.S.; visualization, P.R. and J.D.; supervision, P.R., M.J., and R.S.; project administration, P.R., M.J., and R.S.; funding acquisition, P.R. All authors have read and agreed to the published version of the manuscript.

Funding: European Regional Development Fund (ERDF) and the CDA of La Rochelle through the DYPOMAR research project and the thesis of Julien Duboscq.

Data Availability Statement: The data presented in this study are available on request from the corresponding author. The data are not publicly available because work is still in progress on the subject.

Conflicts of Interest: The authors declare no conflict of interest.

References

1. Refait, P.; Grolleau, A.-M.; Jeannin, M.; Rémazeilles, C.; Sabot, R. Corrosion of carbon steel in marine environments: Role of the corrosion product layer. *Corros. Mater. Degrad.* **2020**, *1*, 198–218. [CrossRef]
2. Lanneluc, I.; Langumier, M.; Sabot, R.; Jeannin, M.; Refait, P.; Sablé, S. On the bacterial communities associated with the corrosion product layer during the early stages of marine corrosion of carbon steel. *Int. Biodeterior. Biodegrad.* **2015**, *99*, 55–65. [CrossRef]
3. Detournay, J.; De Miranda, L.; Dérie, R.; Ghodsi, M. The region of stability of green rust II in the electrochemical potential-pH equilibrium diagram of iron in sulphate medium. *Corros. Sci.* **1975**, *15*, 295–306. [CrossRef]
4. Olowe, A.; Génin, J. The mechanism of oxidation of ferrous hydroxide in sulphated aqueous media: Importance of the initial ratio of the reactants. *Corros. Sci.* **1991**, *32*, 965–984. [CrossRef]
5. Pineau, S.; Sabot, R.; Quillet, L.; Jeannin, M.; Caplat, C.; Dupont-Morral, I.; Refait, P. Formation of the Fe(II-III) hydroxy-sulphate green rust during marine corrosion of steel associated to molecular detection of dissimilatory sulphite-reductase. *Corros. Sci.* **2008**, *50*, 1099–1111. [CrossRef]
6. Duan, J.; Wu, S.; Zhang, X.; Huang, G.; Du, M.; Hou, B. Corrosion of carbon steel influenced by anaerobic biofilm in natural seawater. *Electrochim. Acta* **2008**, *54*, 22–28. [CrossRef]
7. Refait, P.; Grolleau, A.-M.; Jeannin, M.; François, E.; Sabot, R. Localized corrosion of carbon steel in marine media: Galvanic coupling and heterogeneity of the corrosion product layer. *Corros. Sci.* **2016**, *111*, 583–595. [CrossRef]
8. Refait, P.; Jeannin, M.; Sabot, R.; Antony, H.; Pineau, S. Electrochemical formation and transformation of corrosion products on carbon steel under cathodic protection in seawater. *Corros. Sci.* **2013**, *71*, 32–36. [CrossRef]
9. Refait, P.; Jeannin, M.; Sabot, R.; Antony, H.; Pineau, S. Corrosion and cathodic protection of carbon steel in the tidal zone: Products, mechanisms and kinetics. *Corros. Sci.* **2015**, *90*, 375–382. [CrossRef]
10. Riley, J.P.; Chester, R. *Introduction to Marine Chemistry*; Academic Press: London, UK, 1971.
11. Refait, P.; Sabot, R.; Jeannin, M. Role of Al(III) and Cr(III) on the formation and oxidation of the Fe(II-III) hydroxysulfate green rust. *Colloids Surf. A Phys. Eng. Asp.* **2017**, *531*, 203–212. [CrossRef]
12. Boucherit, M.N.; Goff, A.H.-L.; Joiret, S. Raman studies of corrosion films grown on Fe and Fe-6Mo in pitting conditions. *Corros. Sci.* **1991**, *32*, 497–507. [CrossRef]
13. Hansen, H.C.B. Composition, stabilisation, and light absorption of Fe(II)-Fe(III) hydroxycarbonate (green rust). *Clay Miner.* **1989**, *24*, 663–669. [CrossRef]
14. Braithwaite, R.S.W.; Dunn, P.J.; Pritchard, R.G.; Paar, W.H. Iowaite, a re-investigation. *Miner. Mag.* **1994**, *58*, 79–85. [CrossRef]
15. Kohls, D.W.; Rodda, J.L. Iowaite, a new hydrous magnesium hydroxide-ferric oxychloride from the precambrian of iowa. *Am. Miner.* **1967**, *52*, 1261–1271.
16. McGill, I.R.; McEnaney, B.; Smith, D.C. Crystal structure of green rust formed by corrosion of cast iron. *Nat. Cell Biol.* **1976**, *259*, 200–201. [CrossRef]
17. Refait, P.; Abdelmoula, M.; Génin, J.-M. Mechanisms of formation and structure of green rust one in aqueous corrosion of iron in the presence of chloride ions. *Corros. Sci.* **1998**, *40*, 1547–1560. [CrossRef]
18. Rius, J.; Allmann, R. The superstructure of the double layer mineral wermlandite [Mg$_7$(Al$_{0.57}$, Fe$^{3+}_{0.43}$) (OH)$_{18}$]$^{2+}$ [(Ca$_{0.6}$, Mg$_{0.4}$) (SO$_4$)$_2$(H$_2$O)$_{12}$]$^{2-}$. *Z. Für Krist. Cryst. Mater.* **2015**, *168*, 133–144.
19. Bookin, A.S.; Cherkashin, V.I.; Drits, V.A. Polytype diversity of the hydrotalcite-like minerals II: Determination of the polytypes of experimentally studied varieties. *Clays Clay Miner.* **1993**, *41*, 558–564. [CrossRef]
20. Shannon, R.D. Revised effective ionic radii and systematic studies of interatomic distances in halides and chalcogenides. *Acta Cryst.* **1976**, *32*, 751–767. [CrossRef]
21. Grégoire, B.; Ruby, C.; Carteret, C. Structural cohesion of MII-MIII layered double hydroxides crystals: Electrostatic forces and cationic polarizing power. *Cryst. Growth Des.* **2012**, *12*, 4324–4333. [CrossRef]
22. Khaldi, M.; De Roy, A.; Chaouch, M.; Besse, J. New varieties of zinc–chromium–sulfate lamellar double hydroxides. *J. Solid State Chem.* **1997**, *130*, 66–73. [CrossRef]

3. Simon, L.; François, M.; Refait, P.; Renaudin, G.; Lelaurain, L.; Génin, J.M. Structure of the Fe(II-III) layered double hydroxysulphate green rust two from Rietveld analysis. *Solid State Sci.* **2003**, *5*, 327–334. [CrossRef]
4. Bernal, J.D.; Dasgupta, D.R.; Mackay, A.L. The oxides and hydroxides of iron and their structural inter-relationships. *Clay Min. Bull.* **1959**, *4*, 15–30. [CrossRef]
5. Galvão, T.L.P.; Neves, C.S.; Caetano, A.P.F.; Maia, F.; Mata, D.; Malheiro, E.; Ferreira, M.J.; Bastos, A.C.; Salak, A.N.; Gomes, J.R.B.; et al. Control of crystallite and particle size in the synthesis of layered double hydroxides: Macromolecular insights and a complementary modeling tool. *J. Colloid Interface Sci.* **2016**, *468*, 86–94. [CrossRef] [PubMed]

Review

Corrosion Performance of Electrodeposited Zinc and Zinc-Alloy Coatings in Marine Environment

Kranthi Kumar Maniam [1] and Shiladitya Paul [1,2,*]

1 Materials Innovation Centre, School of Engineering, University of Leicester, Leicester LE1 7RH, UK; km508@leicester.ac.uk
2 Materials and Structural Integrity Technology Group, TWI, Cambridge CB21 6AL, UK
* Correspondence: Shiladitya.paul@twi.co.uk

Abstract: Electrodeposited zinc and zinc-alloy coatings have been extensively used in a wide variety of applications such as transport, automotive, marine, and aerospace owing to their good corrosion resistance and the potential to be economically competitive. As a consequence, these coatings have become the industry choice for many applications to protect carbon and low alloy steels against degradation upon their exposure in different corrosive environments such as industrial, marine, coastal, etc. Significant works on the electrodeposition of Zn, Zn-alloys and their composites from conventional chloride, sulfate, aqueous and non-aqueous electrolyte media have been progressed over the past decade. This paper provides a review covering the corrosion performance of the electrodeposited Zn, Zn-alloy and composite with different coating properties that have been developed over the past decade employing low-toxic aqueous and halide-free non-aqueous electrolyte media. The influence of additives, nano-particle addition to the electrolyte media on the morphology, texture in relation to the corrosion performance of coatings with additional functionalities are reviewed in detail. In addition, the review covers the recent developments along with cost considerations and the future scope of Zn and Zn-alloy coatings.

Keywords: corrosion; marine; composites; electrodeposition; superhydrophobic coatings; zinc and zinc-alloys; electroplating; aerospace

Citation: Maniam, K.K.; Paul, S. Corrosion Performance of Electrodeposited Zinc and Zinc-Alloy Coatings in Marine Environment. *Corros. Mater. Degrad.* **2021**, 2, 163–189. https://doi.org/10.3390/cmd2020010

Academic Editor: Henryk Bala

Received: 28 February 2021
Accepted: 14 April 2021
Published: 21 April 2021

Publisher's Note: MDPI stays neutral with regard to jurisdictional claims in published maps and institutional affiliations.

Copyright: © 2021 by the authors. Licensee MDPI, Basel, Switzerland. This article is an open access article distributed under the terms and conditions of the Creative Commons Attribution (CC BY) license (https://creativecommons.org/licenses/by/4.0/).

1. Introduction

Steels are commonly used as structural materials in diverse fields (construction, marine, aerospace, automotive, mining) [1] as they possess interesting engineering properties such as (i) high tensile strength, (ii) melting point, (iii) hardness [2,3]. Among them, mild steel is one of the most widely used materials in various industrial applications such as automotive [4], oil and gas [5], marine (ship hull, naval architecture) [6], owing to its compatible functions and properties with diversified industrial functions. The formation of rust is one of the most widely recognized case of the corrosion, commonly observed with ferrous steel materials such as carbon steel, and can be visualized as a salt of the original metal with different phases (oxides, oxy-hydroxides) in reddish-brown color. The formation of rust scale affects the functional engineering properties of structures, materials besides appearance, strength and liquid, gas permeability (through pores), indicative of material deterioration by corrosion process [7].

Corrosion mitigation is an indispensable challenge, particularly in aggressive environments such as seawater, underground mining, aerospace, automotive to biomedical implants, etc. The annual global cost of corrosion is estimated to be around 2.5 trillion USD, which is ~3.4% of the world's gross domestic product (GDP) [8,9]. Corrosion protection of carbon steel and other low alloy steels has been a topic of interest for many years and are continuously being studied with more emphasis on identifying a suitable alternative to the conventional toxic cadmium coatings [10]. One of the recommended solutions to combat corrosion is to employ a metallic protective coating that can improve the corrosion resistance of ferrous steel materials such as mild steel/carbon steel, low alloy steels, etc. [11] These coatings protect the metal structures by acting either as a physical barrier or as a sacrificial coating [12]. Adherence of coating to the substrate surface and their internal strains are two key parameters that needs to be optimized in order to overcome in-service mechanical stresses such as vibration, friction, etc. [13]. An ideal coating should have higher corrosion resistance and pose minimal environmental threat. Proper selection of metal and its alloys as preferred coating material can be considered as a mitigation technique to combat severe corrosion. In order to choose an ideal metallic protective coating, it is important to consider the intended application and the exposure environment [11].

Considering the intended application and economics, zinc (Zn) is the most commonly used metal that is identified for corrosion protection due to its highly sacrificial nature with electrochemical potential less than that of the ferrous metals such as mild steel/carbon steel [14,15]. Zn is widely used to coat mild steel to prevent corrosion by at least 50%. Additionally, it is the fourth most common metal in use with annual production just below that of iron, aluminum and copper [16]. Moreover, Zn and the corrosion products of Zn are not as toxic as cadmium and are found to be most suitable for corrosion resistant coatings applications. Figure 1 lists the various applications in which zinc and its alloys are employed as protective coatings (either as composite or metallic layers) along with their primary requirements. Overall, zinc and zinc-alloy protective coatings cover a wide range of applications ranging from structural steelwork for buildings, offshore platforms and bridges with flat structures to nuts, bolts, sheet, wire, tubes.

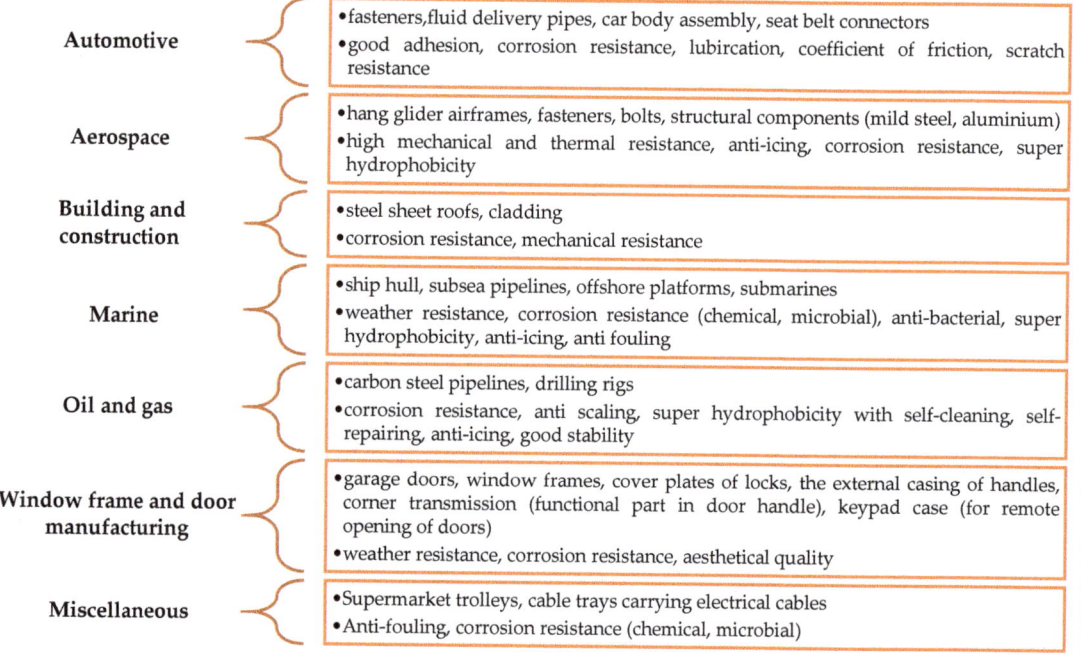

Figure 1. Figure shows the applications of zinc and its alloys along with their functional requirements in various industry sectors.

One of the key benefits in employing a Zn/Zn–alloy coating to protect the metal surfaces is that it offers a cathodic corrosion protection layer which dissolves and significantly delays the time until the substrate material can be attacked by the corrosive environment [17,18]. Generally, corrosion performance of Zn/Zn–alloy coatings are studied under different climatic conditions, regions depending on the nature of corrosivity and test conditions as per different standards. The primary objective of performing corrosion studies is to evaluate the coating durability when exposed to a certain corrosive environment.

When corrosion studies related to the Zn-coated structural materials are performed during their exposure to different corrosive environments, one can expect an initial mass increase. This initial mass increase can be generally ascribed to the corrosion products that are formed on the coating surface as a result of Zn corrosion followed by a decrease in mass indicating the corrosion products separation from coating surface [19]. Such a transition during the initial period of studies (between 1 to 3 years) are reported to be uneven and can occur either sooner or later depending on the presence of certain aggressive constituents (such as carbon dioxide, chloride, sulphates, nitrates), and their relative concentrations. Considering such a scenario, conducting long-term corrosion studies of Zn/Zn–alloy protective coatings under atmospheric conditions deserve significant attention [20] as they provide information on the corrosion products, processes and their formation mechanisms on the coated surface. Studies covering the atmospheric corrosion of zinc in both short and medium term have been published by different groups [21]. A consolidated review on the corrosion performance of the electrodeposited Zn, Zn–alloy coatings performed in different environments such as urban, rural, sea (natural, synthetic), microbial corrosion has been not covered so far. The number of articles that have been published on the zinc-based coatings for different applications in the past 10 years range from 1200–1700 every year (based on the data from scopus), signifying the importance of the field. This review will cover the progress on the recent developments in Zn, Zn–alloy, composite coatings, electrodeposited on different commonly used industrial substrates and their corrosion performance along with future challenges and economics.

2. Corrosion Performance of Zinc and Zinc–Alloy Coatings

2.1. Zn Coatings

Zinc coatings offer flexibility in fabrication and good affordability owing to their sacrificial property [22]. As a consequence, these coatings were fabricated by different methods to protect the bare metallic structures against deterioration and degradation upon their exposure in different corrosive environments. Among the different techniques, electrodeposition is simple, economic and versatile in producing uniform, adherent coatings with variable thickness at processing temperatures <100 °C. On the contrary, other techniques such as hot dip galvanization, ion vapor deposition techniques require high processing temperatures and expensive equipment to produce Zn coatings, and are relatively expensive in electrodeposition besides achieving uniformity in coatings. For instance, the cost to produce a 35-micron thick hot dip galvanized coating is $1.76 /ft^2 in contrast to $0.1/ft^2 [23,24] for the electrodeposited Zn coatings, signifying the techno-economic benefit of electrodeposition. When Zn coatings are exposed to aggressive environments such as coastal, marine which contain rich amounts of chlorides, sulfates, etc., their corrosion resistance is significantly influenced. The factors that influence the corrosion resistance of the zinc coatings obtained via electrodeposition method include: (i) applied current density, (ii) deposition temperature, (iii) electrolyte pH, (iv) mode of current deposition, (v) additives (grain refiners, brightening agents). For instance, Zhang et al. [25] showed that increasing the applied current density to an optimum value during the electrodeposition of zinc increased the nucleation density, cathodic current efficiency and most importantly, improved the grain refinement of the Zn. Grain refinement favors the nucleation while controlling the growth, resulting in a compact deposit. Such features delay the corrosion by reducing the contact area between the corrosive environment and the coating surface [25,26]. The same study has shown a deposit deterioration when the Zn deposition is performed beyond the optimum. Deposition temperature might play a role in (i) controlling the average size of the crystallite, (ii) energy consumption during the process, (iii) current efficiency. Tuaweri et al. [27] reported an increase in current efficiency when employing the acidic sulphate-based electrolyte while achieving a relatively low energy consumption (per unit mass of the deposit) by controlling the temperature between

40–45 °C. Increasing the temperature influenced the rate of deposition and crystal size reduction of the Zn deposit owing to their high cathodic reduction and increased nucleation density while controlling their growth. Increasing the pH of the electrolyte dictates the conductivity which significantly influences the hydrogen ion concentration at the cathode in addition to the electrodeposition of Zn. While lowering the pH favors the conductivity increase and facilitate good deposition, it acidifies the solution below a certain pH. Acidification of the electrolyte solution elevates the hydrogen ion concentration of the cathode and as a result, hydrogen evolution reaction dominates the Zn deposition, thereby affecting the overall deposition process and the corrosion resistance of the Zn deposit As a consequence, significant works were carried out with different types of deposition media with different pH such as acid chloride [28], acid sulphate [29], mixed bath (chloride and sulphate, sulfate–gluconate) [30,31], alkaline zincate baths [32] and acetate baths [33]. Amongst them, acid sulphate was demonstrated to perform better in terms of plating, non-toxic nature and wide operating current density ranges [34].

Many recent studies reported that the modes of deposition, direct current (DC), pulse mode (PC), pulsed cycle reversal mode (PCR), influence the structure and indeed, the corrosion resistance of the Zn deposits. Results from [35–37] showed that deposition of Zn via pulse mode resulted in more compact thinner deposits with (i) less porosity, (ii) better corrosion resistance than the direct current mode, with PCR being predominant. The key advantage with the PCR mode of deposition is that it facilitates the formation of Zn deposits with nano-grains and contributes to better hardness and corrosion resistance than the Zn deposited by other modes of deposition. Wasekar et al. [35] demonstrated this approach by depositing Zn employing different deposition techniques, DC, PC, PCR, and correlated this with the formation of corrosion products on Zn surface. The authors observed that different corrosion products were formed when Zn was deposited using different deposition modes. A compact ZnO was reported to be formed from the corrosion of Zn deposited from PCR. On the contrary, corrosion of the Zn deposited from the other two modes (DC, PC) was shown to form zinc hydroxy-chloride, a highly porous corrosion product. Obtaining Zn deposits with grains in the nanometer range via the PCR mode of deposition was demonstrated to be the key in achieving better corrosion protection properties with high hardness. Such a morphology might facilitate the formation of ZnO film easily via the controlled diffusion mechanism occurring through the grain boundaries.

A different approach that has been identified to improve the corrosion resistance of Zn deposits is the introduction of additives in the electrolyte. Additives can be organic or inorganic and they greatly influence the corrosion resistance of Zn deposits by modifying their structural characteristics, such as (i) surface composition, morphology (microstructure), (ii) grain size, (iii) crystal orientation, texture via controlling the reduction of metal ions [38–40]. They usually get adsorbed to the substrate that is being deposited via the non-bonding electron pairs present in nitrogen, Sulphur, oxygen, hydrophilic groups and (i) enhance the rate of nucleation while controlling the grain growth, (ii) aid the formation of fine, compact, refined deposit. One of the key advantages in obtaining a compact deposit via employing additives is the formation of crystallographic planes with closed packed structure. This contributes to the overall improvement in corrosion resistance of the Zn deposit [41,42]. For instance, Mouanga et al. [43] demonstrated an increase in the intensity of Zn crystal plane (1 1 2) with the addition of urea as an additive in a chloride-based zinc electrolyte. The study focused on the influence of 3 additives: (i) urea, (ii) thiourea, (iii) guanidine (which has same molecular structure but different electron pairing groups: oxygen, Sulphur and nitrogen), and studied the corrosion behavior in relation to the structural characteristics of the deposit. The study concluded that corrosion test results (performed by polarization, weight loss) showed an increase in the corrosion resistance for the Zn deposited in the presence of urea. This was attributed to the presence of oxygen in the molecular structure of urea to function as an effective additive. Though it has been demonstrated that the radical with more free electrons interacted more effectively with the metal substrate in controlling the morphology of the final deposit, the molecular weight of the additive influences its adsorption capability. Ballesteros et al. [44] observed that the molecular weight of an oxygen group containing radical poly-ethylene glycol (PEG) had a significant influence on the final quality of the Zn deposit. When PEG with a molecular weight in the order of $<10^4$ was introduced into the Zn deposition bath, the results showed a greater adsorption of Zn(II) ions with the substrate than the ones containing the higher molecular weight PEG ($>10^4$). Issues were shown to occur on the addition of PEG with molecular weights $>10^4$, which decreased the number of oxygen pair electrons that can form an effective bond with the additive and affected the adsorption characteristics of the Zn(II) ions. The influence of additives towards improving the corrosion resistance properties of the deposits could be correlated with their ability to increase nucleation rate while retarding growth. Employing an additive may result in a higher cathodic overpotential than the non-additive containing electrolytes. High cathodic overpotential tends to increase the formation of new nuclei, increasing its nucleation rate, utilizing the free energy, thereby inhibiting the growth of Zn [45]. In general, the contribution from adding an additive (mostly organic) towards improvement in the corrosion resistance property of electrodeposited Zn coatings can be related to either of the following or their combination:

- texture
- composition
- morphology
- grain size

Electrodeposited Zn is composed of Zn with a hexagonal closed packed (hcp) structure with different crystallographic orientations representing different planes: basal, pyramidal, prismatic. These planes differ in terms of their packing density and significantly influence the corrosion rate. Zn crystals possessing low-index basal plane texture such as (0 0 1) possess high packing density and were reported to be significantly corrosion resistant relative to other orientations and different planes [46]. Based on the published literature, it was identified that promoting the presence of (0 0 2) basal plane via the additives contributed to the corrosion resistance property more effectively than the other crystal planes. For instance, Chandrasekar et al. [37] obtained a more compact Zn deposit with (0 0 2) as the dominant facet by employing polyvinyl alcohol (PVA) as the additive, and demonstrated a significant increase in the corrosion resistance. In this context, it is important to consider the influence of surface roughness over the crystal plane texture. Lowering the surface roughness results in a deposit with fine grain size which lowers the corrosion rate by providing a lower contact area between the deposit surface and the corrosive environment, indicating the predominant influence of grain refining over the crystal orientation/texture. Grain refining achieved via the addition of additives will produce a coating that accelerate the formation of ZnO passive films via the diffusional mechanism and elevate the corrosion resistance. Table 1 lists the most commonly used organic additives that are employed during the Zn deposition in different deposition media along with their functional role. These additives were demonstrated to be contributing towards the enhancement of corrosion protection by imparting additional functionalities to the deposit.

Besides many functions, additives such as thiourea [43] can also influence the compositional change in the Zn deposit with fine grains when added to the electrolyte. Despite its attractive grain refining property, such an addition incorporates sulfur in the deposit which made the neighboring regions anodic and decreased the corrosion resistance. Almeida et al. [47] performed a detailed investigation by studying the influence of glycerol on the corrosion resistance of the electrodeposited zinc obtained via the galvanostatic mode. Glycerol exhibit similar characteristics to urea, coumarin wherein the oxygen atoms double bonded with carbon act as radicals (free unpaired electrons) and favor the adsorption of the organic additive in the Zn deposit. Physical characterization revealed that addition of glycerol played the role of a grain refiner but decreased the intensity of (0 0 2) basal planes similar to the observations made by Chandrasekar et al. [37] and Nayana et al. [48] when the combinations of piperonal +PVA [37], cetyltrimethyl ammonium bromide (CTAB) + veratraldehyde (VV), formic acid (FA) + cyclohexylamine (CHA) [45] are employed as additives. However, the electrochemical test results showed that these coatings possessed the best corrosion resistance. The authors ascribed this to the predominance of grain size over the texture by demonstrating the results from microhardness, surface measurements. An increase in compactness due to the grain refining was shown to exhibit better corrosion resistance despite the decrease in (0 0 2) basal plane.

Table 1. Table listing the additives that have been employed to improve the corrosion resistance property and impart addition functional properties to the Zn deposit.

System	Substrate	Additive	Functional Role [1]	References
Alkaline zincate	mild steel	Poly vinyl alcohol (PVA)	Texture	[37]
Alkaline zincate	mild steel	(PVA) + piperonal	grain refiner	[37]
Acidic sulphate	steel sheet	Gelatin	grain refiner, lowering the surface roughness	[49]
Acidic sulphate	steel sheet	polyethylene glycol (PEG)	grain refiner, lowering the surface roughness	[49]
Acidic sulphate	steel sheet	Saccharin	grain refiner, lowering the surface roughness	[49]
Acidic sulphate	steel sheet	tetrabutylammonium chloride	grain refiner, lowering the surface roughness	[49]
Acidic sulphate	steel sheet	sodium lauryl sulfate	grain refiner, lowering the surface roughness	[49]
Acidic sulphate	mild steel	cetyltrimethyl ammonium bromide (CTAB) + ethyl vanillin	grain refiner	[50]
Acidic chloride	carbon steel	Sodium benzoate	grain refiner	[51]
Alkaline zincate	carbon steel	trisodium nitrilotriacetic (NTA)	complexing agent	[52]
Acidic sulphate	mild steel	(CTAB) + veratraldehyde (VV)	grain refiner, texture, morphology	[48]
Acidic sulphate	glassy carbon	[3-(2-furyl) acrolein]	grain refiner	[29]
Acidic sulphate	mild steel	PEG	grain refiner, texture	[31]
Acidic sulphate	mild steel	CTAB	grain refiner, texture	[31]
Acidic sulphate	mild steel	Thiourea	grain refiner, texture	[31]
Acidic sulphate + gluconate	mild steel	PEG	grain refiner, texture	[31]
Acidic sulphate + gluconate	mild steel	CTAB	grain refiner, texture	[31]
Acidic sulphate + gluconate	mild steel	Thiourea	grain refiner, texture	[31]
Acidic sulphate	mild steel	Polyacrylamide	grain refiner	[36]
Acidic chloride	mild steel	(PEG) and syringaldehyde (SGA)	grain refiner, texture	[28]
Acidic chloride	carbon steel	Formic acid (FA) + cyclohexylamine (CHA)	Texture	[45]

[1] Functional roles are listed based on the conclusions reported by the references mentioned in the table.

2.2. Zn-Alloy Coatings

Though zinc coatings have proven to be acting as a sacrificial layer to protect the ferrous substrates from corrosion, they readily undergo rapid corrosion within a short period of time which significantly impact the overall performance and durable life of the coatings over the period of time depending on its interaction with the type of environment. To enhance the corrosion performance in a harsh environment such as marine, Zn is alloyed with iron group metals, namely cobalt (Co), nickel (Ni), iron (Fe), introduced during the last three decades, with an intention to impart additional functional properties and match the industry market requirements [4,18,53]. Few of these include hardness, uniformity, deformability, weldability, paintability, corrosion and wear resistance. With the ever changing demands from automotive, aerospace, fastener, building and frame and marine industry, active research in the field is being pursued [54]. An exhaustive research has been conducted for many years to explore the possibility of replacing the toxic cadmium coatings with similar corrosion resistant zinc–nickel alloy coatings [10,55–58]. It was demonstrated that Zn–Ni alloys with Ni content of 12 to 15 wt.% possessed excellent corrosion resistance properties with longer corrosion protection life, reduced corrosion rate while retaining the primary sacrificial anodic behavior. Numerous studies were conducted to support the fact that incorporating Ni in the Zn–Ni alloys enhances the corrosion resistance of the overall coatings [53,59–61]. Besides, studies with varying Ni contents concluded that Zn–Ni alloys tend to become nobler with increasing Ni content and tend to lose their sacrificial property (with respect to steel) when the deposit contains above 30 wt.% Ni. Such Zn–Ni coatings transit from active to passive owing to their increasing nobler character, show cathodic behavior and favor the corrosion of bare ferrous steel substrates. Incorporation of Ni could slow down the dissolution rate of Zn when present in the range of 12 to 15 wt.%, retarding the dissolution of zinc and delaying the corrosion of bare ferrous steel substrate. Zn–Ni coatings with 12–15 wt.% Ni are known as "γ"-phase coatings and exhibit the best corrosion resistance [62]. Despite their excellent corrosion resistance, Zn–Ni alloy coatings lack two properties: (i) phosphatability and (ii) paintability, rendering them weak in coating applications. As a consequence, zinc–iron (Zn–Fe) alloys were introduced and studied extensively on the deposition from chloride, sulfate (with moderate pH) alkaline baths [54,63] and extended to Zn–Co coatings. While electrodeposited Zn–X (X: Ni, Co, Fe, Mn) have gained significant attention, development of Zn–Mn alloys with Mn contents varying from 10 to 40 wt.% paced up rapidly. Alloying Zn with Mn (10 to 40 wt.%) could facilitate the formation of an insoluble passive barrier layer, which enhances the protective ability of the coatings, and impart better corrosion properties [64]. However, Zn–Ni alloy coatings are reported to be corrosion resistant amongst the other alloy coatings such as Zn–Fe, Zn–Co, Zn–Mn in a marine environment, with good mechanical properties, and considered as a potential alternative to toxic Cd coatings [65]. As the industry interest is shifting towards the development of lightweight materials, automotive industries shed some light on the development of Zn–Mn electrodeposits on base substrates such as aluminum (Al), magnesium (Mg). As the potentials of electrodeposited Zn–Mn alloys are in close proximity with reactive substrates: Al, Mg, they tend to serve more actively as a sacrificial anode and justify their ability to protect the surface from corrosion. Zn–Mn coatings offer excellent steel corrosion protection due to their good synergy, passive corrosion product layer, that are formed in corrosive environments [66] despite the fact that Mn is a thermodynamically less noble character than Zn. The synergistic effects can be attributed to the protective ability of Zn–Mn alloy deposits combined with the insoluble passive corrosion product layer. Obtaining Zn–Mn alloys by electrodeposition needs complexation because zinc and manganese have reversible potentials different by more than 0.4 V [67]. This motivated the scientific community to study Zn–Mn alloy electrodeposition, and previous results have shown that the coatings with increased Mn content offer salient benefits such as: (i) passive layer formation comprising oxides of Mn and Zn salts, (ii) monophasic structure. The formation of a compact, insoluble passive layer will not only control the anodic dissolution [68], but also favor the inhibition of dissolved oxygen reduction at the cathode [69].

Zn–Mn alloys with monophasic structure are reported to hinder the local corrosion cell formation that generally originates in dual phase structure, indicative of better corrosion resistance in the former [37]. Fashu et al. [70] demonstrated that crystal size influences the behavior of Zn–Mn alloy deposits during corrosion, with a smaller size showing the best results. Claudel et al. [71] demonstrated that Zn-Mn alloys with Mn contents up to 30 wt.% could be achieved on steel substrates by pulse plating with a faradaic efficiency up to 90% in contrast to 65% efficiency by direct current. Additionally, the deposits obtained were pore-free and homogeneous when pulse plating was employed. Obtaining a small crystal size with high Mn content is difficult to achieve, as increasing the Mn content could aid the increase in crystal size of the monophasic Zn–Mn alloys and affect its corrosion resistance. Bucko et al. [72] observed such a phenomenon while depositing Zn–Mn alloys and concluded that incorporation of a high amount of Mn in the Zn–Mn alloy and monophasic structure are not the only conditions that enhance corrosion resistance. There are certain factors that affect the corrosion behavior of Zn–alloy coatings. Deposition temperature is a parameter which influences the metal–alloy electrodeposition process, and has the capability to tailor the corrosion resistance, structural characteristics (micro/nano), mechanical properties and alloy composition of Zn–alloy coatings. Beheshti et al. [73] conducted an experimental study on the effect of deposition temperature in relation to the structural properties, phase composition and corrosion behavior of Zn–Ni alloy electrodeposits on API 5L X52 low carbon steel using an aqueous chloride bath. The deposition temperature was varied from 25–70 °C and the corrosion behavioral study was conducted via electrochemical characterization techniques: linear polarization resistance, immersion method using 3.5 wt.% NaCl solution, analyzed in relation to the surface morphology. The study demonstrated that the Zn–Ni alloy electrodeposited via chronopotentiometric (constant current) method at 25 °C exhibited a compact and dense morphology with good uniformity, less crack and highest corrosion resistance. Additionally, Ni content was reported to be within the range of 12 to 15 wt.%. Increasing the temperature beyond 25 °C resulted in an increase in Ni content, decreased the uniformity, compactness of the deposits and the corrosion resistance. This was attributed to the formation of more cracks in the Zn–Ni coatings with increasing temperature due to the internal stress resulting from hydrogen embrittlement, indicative of predominant hydrogen evolution reaction. Hydrogen evolution reaction is a cathodic reaction commonly observed in aqueous electrolyte media which competes with the electrochemical reduction reaction between Zn/Ni, facilitates the hydrogen to diffuse inside the coatings, resulting in a brittle deposit inducing crack. Additionally, deposition of Zn–Ni at higher temperatures shall increase the Ni content in the alloy, making the deposit nobler than the ferrous steel substrate. Zn–Ni deposits shall then lose their sacrificial ability, thereby accelerating the corrosion of the underlying less noble ferrous steel substrate. Therefore, optimizing the deposition temperature was shown to be an important parameter in improving the properties of Zn–Ni alloys in aqueous solutions such as (i) corrosion resistance, (ii) mechanical (crack formation control), (iii) phase composition, (iv) structural (uniformity, compactness).

Alloying Zn with cobalt (Co) in low contents (<3 wt.%) are considered a potential alternative to the conventional Zn–Ni systems owing to their (i) less noble character than steel, (ii) better corrosion protection properties than Zn coatings [74,75]. In addition, their possibility to achieve the desirable surface finishing properties such as brightness, decorative aspects with low Co contents (1–3 wt.%) in contrast to high Ni wt.% (12–18 wt.%) in Zn-Ni makes it an economically viable candidate to replace the toxic Cd coatings. Significant works has been reported from past 2–3 decades from different electrolytes such as (i) acidic-chloride, (ii) alkaline-sulfate, (iii) cyanide, and shown that the deposition follows an anomalous type similar to Zn–Ni. Among them, Zn–Co with Co content in the range of 1 wt.% was shown to exhibit superior corrosion resistance and is widely accepted by the various industry segments (automotive, marine, sanitary) [76,77]. One of the major hurdles with the current chemistries is the presence of carcinogenic compounds as cyanides, complexing agents which pose human threats, environmental challenges.

Replacing the electrolyte with acetate ones has shed some light on these alloy coatings and is shown to produce Zn–Co alloys with good corrosion protection properties. Selvaraju and Thangaraj [78] fabricated Zn–Co alloys via direct current electrodeposition on mild steel substrates and studied the influence of current density in relation to the corrosion resistance of the Zn–Co coating. It was demonstrated that Zn–Co deposited at 4 A dm^{-2} exhibited (i) better coverage with good throwing power, (ii) hardness with high corrosion resistance and (iii) reduced corrosion rate. The authors attributed the enhancement in corrosion resistance to the texture, morphology obtained with the acetate-based electrolyte and demonstrated its techno-commercial capability to replace the currently used electrolytes.

2.3. Zn and Zn–Alloy Composite Coatings

To enhance the strength, durability of zinc-based coatings for their application in harsh conditions, metal nanoparticles with better chemical stability than the matrix are often incorporated. These additions promote the development of microstructures with a uniform lower number of surface defects, facilitate the formation of stable passivation film with good adherence and resist further corrosion attack. Such coatings are referred to as composite coatings. Unlike alloys, composites are made from two or more different materials, which are physically distinct from each other by certain boundary/interface and contain 2 phases: (i) a continuous matrix phase, (ii) an insoluble reinforcement phase, bonded in such a way as to form a solid material. Alloys are obtained through a combination of two or more materials (metallic, non-metallic) which form a homogeneous solid solution at a certain temperature. Composites are reinforced materials that are tailored to either enhance the existing properties of a coating or impart additional functional properties that might be required for a specific application. Studies of composite coatings are shown to possess improved corrosion, mechanical properties than the traditional metallic coatings [79,80]. Therefore incorporating the metal nanoparticles into the metallic coatings broadens their range of applications and is reported to perform good while minimizing the addition of hazardous chemical agents (complexing, organic chelating) with elimination of chrome passivation. This could reduce (i) the environmental impact, (ii) economics, and as a result, these composite coatings are encouraged to substitute for the cyanide-based aqueous Zn/Zn-alloy coatings [81–83]. The development of Zn, Zn–alloy composite coatings by electrodeposition is motivated by the sacrificial ability of Zn in protecting bare steel against corrosion, thereby making it attractive to fabricate advanced novel matrix composite coatings with improved surface properties in oil and gas, marine, automotive, aerospace, etc. [84,85]. These applications are quite demanding with ever changing market dynamics, and hence, the Zn-based composite coatings technology attracts significant interest.

Zn-composites have been demonstrated as technically competitive in comparison to the Zn coatings in harsh corrosive environments such as marine, coastal, their overall corrosion protection life is mitigated by the early formation of their corrosion products [80]. It is important to optimize the conditions to obtain a composite coating with improved particle dispersion and microstructure as the quality of the composite based deposit is dependent on the deposition conditions besides particle loading, concentration and the way of particle incorporation [86–88]. Tuaweri et al. [88] studied the influence of applied current density, deposition time, particle concentration, agitation in relation to the current efficiency, deposit characteristics of Zn–SiO$_2$ composite coatings. The results showed that Zn–SiO$_2$ composite coatings displayed a higher cathode current efficiency at low current densities, SiO$_2$ concentration of 26 g L^{-1} under an agitated condition. With a further increase in time, Zn dendrites were shown to face certain struggle in building up through the dense SiO$_2$ layer, indicative of predominant dense SiO$_2$ as the top layer. Tuaweri and Ohgai [88–90] investigated the effect of time, current density on the composite growth, thickness and studied in relation to the increase in weight, thickness, microstructural characteristic of the Zn-SiO$_2$ deposit. It was shown that the composite thickness and its growth was not significantly affected on varying the current density. Though the coating became thicker with deposition time, cracks were reported to be growing with time. Such a composite is prone to rapid corrosion owing to the rapid transport of corrosive species through the cracks formed at the surface. In order to achieve good composite coatings with enhanced properties, it is necessary to optimize not only the deposition conditions but also control the particle dispersion and distribution. Incorporation of particles (metal oxides, ceramics, borides, nitrides, carbides, etc.) into the matrix might tend to impede the grain growth, structural characteristics which subsequently shall result in the formation of small-sized crystals containing the microstructures [91–93].

By dispersing them in the Zn matrix, the defect prone regions of the composite coatings such as pores, gaps, microholes, crevices, etc., which represent the corrosion active defective sites, get covered up and form a compact layer, acting like a physical barrier in separating the corrosive species from the metal matrix [92,94]. Praveen et al. [95], Punith et al. [96] and Rekha et al. [97] reported on the corrosion performance of Zn composites containing nano-sized carbon particles. The test results conducted employing 3.5 wt.% NaCl electrolyte solution as the corrosive media showed that the Zn metal dissolution in the matrix took place at a steady rate in comparison to the Zn metal coatings and at higher anodic potentials. Zirconia (ZrO_2) is reported to exhibit high hardness and thermal stability with excellent wear resistance and a similar coefficient of thermal expansion to that of iron [98]. Considering the advantages of ZrO_2, Vathsala et al. [91] and Setiawan et al. [99] studied the influence on the corrosion resistance of Zn by incorporating ZrO_2 nanoparticles in the Zn metal matrix. The results demonstrated that incorporating ZrO_2 (i) influenced the kinetics of the electrode reactions, (ii) favored the formation of a stable passivation layer, (iii) enhanced the corrosion protection of the composite coatings. To improve the coatings' corrosion performance, it is necessary to optimize and establish the particle loading/incorporation. For instance, Malatji et al. [100] demonstrated that addition of Al_2O_3, SiO_2 to the Zn metal matrix beyond an optimum concentration of 5 g L^{-1}, resulted in the formation of agglomerates, decreasing the corrosion performance of the composite coatings significantly. Incorporating the agglomerates in the metal matrix could promote the initiation of surface defect sites, chemical heterogeneities in the final composite coatings that will directly (or indirectly) contribute to the overall corrosion degradation performance.

There are significant works reported on the Zn-alloy composite coatings: composite coatings prepared from Zn–Co [101,102], Zn–Ni [63,80], Zn–Fe [103] were identified to be excellent candidates for corrosion protection. Among them, Zn–Ni composite coatings attracted predominant interest owing to the chemical stability of Ni and mechanical properties [63]. Zn–Ni composite coatings incorporated with metal oxides (Al_2O_3, SiO_2, TiO_2, ZrO_2) and carbides (SiC) were formulated to enhance (i) corrosion resistance, (ii) good adhesion, (iii) hardness, (iv) wear resistance, (v) crack free surface [81,85]. For instance, incorporating Al_2O_3 in the Zn–Ni matrix with uniform distribution was shown to (i) minimize surface defects, (ii) achieve smaller crystallites, (iii) improve the grain growth, compactness in the final deposit [104]. It was also demonstrated that incorporating the metal oxide particles along with its size, influenced the crystallite size of the deposited Zn–Ni composite. Besides, the addition of second phase metal oxide particles shall also influence the Ni content in the electrodeposited Zn–Ni composite. Works from [104,105] demonstrated the addition of metal oxide particles: Al_2O_3, SiO_2 influenced the Ni content which increased up to 12.3 wt.% with a significant decrease in Zn up to 87.7 wt.% in the final Zn-Ni composite. Furthermore, inclusion of metal oxide particles into the Zn-Ni matrix shall influence the morphological features of the resultant composite coating. [106] reported on the morphological transitions of the Zn–Ni composite coating from spherical nodular like to cauliflower type morphology when Al_2O_3 was added into the matrix. Corrosion test results from [107] showed that the addition of Al_2O_3 particles in the concentration range of 5 g L^{-1} to the deposition electrolyte solution yielded a deposit which displayed (i) reduced corrosion currents, (ii) increased polarization resistance. These results in combination with the data obtained by [83,108] conclude in the fact that the Al_2O_3 imparts a corrosion inhibiting effect on the corrosion of the composite matrix owing to low electronic conductivity, thereby perturbing the corrosion current when present. Similarly, the Zn–Ni composite matrix consisting of SiO_2 nano-sized particles was reported to be possessing excellent corrosion resistance when tested in 3 wt.% NaCl solution [100,105]. Table 2 shows the list of Zn and Zn-alloy composites that showed significant progressive developments in the past 10 years and have been focused upon in the recent reviews [109].

Table 2. Table listing the Zn and Zn–alloy composites that have been developed in the past 10 years.

Zn/Zn-X	Second Phase	Substrate	System	Mode of Deposition	E_{Corr}, V (SCE)	i_{Corr}, µA cm^{-2}	References
Zn	CeO_2	mild steel	chloride	direct current	−1.127	3.56	[110]
				pulse current	−1.147	0.69	
Zn	$TiO_{2.5}$	steel	sulfate	direct current	−1.052	2.7	[111]
				pulse current	−1.118	15.1	
Zn	SiO_2	mild steel	chloride	galvanostatic	−1.127	~1	[100]
Zn	Al_2O_3	mild steel	chloride	galvanostatic	−1.282	~1	[100]
Zn	ZrO_2	mild steel	sulfate	direct current	−1.034	4.45	[91]
Zn	SiC	mild steel	sulfate	direct current	−1.100	2.090	[112]
Zn	graphene oxide	mild steel	sulfate	direct current	−1.131	4.1	[113]
Zn–Ni	TiO_2	steel	citrate	galvanostatic	−0.90	176	[82]
Zn–Ni	Fe_2O_3	mild steel	sulfate	direct current	−1.1991	0.682	[114]
Zn–Ni	CeO_2	mild steel	chloride	reverse pulse current	−0.78	28	[115]
Zn–Fe	graphene	mild steel	sulfate	direct current	−1.087	19.20	[103]
Zn–Co	CNTs	mild steel	sulfate	direct current	−0.901	0.156	[102]

Considering the significant advances in the utilization of TiO_2 for development of Zn–Ni–TiO_2 composites to achieve better corrosion resistant, mechanical properties, Anwar et al. [82] studied the corrosion behavior analysis of the Zn–Ni–TiO_2 composite deposited via galvanostatic mode on mild steel substrates. Deposition was performed from citrate-based baths containing nano-sized TiO_2 as these baths are identified to be stable in nature and comparison was made with chloride (non-citrate)-based bath. The authors observed that the Zn–Ni–TiO_2 deposits from citrate-based electrolyte yielded the following: (i) formation of compact coatings, (ii) small sized crystals, (iii) uniform texture, (iv) reduced hydrogen evolution, (v) good corrosion resistance. In addition, they revealed the corrosion products that are formed on the γ-phase Zn–Ni composites upon exposure to seawater environment when conducted in laboratory. Their study demonstrated that ZnO (zincite), $Zn(OH)_2$ (Wulfingite), $Zn_5(CO_3)_2(OH)_6$ (hydrozincite), $Zn_5(OH)_8Cl_2$ (simonkolleite) are the predominant products that are formed due to the corrosion which is aligned with the data proposed by Leygraf et al. [21]. Figure 2 shows the sequence of corrosion products that are formed on the Zn–Ni–TiO_2 composite surface upon exposure to corrosive media over a period of time recorded up to 72 h. It was shown that the immersion time had significantly influenced the composition of the corrosion products with simokolleite being the predominant. Though the initial corrosion resistance was shown to be lower, there was a significant increase in corrosion resistance on increasing the exposure time beyond 24 h owing to the formation of the robust, compact corrosion product layers (hydrozincite, simonkolleite). The authors observed the conversion of simonkolleite back to hydrozincite, and attributed this to the unstable nature of hydrozincite and demonstrated its conversion back to simonkolleite, which is represented as a reversible loop between hydrozincite and simonkolleite in Figure 2.

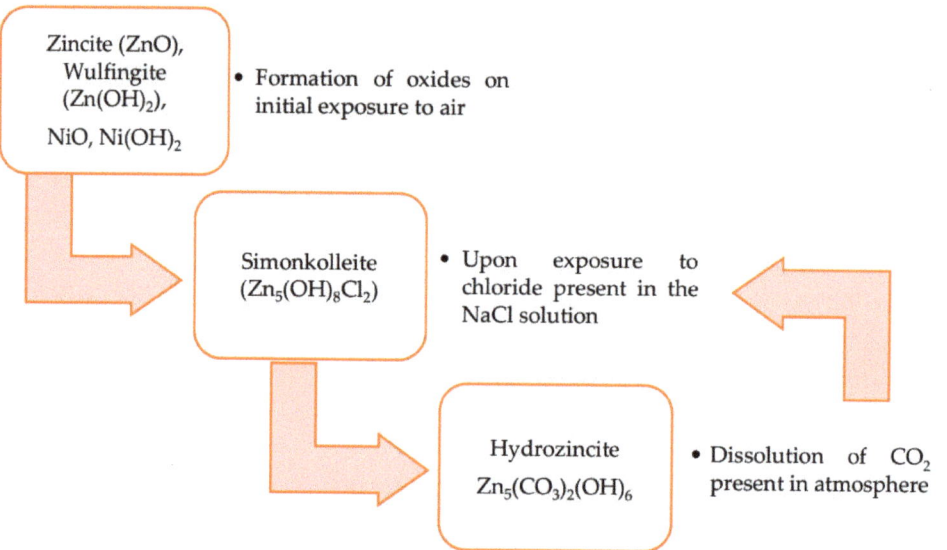

Figure 2. Figure depicting the schematic sequence of corrosion products that are formed on the surface of Zn-Ni-TiO$_2$, as reported in [82].

3. Recent Developments
3.1. Zn and Zn–Alloy Deposition in Ionic Liquids

Though there are significant works on the deposition of Zn and Zn-alloys on various substrates from aqueous solutions, there are certain challenging issues that remain to be critically solved. All the zinc alloys with an alloying metal from the iron group (nickel, iron, cobalt) are obtained under so called anomalous codeposition; that is, with preferential deposition of the less noble zinc. One of the possible reasons for the anomalous codeposition is the formation of zinc hydroxide followed by its adsorption on the surface during the hydrogen evolution reaction. This will hinder the reduction of respective alloying metal ions (Ni^{2+}) and control the overall alloy (Zn-Ni) composition as the high surface activity of zinc ions facilitates the easy replacement, inhibition of Ni ions and its nucleation, growth in the case of Zn–Ni alloy deposition. This combined with the hydrogen evolution affected the quality of the deposit drastically in terms of (i) visual appearance, (ii) crack formation, (iii) adhesion, (iv) brittleness, (v) throwing power, (vi) structural properties, (vii) corrosion resistance behavior. Besides, the optimization of the deposition parameters (current density, temperature, mode of deposition, bath agitation), electrolyte conditions (pH, concentration), additives are added either as complexing agents or levelers or brighteners or their combination which favor the anomalous deposition, formation of passivation layers (resulting due to corrosion). These shall circumvent the brittleness of the deposit by controlling the hydrogen embrittlement due to the evolution reaction and reduce the crack formation, thereby delaying the corrosion. Though there are significant research works carried out on the structural-property relation of the Zn–alloys, scientific understanding of the structural features (morphology, crystal size and orientation, alloy composition) in relation to their corrosion behavior has not been fully established. For instance, it is known that the rate of Ni deposition in the Zn–Ni alloy is hindered by the formation of zinc hydroxide. From the studies of [60,116], it was shown that the pH value measured near to the cathode surface doesn't form zinc hydroxide, indicating that the hindrance in the electroreduction of Ni ions did not occur via the hydroxide formation mechanism. The combination of Zn deposition and monolayer formation during the underpotential deposition and high overpotential of Ni resulted in anomalous Zn–Ni deposition with controlled inhibition of the Ni ions' nucleation and growth process. Due to the ever-growing demand across a range of engineering and structural market applications, in deposition of Zn and Zn–alloys in non-aqueous electrolyte media, (i) ionic liquids and (ii) deep eutectic solvents were identified as an alternative technical competitive approach. Ionic liquids (ILs) are composed of single organic cation and an inorganic/organic anion while deep eutectic solvents (DESs) contain a combination of cations, anions. These media exhibit similar physical properties but differ in terms of synthesis, chemical properties. By employing ILs, it is possible to (i) eliminate the hydrogen gas liberation, (ii) tailor the redox properties, (iii) achieve the desired physical, chemical properties, (iv) control nucleation characteristics [117–121]. Their large electrochemical windows in combination with their good physico-chemical properties, thermal stabilities, low vapor pressures make them versatile for electrodeposition of Zn and Zn-alloys, enhance the coatings' corrosion resistance performance [122]. ILs offer an ideal alternative for the electrodeposition of Zn and its alloys such as Zn–Ni, Zn–Fe, Zn–Co, Zn–Mn in two ways. First is the hydroxide suppression mechanism that is responsible for the formation of anomalous deposits can be eliminated, and second is the elimination of hydrogen liberation owing to the absence of water in the non-aqueous bath [123]. The motivation for using ILs in Zn-alloy deposition such as Zn–Mn is due to (i) solution instability in aqueous media, (ii) low current efficiency, (iii) poor deposit morphology. Poor quality deposits and low current efficiencies arise in the case of Zn–Mn alloy coatings because these require higher negative potentials to reduce Mn, which results in drastic hydrogen gas liberation at the cathode [124]. One key benefit in using ILs is their ability to tune redox potentials via the metal speciation and promote better co-deposition of metal alloys: Zn–X (X: Ni, Co, Fe, Mn) without the need for a complexing agent, unlike aqueous electrolytes [125–128]. Since Zn and Mn co-deposits have a large difference in their redox potentials, employing an IL with a high electrochemical window shall favor the metals' co-deposition owing to their better tailoring properties. Table 3 shows the corrosion parameters obtained from the electrochemical characterization of the Zn and Zn–alloys deposited from ionic liquids.

Table 3. Table showing the corrosion parameters for the deposition of Zn and Zn-X alloys (X = Ni, Mn etc.) in ionic liquids (ILs).

System	Coating	Substrate	Mode of Deposition	Corrosion Test Method	E_{Corr}, V (vs Pt/SCE)	I_{Corr}, $\mu A\ cm^{-2}$	References
ChCl–Urea	Zn	Carbon steel	Potentiostatic	LPP	−0.289 [1]	0.68	[129]
	Zn–Mn (0.4–0.7) [3]	Copper	Potentiostatic	LPP, EIS	−1.021 [1]	1.075	[70]
	Zn–Mn (0.4–1.0) [3]	Copper	Potentiostatic	LPP, EIS	−1.054 [1]	0.917	
	Zn–Mn (0.4–1.4) [3]	Copper	Potentiostatic	LPP, EIS	−1.098 [1]	1.175	
	Zn–Mn (0.4–0.7) [3]	Copper	Potentiostatic	LPP, EIS	−1.062 [1]	0.989	
	Zn–Mn (0.4–1.0) [3]	Copper	Potentiostatic	LPP, EIS	−1.079 [1]	0.875	
	Zn–Mn (0.4–1.4) [3]	Copper	Potentiostatic	LPP, EIS	−1.109 [1]	1.251	
ChCl–Urea (1 wt.% H_2O)	Zn–Ni	Carbon steel	Potentiostatic	LPP	−0.414 [1]	0.82	[129]
ChCl–Urea (3 wt.% H_2O)	Zn–Ni	Carbon steel	Potentiostatic	LPP	−0.478 [1]	1.3	
ChCl–Urea (5 wt.% H_2O)	Zn–Ni	Carbon steel	Potentiostatic	LPP	−0.801 [1]	2.1	
ChCl–Urea (7 wt.% H_2O)	Zn–Ni	Carbon steel	Potentiostatic	LPP	−0.931 [1]	5.6	
ChCl –EG	Zn	Mild steel (AISI 304)	Potentiostatic	LPP, EIS	−1.040 [1]	6.57	[130]
[EMIm][Tf_2N]-	Zn–Mn	DP-1000 steel	Potentiostatic	LPP	−1.016 [1]	0.0119	[131]
Zn[Tf_2N]	Zn–Mn	DP-1000 steel	Potentiostatic	LPP	−0.776 [1]	0.0112	[131]
ChCl–Urea	Zn	WE43-T6 Mg alloy	galvanostatic	LPP	−1.420 [1]	38.68	[132]
ChCl–Urea	Zn–Mn (1–1) [4]	Steel	galvanostatic	LPP	1.110	1.06	[128]
ChCl–Urea	Zn–Mn (1–1) [4]	Steel	galvanostatic	LPP	1.040	3.2	[128]
ChCl–Urea	Zn–Mn (1–1) [4]	Steel	galvanostatic	LPP	1.045	3.6	[128]
ChCl–Urea	Zn–Mn (1–3) [4]	Steel	galvanostatic	LPP	1.130	0.90	[128]
ChCl–Urea	Zn–Mn (1–3) [4]	Steel	galvanostatic	LPP	1.040	0.82	[128]
ChCl–Urea	Zn–Mn (1–3) [4]	Steel	galvanostatic	LPP	1.046	5.3	[128]
ChCl –EG	Zn	Copper	galvanostatic	LPP	−1.197 [2]	7.987	[133]
NaOAc: EG[2]	Zn	Mild steel	galvanostatic	LPP	−1.066 [2]	1.01	[134]

[1] Linear potentiodynamic polarization (LPP) conducted in 0.1 M $NaNO_3$ (or) 3 wt % NaCl solution; EIS: Electrochemical impedance spectroscopy. [2] NaOAc: EG–Sodium Acetate: Ethylene Glycol. [3] Zn–Mn(0.4–X) indicate 0.4 M of $ZnCl_2$ + X M $MnCl_2 \cdot 4H_2O$ in the electrolyte solution containing choline chloride: Urea in the molar ratio 1:2. X: 0.7–1.4 [70]. [4] Zn–Mn(1–Y) indicate 0.1 M of $ZnCl_2$ + Y M $MnCl_2 \cdot 4H_2O$ in the electrolyte solution containing choline chloride: Urea in the molar ratio 1:2. Y: 0.1; 0.3 [128].

3.2. Superhydrophobic Zn and Zn–Alloy Coatings

In recent times, superhydrophobic coatings are considered a beneficial approach for corrosion protection of metallic structures for a variety of applications such as aerospace, marine, oil and gas and so on. Superhydrophobic surfaces are usually formed with a combination of low surface energy materials and rough microstructures. To create superhydrophobic surfaces to resist against corrosion, it is important to create rough microstructures [135,136]. On one hand, the rough microstructure surfaces trap the air within them when they are in contact with water, acting like an additional barrier and retard the corrosion rate on aircraft and ship surfaces. On the other hand, they exhibit self-cleaning, anti-fouling, anti-icing/de-icing properties which enable them to be suitable potential candidates for protecting pipelines and other surfaces that are exposed to the marine environment besides corrosion [137]. Low surface energy substances (generally organic-based) are often added directly to the electrolyte solutions to achieve superhydrophobic coatings. With the addition of low surface energy materials to the electrolyte solution containing the metal ions, they tend to react with the functional groups of these substances during electrodeposition and form a coating with low surface energy and high-water angle on the cathode surface. The key advantage of such an addition is that superhydrophobicity can be obtained without the need for any surface modification after electrodeposition [138]. The high-water contact angle will directly influence the reaction between the corrosive species and the bare metallic substrates (generally mild steel) and prolong the life of the coatings by lessening their reaction time. On increasing surface hydrophobicity, it is possible to limit the metals' interaction with corrosive species, such as water and other ions such as Cl^-, SO_4^{2-}, CO_2, etc., and reduce the corrosion rate of the coatings deposited. For organic anticorrosive coatings, incorporating a superhydrophobicity property would impede the diffusive mass transport of water molecules and enhance the coating's protectiveness against corrosion of underlying metallic structures for longer periods [139]. In cases such as oil and gas, these coatings seem to be an economical solution to control the corrosion and fouling in pipelines for transporting oil and gas related products such as natural gas liquid products and liquid propane via subsea, and also, they have a high tendency to be used over different substrates [140].

Considering the likelihood of obtaining several surface morphologies with varied roughness and different microstructures, electrochemical deposition is considered to be the most versatile in terms of simplicity, scalability and cost effectiveness. Table 4 lists the Zn coatings prepared by electrodeposition that exhibit superhydrophobic properties. The most widely preferred mode to obtain a metallic coating is via the electrochemical deposition at the cathode. It is also possible to achieve coatings with superhydrophobic, corrosion resistance properties by anodic electrodeposition. For instance, Wang et al. [141] performed anodic electrodeposition and obtained a superhydrophobic coating on metallic zinc anode surfaces from the solution containing zinc tetradecanoate with platinum as the cathode. A corrosion resistant superhydrophobic Zn layer was formed on the zinc anode substrate by one-step potentiostatic deposition at 30 V for 2 h and room temperature. The authors demonstrated the possibility of obtaining the superhydrophobic coatings by oxidizing the Zn to Zn^{2+} initially, which resulted in the formation of a superhydrophobic Zn deposit film by combining with tetradeconate on the anode surface. Corrosion test results of the superhydrophobic Zn coatings showed an enhancement in corrosion protection of the substrate. The behavior of the air medium that is trapped between the pockets of the superhydrophobic surface was shown to be similar in the action of a dielectric film in a parallel plate type pure capacitor. Such a configuration would improve the corrosion resistance life of the substrate through circumventing the metallic pathway between the substrate and the electrolyte.

Table 4. Table showing the list of Zn coatings with superhydrophobic properties prepared by electrodeposition.

Coating	Substrate	System/Bath	Surface Energy Reducer Agent	CA°	Reference
Zn	steel	chloride	vulcanized silicone polymer	155 ± 1	[142]
Zn	X65 steel	sulphate	stearic acid	158.4 ± 1.5	[143]
Zn	X90 steel	sulphate	perfluoro octanoic acid	154.21	[144]
Zn	carbon steel	Sulfate-acetate	stearic acid	153	[145]
Zn	carbon steel	alkaline	stearic acid	158.7	[146]
Zn	copper	DES [1]	stearic acid	164.8 ± 0.6	[147]

[1] DES: Deep eutectic solvent consisting of chloine chloride:ethylene glycol (1:2).

In a study by Wang et al. [148], the zinc-laurylamine superhydrophobic complex film with corrosion resistant properties was obtained on a zinc substrate via the same anodic electrodeposition route. The corrosion resistance of the deposited film was investigated in a simulated marine environment. The results showed that the superhydrophobic film coating was corrosion resistant with a protection efficiency of ≥99% [149]. Obtaining structures similar to Micropogonias Undulatus scales on the coatings via electrodeposition could result in micro patterns with superhydrophobicity. Such micro patterns exhibit the similar skin surface topographical features that are observed with marine creatures (sharks and fish) [150]. Considering the advantages with electrodeposition in obtaining structures on various geometries from simple to complex, it can be assumed that such a pattern is achievable. Inclusion of micropatterns similar to the topographical features of marine creatures (e.g., Micropogonias Undulatus-like scales) is expected to boost the physical properties and contribute to the enhancement in the corrosion resistance of the mild steels [151]. A number of scientists and researchers have leveraged the benefits of zinc coatings fabricated by electrodeposition in improving the corrosion resistance [11]. Li et al. [144] fabricated a crater-like Zn structure on an X90 steel pipe surface with superhydrophobic coating via 2 steps: (i) galvanostatic electrodeposition in sulfate electrolyte followed by (ii) chemical modification using perfluorooctanoic acid (PFOA). Contact angle measurement data showed a stable value of ~150 °C even after exposure to air for 80 days and the superhydrophobic coatings demonstrated good quality with self-cleaning properties and air stability. In addition, these coatings were shown to play a dual role acting as self-cleaning coatings on the one hand and exhibiting cathodic protection on the other hand, thereby enabling a double protection to the bare metal substrate. Imparting superhydrophobic properties to Zn coatings shall overcome the limitations of short corrosion life that are commonly observed with conventional Zn coatings under high humid conditions such as coastal and marine environments. Such a surface can resist the formation of a moisture film owing to its small tilt angle or high-water contact angle, which makes it difficult to hold the water molecules. Polyakov et al. [145] aimed at investigating the possibility of forming superhydrophobic Zn coatings and estimating their corrosion protection ability under salt spray chamber conditions, using 0.5 M NaCl test solution. Attempts were made by modifying the electrochemical pretreatment of carbon steel surface prior to deposition followed by 2-stage treatment in obtaining the Zn coatings with superhydrophobic properties. The 2-stage treatment involved the potentiostatic deposition of Zn dendrites from sulfate–acetate-based electrolytic solution followed by treatment with stearic acid (hydrophibising/surface energy reducing agent). The results showed that employing such an electrochemical pretreatment will play a vital role in preserving the superhydrophobic properties of the obtained coatings as the pretreated surface via galvanostatic method provides a polymodal surface with adequate roughness for creating an anti-wetting surface. Additionally, corrosion test results from salt spray, 0.5 M NaCl confirmed that the coatings can withstand severe corrosion owing to the formation of a gas interlayer on the superhydrophobic coating surface which acted like an insulator or dielectric film, thereby preventing the Zn dissolution. This was justified by evaluating the average value of the wetting angle for the superhydrophobic coated with Zn/Stearic acid that was shown to be $\geq 151°$ after 148 h of exposure in the salt spray chamber. The authors also identified that ultrasonication of Zn coatings with stearic acid specimens had a positive influence on improving the superhydrophobic properties, preserving it for a long duration while possessing excellent corrosion resistance.

Subsequently, research has shifted towards the deposition of Zn, Zn–alloys with ferromagnetic iron group metals such as Fe, Co from ionic liquids resulting in the formation of superhydrophobic coatings. For instance, Li et al. [147] utilized the advantages associated with DESs, a class of ILs, in obtaining nanostructured deposits and synthesized hierarchical Zn structures via two-step electrodeposition from choline chloride: ethylene glycol-based DESs on copper-based substrates. It was observed and shown that the Zn structural coating was mainly composed of a combination of micro-slices containing pure, uniform, dense nano-concaves of Zn and zinc–stearate. Designing a superhydrophobic coating with micro- and nano-structural combinations was demonstrated to be highly adherent to the substrate and a promising potential solution. While nano-concaves generate van der Waals' forces and strong negative pressure, micro-slices control the surface wettability and the degree of super hydrophobicity. Development of such a unique structure shall not only endow the Zn-based coatings with high surface roughness but also with low surface energy and can be employed for applications such as self-cleaning, anti-icing and so on. Chu et al. [152] demonstrated the formation of Zn–Co alloy coating with superhydrophobic properties on AM60B magnesium alloy via electrodeposition from choline chloride-based ionic liquid and subsequent surface modification employing stearic acid as the surface energy reducer. The coating so obtained displayed improved corrosion resistance behavior and immersion test results. Additionally, the superhydrophobic coating exhibited high stability in aqueous solution and could maintain the rough surface textures even after mechanical destruction, indicative of mechanical scratch resistance. Development of superhydrophobic surfaces on lightweight metal alloy substrates such as Al, Mg provide a water-repellent surface and prevent the permeation of water into the substrate, thereby enhancing the corrosion performance of the coatings. Additionally, the scanning electrodeposition technique was developed recently, where the electrodeposition process takes place by holding the substrate stationary while the anode nozzle is kept in motion [153]. Such a technique shall overcome the difficulty associated with the plating complex shaped part such as cargo restraints (marine), propeller shaft housings (marine), wing flap bearings (aerospace). The unique structure and surface composition are expected to bestow the resulting Zn-based coatings on lightweight materials such as Al and Mg with several desirable properties. These include: (i) high surface roughness, (ii) low surface energy, (iii) reduced water-contact surface, (iv) flexibility for use in various applications, and they show a great potential in developing smart materials for corrosion protection of metallic parts in marine, aerospace, oil and gas subsea lines against chemical, mechanical, biological, physical corrosion causing agents.

4. Cost Considerations and Future Challenges

4.1. Economic Aspects

Electrodeposited Zn and Zn–alloy coatings have been demonstrated as low cost, scalable solutions to minimize the surface degradation of mild steel parts, structures (of reasonable size) in the marine environment. Their corrosion protection abilities combined with additional functionalities gained by incorporation of nanoparticles make them techno-commercially viable for such applications. While there were potential developments in the Zn, Zn–alloy coatings and composite coatings over the past decade, it is important to get an appreciation of the costs that are required for the fabrication of the material system. Figure 3 shows the cost comparison of the electrodeposited Zn, Zn–alloy and composite coatings produced by different electrolytes: less toxic citrate, acetate; non aqueous ILs [33,82,127,134,154], which were selected based on their significant developments over the past decade, the number of works published by scientific community, and that were demonstrated to be techno-commercially viable solutions as corrosion resistant coatings. As can be seen, the Zn, Zn–alloy deposition from ILs seems to be more expensive while the Zn, Zn–alloy composites deposited from the aqueous system seem to be economical and are in line with the conventional Zn, Zn-alloy deposits. In addition, certain relatively low-toxic aqueous electrolytes such as citrate, acetate are less expensive than the sulfate. Hopefully, these electrolytes could live up to their primary aptitude and offer real, ecofriendly solutions to numerous technical challenges that are currently being faced with the conventional chloride, sulfate systems. With the rapid advancement in the applications of ILs at a relatively large scale, it becomes obvious that the bulk production of ILs may increase and reduce the costs by up to ~1 USD per kg. Additionally, certain advantages such as recycling large fractions of plating solutions combined with the possibility of bulk scale production illustrate their economic competitiveness in producing high quality Zn, Zn–alloy deposits and are expected to make potential savings in costs [155]. With ever-growing composites, it becomes more important to understand the composite properties, its intended application and corrosion behavior in a particular service condition. Therefore, judicious selection of the electrolyte combined with the properties and the costs required must be put into consideration. Development of Zn and Zn–alloy composites from aqueous media have been demonstrated not only to be technically competitive but also economically viable and are expected to overcome the challenges associated with marine applications such as (i) corrosion resistance, (ii) wear resistance, (iii) frictional drag, (iv) fouling. Minimizing the costs of the coatings developed while meeting the performance requirements for marine applications remains of significant interest while their service life remains a great challenge.

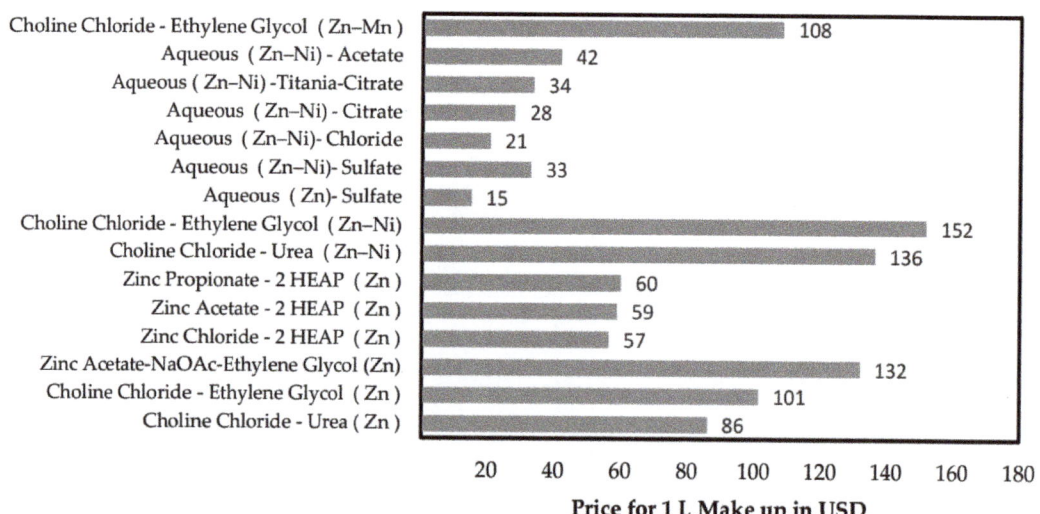

Figure 3. Plot showing the cost comparison of different electrolytes that were developed for the deposition of Zn, Zn-alloys and Zn, Zn-alloy composites considering the initial make up of 1 L solution [33,82,127,134,154]. NaOAc-sodium acetate; HEAP-2-hydroxyethyl ammonium propionate.

4.2. Future Challenges

Electrodeposition of Zn and its alloys, and composites, have significantly progressed, employing low toxicity aqueous electrolytes (e.g., citrate, acetate) and ILs over the past decade. These have shown promising applications with additional properties such as wear resistance, superhydrophobicity and hardness in addition to corrosion resistance [63,78,81,119]. However, the globe continues to face challenges that are critical in materials deterioration, often resulting in failures of components during service. The conflicting choice of a compatible material envisioned for an explicit use and environmental hazards' control will be one of the key challenges that remains to be critically addressed by the researchers, scientists, industrial experts, etc., in the field. One of the cost-effective approaches to protect the metal parts that are being used in various applications ranging from automotive to marine is by choosing the right material and method. Electrodeposition of Zn and Zn-alloys have been demonstrated to fit the requirement owing to the sacrificial property offered by a relatively cheap Zn metal combined with the simple, cost effective method of fabrication. Electrodeposited Zn, Zn–alloy coatings combined with incorporation of nano-particles could not only protect surface degradation of mild steel in the marine environment but also impart the necessary additional properties such as superhydrophobicity, improved hardness, wear resistance, fouling resistance of mild steel components. As a consequence, these become the foundation for surface adhesion and corrosion property enhancement [11].

It is well known that Zn coatings are prone to atmospheric corrosion and result in the formation of different corrosion products depending on the nature of environment they are exposed to. The evolution of corrosion products on zinc surfaces when exposed to either sulfur-dominated or chloride-dominated environments are most commonly observed. Exposure to a sulfur-based environment might result in the formation of zinc hydroxysulfate ($Zn_4SO_4(OH)_6 \cdot nH_2O$) with different amounts of water depending on the moisture it is exposed to, followed by the formation of $Zn_4Cl_2(OH)_4SO_4 \cdot 5H_2O$. In aerosol dominated environments such as marine and offshore industries, the presence of high concentrations of NaCl or other chlorides favors the formation of Simonkolleite $Zn_5Cl_2(OH)_8 \cdot H_2O$ with Gordaite ($NaZn_4Cl(OH)_6SO_4 \cdot 6H_2O$) as the final product. The sequence of different corrosion products that are formed due to exposure to sulfur-rich and chloride-rich environment over the period is represented in the Figure 4 as shown below. In chloride rich environments such as marine, the aerosols are expected to have a high concentration of NaCl which initially form $Zn_5Cl_2(OH)_8 \cdot H_2O$. These will tend to form gordaite ($NaZn_4Cl(OH)_6SO_4 \cdot 6H_2O$) by interacting with sulfate, whose emission comes from the biological activity (microorganisms, sulfate producing bacteria), droplets containing sulphate. This is reported by many works as the main reason for the frequent detection of gordaite as the final corrosion product from a chloride rich environment [156,157].

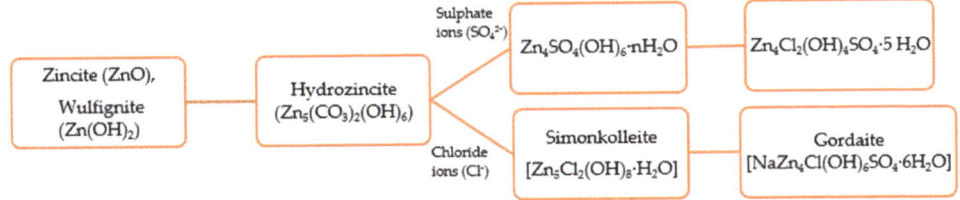

Figure 4. Figure showing the sequence of corrosion products that are formed on the Zn surface exposed to chloride and sulfate-based environment [21].

Considering the recent changes in the world's climatic conditions, it has been observed that the chloride-based environment is predominantly increasing relative to the sulfur-based environment [21]. Corrosion studies on Zn and Zn–alloy composites showed that the nanoparticle incorporation tend to lower the corrosion current, signifying a reduction in the corrosion rate for the coatings. The nanoparticles are thought to be capable of blocking the penetration path of the corrosive medium through the voids, gaps, crevices and holes within and on the surface of the coatings, besides incorporating additional functionalities such as brightness, uniformity, etc. In such cases, the transport of corrosive species to reach the metal substrate is further delayed, indicating improved corrosion resistance. Most of the reports signify the corrosion protection offered by the Zn, Zn–alloy nanocomposite coatings was improved as compared to pure Zn, Zn–alloy coatings [80,109]. Though there are significant studies that reported on the atmospheric corrosion of Zn, especially on the development of corrosion products on their surface, such studies on the electrodeposited Zn, Zn–alloys from aqueous and non-aqueous media (such as ILs) remain scarce. Hence, it becomes very important to understand not only the chemical speciation of electrodeposited Zn, Zn–alloys surfaces developed from different electrolyte media such as aqueous, non-aqueous (ILs) in different environmental settings, but also from potential Zn–composite surfaces.

Different environments have different chemistries and include a variety of combination of anions such as Cl^-, OH^- and others found in marine environment. Understanding the nature of the zinc corrosion platina layer at different circumstances might require primary attention. The formation of porous or a compact ZnO layer either at the bottom or intermediate remains critical for the electrodeposited Zn, Zn-alloy metal structures and their composites fabricated using the less toxic citrate, acetate baths and ILs used in a marine environment. Surface methods such as scanning electron microscope (SEM) and X-ray diffraction (XRD) that are utilized for identification of corrosion products formed may not be a reliable indicator in determining the relationship between the corrosion performance of the layer formed and its full structural assembly. A prodigious deal of work is required in this area to produce a much superior understanding on the (i) formation of layered structures, (ii) conditions under which the various elements interact on the surface of electrodeposited Zn, Zn–alloys fabricated using the newly developed electrolytes. Finally, the role of each corrosion product layer towards the formation of a corrosion barrier needs to be elucidated with electrodeposited Zn, Zn–alloy and their composite coatings' surfaces [157]. This shall help to identify the environmental risks, service life of the coatings and contribute significantly when performing an assessment study such as life cycle assessment, atmospheric corrosion.

Superhydrophobic Zn coatings are considered as potential corrosion resistant coatings for marine applications and subsea pipelines in oil and gas, and incorporate several additional functions such as (i) self-healing, (ii) anti fouling, (iii) nominal involvement from external agents (e.g., UV light). There are limited studies on the Zn-based superhydrophobic coatings synthesized by electrodeposition and the corrosion performance of these coatings has only been evaluated for shorter periods in laboratory. Hence, there is an imminent necessity to fabricate Zn-based superhydrophobic coatings that maintain corrosion protection capabilities and other properties such as wear, self-cleaning for a long period. Long term exposure to severe corrosive conditions would represent the marine environment where structures, subsea pipelines (e.g., X60, X80, X90 pipelines) and parts coated with Zn (combined with a polymer) are mostly used. In addition, the development of Zn superhydrophobic coatings could be extended to protect the industry grade carbon steel substrates that are generally used for transporting oil, gas through the pipelines. With respect to the corrosion studies, validation of the indirect lab corrosion measurements via the electrochemical characterizations, weight loss measurements while considering the factors that control the electrochemical corrosion rate become critically important. There is a wide growing interest in the development of Zn, Zn–alloy composites from the newly developed electrolytes with different surface structures that offer superhydrophobic properties combined with tribological properties for marine applications [158]. Studies on the evolution of corrosion products of such surfaces on exposure to simulated marine environment could provide an understanding of the service life of the coating and help the engineers, scientists, manufacturers to decide the service life of the ship hull.

5. Conclusions and Outlook

This review covered the developments in the electrodeposition of Zn, Zn–alloys and their composite coatings produced from different electrolyte media ranging from low toxicity aqueous citrate to halide free acetate- based ionic liquids. These electrolyte media have shown a promising potential to develop coatings with improvement in their texture, morphology, phases in the case of Zn–alloys which showed good results in terms of corrosion resistance. Electrodeposited Zn, Zn–alloy composite coatings have been encouraged by the availability of newer and nanostructured materials such as TiO_2, Al_2O_3, CeO_2, etc., and have seen major progress for potential application in marine, automotive aerospace, heavy duty engineering and so on.

Some recent works on forthcoming applications such as Zn-superhydrophobic coatings offering improved corrosion resistance with mechanical and tribological properties have been reviewed. The review also covered the works that have been conducted to investigate the corrosion behavior of electrodeposited Zn and Zn–alloys in controlled laboratory environments. Recent developments in the electrodeposition of Zn and Zn–alloys using ionic liquids and composites with different nano-particle incorporations have been discussed. Despite the significant developments of the electrodeposition of Zn, Zn–alloys and Zn-composites using low toxicity aqueous electrolytes, halide free ILs, the evolution and roles of different corrosion products formed on such deposited surfaces in tropical marine environments still need a detailed investigation. Corrosion studies of the Zn–coatings produced by newly developed electrolytes in the tropical marine atmosphere could help the marine industry understand the challenges associated with the coatings. Understanding the relation between the tests performed at lab scale combined with the corrosion kinetics and chloride salt deposition rate remains of critical importance for predicting the corrosion behavior of the electrodeposited Zn-based coatings. Development of zinc-manganese (Zn–Mn) alloys have started receiving primary attention for their high corrosion resistance and formation of denser corrosion products and can be potential alternatives to the Zn–Ni, Co alloys. However, certain challenges associated with Zn–Mn electroplating such as the need for complexing agents and their corrosion behavior in real environments still require detailed investigation. In addition, studies of the formation of layered structures representing different corrosion products on the progressively developed Zn–alloy composites, Zn–alloys from ionic liquids and Zn-superhydrophobic coatings will be of greater significance. Considering future trends towards the development of superhydrophobic coatings, the combination of corrosion protection abilities along with mechanical, tribological studies should also be explored. To conclude, the role of each layer in creating a corrosion barrier needs to be explicated for the rapid pace of industrialization.

Author Contributions: K.K.M. wrote the draft paper and S.P. reviewed the draft and provided supervision. All authors have read and agreed to the published version of the manuscript.

Funding: This research has received funding from the European Union's Horizon 2020 research and innovation program under the Marie Sklodowska-Curie grant agreement No.885793.

Institutional Review Board Statement: Not applicable.

Informed Consent Statement: Not applicable.

Data Availability Statement: Not applicable.

Conflicts of Interest: The authors declare no conflict of interest.

References

1. Yang, H.-Q.; Zhang, Q.; Tu, S.-S.; Wang, Y.; Li, Y.-M.; Huang, Y. A study on time-variant corrosion model for immersed steel plate elements considering the effect of mechanical stress. *Ocean Eng.* **2016**, *125*, 134–146. [CrossRef]
2. Deepa, M.J.; Arunima, S.R.; Riswana, G.; Riyas, A.H.; Sha, M.A.; Suneesh, C.V.; Shibli, S.M.A. Exploration of Mo incorporated TiO_2 composite for sustained biocorrosion control on zinc coating. *Appl. Surf. Sci.* **2019**, *494*, 361–376. [CrossRef]
3. Shibli, S.M.A.; Meena, B.N.; Remya, R. A review on recent approaches in the field of hot dip zinc galvanizing process. *Surf. Coat. Technol.* **2015**, *262*, 210–215. [CrossRef]
4. Winand, R. Electrodeposition of Zinc and Zinc Alloys. In *Modern Electroplating*, 5th ed.; John Wiley & Sons, Inc.: Hoboken, NJ, USA, 2011; pp. 285–307. [CrossRef]
5. Askari, M.; Aliofkhazraei, M.; Afroukhteh, S. A comprehensive review on internal corrosion and cracking of oil and gas pipelines. *J. Nat. Gas Sci. Eng.* **2019**, *71*, 102971. [CrossRef]

6. Weng, T.Y.; Kong, G.; Che, C.S.; Wang, Y.Q. Corrosion behaviour of a Zn/Zn-Al double coating in 5% NaCl solution. *Int. J. Electrochem. Sci.* **2018**, *13*, 11882–11894. [CrossRef]
7. Noor, E.A.; Al-Moubaraki, A.H. Corrosion behavior of mild steel in hydrochloric acid solutions. *Int. J. Electrochem. Sci.* **2008**, *3*, 806–818.
8. Kinugasa, J.; Yuse, F.; Tsunezawa, M.; Nakaya, M. Effect of Corrosion Resistance and Rust Characterization for Hydrogen Absorption into Steel under an Atmospheric Corrosion Condition. *ISIJ Int.* **2016**, *56*, 459–464. [CrossRef]
9. Panagopoulos, C.N.; Tsoutsouva, M.G. Cathodic electrolytic deposition of ZnO on mild steel. *Corros. Eng. Sci. Technol.* **2011**, *46*, 513–516. [CrossRef]
10. Sriraman, K.R.; Strauss, H.W.; Brahimi, S.; Chromik, R.R.; Szpunar, J.A.; Osborne, J.H.; Yue, S. Tribological behavior of electrodeposited Zn, Zn-Ni, Cd and Cd-Ti coatings on low carbon steel substrates. *Tribol. Int.* **2012**, *56*, 107–120. [CrossRef]
11. Abioye, O.P.; Musa, A.J.; Loto, C.A.; Fayomi, O.S.I.; Gaiya, G.P. Evaluation of Corrosive Behavior of Zinc Composite Coating on Mild Steel for Marine Applications. *J. Phys. Conf. Ser.* **2019**, *1378*. [CrossRef]
12. Lopez-Ortega, A.; Bayón, R.; Arana, J.L. Evaluation of protective coatings for offshore applications. Corrosion and tribocorrosion behavior in synthetic seawater. *Surf. Coat. Technol.* **2018**, *349*, 1083–1097. [CrossRef]
13. Croll, S.G. Surface roughness profile and its effect on coating adhesion and corrosion protection: A review. *Prog. Org. Coat.* **2020**, *148*, 105847. [CrossRef]
14. Doerre, M.; Hibbitts, L.; Patrick, G.; Akafuah, N.K. Advances in automotive conversion coatings during pretreatment of the body structure: A review. *Coatings* **2018**, *8*, 405. [CrossRef]
15. Sorour, N.; Zhang, W.; Ghali, E.; Houlachi, G. A review of organic additives in zinc electrodeposition process (performance and evaluation). *Hydrometallurgy* **2017**, *171*, 320–332. [CrossRef]
16. International Zinc Association. Available online: https://www.zinc.org/ (accessed on 18 January 2021).
17. Chatterjee, B. Science and Industry of Processes for Zinc-based Coatings with Improved Properties. *Jahrb. Oberfl.* **2017**, *72*, 1–34.
18. Pushpavanam, M. Critical review on alloy plating: A viable alternative to conventional plating. *Bull. Electrochem.* **2000**, *16*, 559–566.
19. De la Fuente, D.; Castaño, J.G.; Morcillo, M. Long-term atmospheric corrosion of zinc. *Corros. Sci.* **2007**, *49*, 1420–1436. [CrossRef]
20. Narkevicius, A.; Bučinskienė, D.; Ručinskienė, A.; Pakštas, V.; Bikulčius, G. Study on long term atmospheric corrosion of electrodeposited zinc and zinc alloys. *Trans. IMF* **2013**, *91*, 68–73. [CrossRef]
21. Odnevall, W.I.; Leygraf, C. A critical review on corrosion and runoff from zinc and zinc-based alloys in atmospheric environments. *Corrosion* **2017**, *73*, 1016–1077. [CrossRef]
22. Ranganatha, S.; Venkatesha, T.V.; Vathsala, K.; Kumar, M.K.P. Electrochemical studies on Zn/nano-CeO$_2$ electrodeposited composite coatings. *Surf. Coat. Technol.* **2012**, *208*, 64–72. [CrossRef]
23. American Tinning & Galvanizing Company. Corrosion Control & Metal Finishing. Available online: http://www.galvanizeit.com/ (accessed on 11 April 2021).
24. Chung, P.P.; Wang, J.; Durandet, Y. Deposition processes and properties of coatings on steel fasteners—A review. *Friction* **2019**, *7*, 389–416. [CrossRef]
25. Zhang, Q.B.; Hua, Y.X.; Dong, T.G.; Zhou, D.G. Effects of temperature and current density on zinc electrodeposition from acidic sulfate electrolyte with [BMIM]HSO$_4$ as additive. *J. Appl. Electrochem.* **2009**, *39*, 1207–1216. [CrossRef]
26. Shivakumara, S.; Manohar, U.; Arthoba Naik, Y.; Venkatesha, T.V. Effect of condensation product on electrodeposition of zinc on mild steel. *Bull. Mater. Sci.* **2007**, *30*, 463–468. [CrossRef]
27. Tuaweri, T.J.; Adigio, E.M.; Jombo, P.P. A Study of Process Parameters for Zinc Electrodeposition from a Sulphate Bath. *Int. J. Eng. Sci.* **2013**, *2*, 2319–6734.
28. Onkarappa, N.K.; Adarakatti, P.S.; Malingappa, P. A Study on the Effect of Additive Combination on Improving Anticorrosion Property of Zinc Electrodeposit from Acid Chloride Bath. *Ind. Eng. Chem. Res.* **2017**, *56*, 5284–5295. [CrossRef]
29. Yang, Y.; Liu, S.; Yu, X.; Huang, C.; Chen, S.; Chen, G.; Wu, Q.H. Effect of additive on zinc electrodeposition in acidic bath. *Surf. Eng.* **2015**, *31*, 446–451. [CrossRef]
30. Achary, G.; Sachin, H.P.; Naik, Y.A.; Venkatesha, T.V. Effect of a new condensation product on electrodeposition of zinc from a non-cyanide bath. *Bull. Mater. Sci.* **2007**, *30*, 219–224. [CrossRef]
31. Esfahani, M.; Zhang, J.; Durandet, Y.; Wang, J.; Wong, Y.C. Electrodeposition of Nanocrystalline Zinc from Sulfate and Sulfate-Gluconate Electrolytes in the Presence of Additives. *J. Electrochem. Soc.* **2016**, *163*, D476–D484. [CrossRef]
32. Scott, M.; Moats, M. Optimizing Additive Ratios in Alkaline Zincate Electrodeposition. In *PbZn 2020: 9th International Symposium on Lead and Zinc Processing*; Springer: Cham, Switzerland, 2020; pp. 123–131.
33. Selvaraju, V.; Thangaraj, V. Influence of γ-phase on corrosion resistance of Zn–Ni alloy electrodeposition from acetate electrolytic bath. *Mater. Res. Express* **2018**, *5*, 056502. [CrossRef]
34. Onkarappa, N.K.; Satyanarayana, J.C.A.; Suresh, H.; Malingappa, P. Influence of additives on morphology, orientation and anti-corrosion property of bright zinc electrodeposit. *Surf. Coat. Technol.* **2020**, *397*, 126062. [CrossRef]
35. Wasekar, N.P.; Jyothirmayi, A.; Hebalkar, N.; Sundararajan, G. Influence of pulsed current on the aqueous corrosion resistance of electrodeposited zinc. *Surf. Coat. Technol.* **2015**, *272*, 373–379. [CrossRef]
36. Li, Q.; Lu, H.; Cui, J.; An, M.; Li, D. Electrodeposition of nanocrystalline zinc on steel for enhanced resistance to corrosive wear. *Surf. Coat. Technol.* **2016**, *304*, 567–573. [CrossRef]

57. Chandrasekar, M.S.; Malathy, P. Synergetic effects of pulse constraints and additives in electrodeposition of nanocrystalline zinc: Corrosion, structural and textural characterization. *Mater. Chem. Phys.* **2010**, *124*, 516–528. [CrossRef]
58. Raeissi, K.; Saatchi, A.; Golozar, M.A.; Szpunar, J.A. Texture and surface morphology in zinc electrodeposits. *J. Appl. Electrochem.* **2004**, *34*, 1249–1258. [CrossRef]
59. Khorsand, S.; Raeissi, K.; Golozar, M.A. An investigation on the role of texture and surface morphology in the corrosion resistance of zinc electrodeposits. *Corros. Sci.* **2011**, *53*, 2676–2678. [CrossRef]
60. Jantaping, N.; Schuh, C.A.; Boonyongmaneerat, Y. Influences of crystallographic texture and nanostructural features on corrosion properties of electrogalvanized and chromate conversion coatings. *Surf. Coat. Technol.* **2017**, *329*, 120–130. [CrossRef]
61. Park, H.; Szpunar, J.A. The role of texture and morphology in optimizing the corrosion resistance of zinc-based electrogalvanized coatings. *Corros. Sci.* **1998**, *40*, 525–545. [CrossRef]
62. Zor, S.; Erten, Ü.; Bingöl, D. Investigation of the effect of physical conditions of a coating bath on the corrosion behavior of zinc coating using response surface methodology. *Prot. Met. Phys. Chem. Surf.* **2015**, *51*, 304–309. [CrossRef]
63. Mouanga, M.; Ricq, L.; Douglade, J.; Berçot, P. Effects of some additives on the corrosion behaviour and preferred orientations of zinc obtained by continuous current deposition. *J. Appl. Electrochem.* **2007**, *37*, 283–289. [CrossRef]
64. Ballesteros, J.C.; Díaz-Arista, P.; Meas, Y.; Ortega, R.; Trejo, G. Zinc electrodeposition in the presence of polyethylene glycol 20,000. *Electrochim. Acta* **2007**, *52*, 3686–3696. [CrossRef]
65. Da Lopes, C.S.; de Santana, P.M.; Rocha, C.L.F.; de Souza, C.A.C. Evaluation of Formic Acid and Cyclohexylamine as Additives in Electrodeposition of Zn Coating. *Mater. Res.* **2020**, *23*. [CrossRef]
66. Youssef, K.M.; Koch, C.C.; Fedkiw, P.S. Influence of pulse plating parameters on the synthesis and preferred orientation of nanocrystalline zinc from zinc sulfate electrolytes. *Electrochim. Acta* **2008**, *54*, 677–683. [CrossRef]
67. De Jesus Almeida, M.D.; Rovere, C.A.D.; de AndradeLima, L.R.P.; Ribeiro, D.V.; De Souza, C.A.C. Glycerol effect on the corrosion resistance and electrodeposition conditions in a zinc electroplating process. *Mater. Res.* **2019**, *22*. [CrossRef]
68. Nayana, K.O.; Venkatesha, T.V. Bright zinc electrodeposition and study of influence of synergistic interaction of additives on coating properties. *J. Ind. Eng. Chem.* **2015**, *26*, 107–115. [CrossRef]
69. Nakano, H.; Ura, T.; Oue, S.; Kobayashi, S. Effect of preadsorption of organic additives on the appearance and morphology of electrogalvanized steel sheets. *ISIJ Int.* **2014**, *54*, 1653–1660. [CrossRef]
70. Nayana, K.O.; Venkatesha, T.V.; Praveen, B.M.; Vathsala, K. Synergistic effect of additives on bright nanocrystalline zinc electrodeposition. *J. Appl. Electrochem.* **2011**, *41*, 39–49. [CrossRef]
71. Mo, Y.; Huang, Q.; Li, W.; Hu, S.; Huang, M.; Huang, Y. Effect of sodium benzoate on zinc electrodeposition in chloride solution. *J. Appl. Electrochem.* **2011**, *41*, 859–865. [CrossRef]
72. De Carvalho, M.F.; Carlos, I.A. Zinc electrodeposition from alkaline solution containing trisodium nitrilotriacetic added. *Electrochim. Acta* **2013**, *113*, 229–239. [CrossRef]
73. Lotfi, N.; Aliofkhazraei, M.; Rahmani, H.; Darband, G.B. Zinc–Nickel Alloy Electrodeposition: Characterization, Properties, Multilayers and Composites. *Prot. Met. Phys. Chem. Surf.* **2018**, *54*, 1102–1140. [CrossRef]
74. Sironi, L. Plating of Zn-Ni Alloy from Acidic Electrolytes for Corrosion Protection. Master's Thesis, Polytechnic University of Milan, Milan, Italy, 2016. Available online: https://www.lagalvanotecnica.com/images/ZnNi_tesi.pdf (accessed on 20 January 2021).
75. Sriraman, K.R.; Brahimi, S.; Szpunar, J.A.; Osborne, J.H.; Yue, S. Characterization of corrosion resistance of electrodeposited Zn-Ni Zn and Cd coatings. *Electrochim. Acta* **2013**, *105*, 314–323. [CrossRef]
76. Bhat, R.S.; Shet, V.B. Development and characterization of Zn–Ni, Zn–Co and Zn–Ni–Co coatings. *Surf. Eng.* **2020**, *36*, 429–437. [CrossRef]
77. Ferreira Fernandes, M.; dos Santos, J.R.M.; de Oliveira Velloso, V.M.; Voorwald, H.J.C. AISI 4140 Steel Fatigue Performance: Cd Replacement by Electroplated Zn-Ni Alloy Coating. *J. Mater. Eng. Perform.* **2020**, *29*, 1567–1578. [CrossRef]
78. Faid, H.; Mentar, L.; Khelladi, M.R.; Azizi, A. Deposition potential effect on surface properties of Zn–Ni coatings. *Surf. Eng.* **2017**, *33*, 529–535. [CrossRef]
79. Shanmugasigamani, S.; Pushpavanam, M. Bright zinc-nickel alloy deposition from alkaline non-cyanide bath. *Trans. Inst. Met. Finish.* **2008**, *86*, 122–128. [CrossRef]
80. Conde, A.; Arenas, M.A.; de Damborenea, J.J. Electrodeposition of Zn-Ni coatings as Cd replacement for corrosion protection of high strength steel. *Corros. Sci.* **2011**, *53*, 1489–1497. [CrossRef]
81. Ghaziof, S.; Gao, W. Electrodeposition of single gamma phased Zn-Ni alloy coatings from additive-free acidic bath. *Appl. Surf. Sci.* **2014**, *311*, 635–642. [CrossRef]
82. Mohan, G.N.N. Influence of Deposition Temperature on the Corrosion Resistance of Electrodeposited Zinc-Nickel Alloy Coatings. Bachelor's Thesis, Universiti Teknologi Petronas, Seri Iskandar, Malaysia, 2017. Available online: http://utpedia.utp.edu.my/17908/ (accessed on 25 December 2020).
83. Fashu, S.; Khan, R. Recent work on electrochemical deposition of Zn-Ni (-X) alloys for corrosion protection of steel. *Anti-Corros. Methods Mater.* **2019**, *66*, 45–60. [CrossRef]
84. Loukil, N.; Feki, M. Review—Zn–Mn Electrodeposition: A Literature Review. *J. Electrochem. Soc.* **2020**, *167*, 022503. [CrossRef]
85. Zhai, X.; Ren, Y.; Wang, N.; Guan, F.; Agievich, M.; Duan, J.; Hou, B. Microbial Corrosion Resistance and Antibacterial Property of Electrodeposited Zn–Ni–Chitosan Coatings. *Molecules* **2019**, *24*, 1974. [CrossRef]

66. Loukil, N.; Feki, M. Zn–Mn alloy coatings from acidic chloride bath: Effect of deposition conditions on the Zn–Mn electrodeposition-morphological and structural characterization. *Appl. Surf. Sci.* **2017**, *410*, 574–584. [CrossRef]
67. Cavallotti, P.L.; Nobili, L.; Vicenzo, A. Phase structure of electrodeposited alloys. *Electrochim. Acta* **2005**, *50*, 4557–4565. [CrossRef]
68. Boshkov, N.; Petrov, K.; Kovacheva, D.; Vitkova, S.; Nemska, S. Influence of the alloying component on the protective ability of some zinc galvanic coatings. *Electrochim. Acta* **2005**, *51*, 77–84. [CrossRef]
69. Ballote, L.D.; Ramanauskas, R.; Bartolo-Perez, P. Mn Oxide Film As Corrosion Inhibitor of Zn-Mn Coatings. *Corros. Rev.* **2000**, *18*, 41–52. [CrossRef]
70. Fashu, S.; Gu, C.D.; Zhang, J.L.; Zheng, H.; Wang, X.L.; Tu, J.P. Electrodeposition, Morphology, Composition, and Corrosion Performance of Zn-Mn Coatings from a Deep Eutectic Solvent. *J. Mater. Eng. Perform.* **2015**, *24*, 434–444. [CrossRef]
71. Claudel, F.; Stein, N.; Allain, N.; Tidu, A.; Hajczak, N.; Lallement, R.; Close, D. Pulse electrodeposition and characterization of Zn–Mn coatings deposited from additive-free chloride electrolytes. *J. Appl. Electrochem.* **2019**, *49*, 399–411. [CrossRef]
72. Bucko, M.; Rogan, J.; Stevanovic, S.I.; Stankovic, S.; Bajat, J.B. The influence of anion type in electrolyte on the properties of electrodeposited Zn Mn alloy coatings. *Surf. Coat. Technol.* **2013**, *228*, 221–228. [CrossRef]
73. Beheshti, M.; Ismail, M.C.; Kakooei, S.; Shahrestani, S.; Mohan, G.; Zabihiazadboni, M. Influence of deposition temperature on the corrosion resistance of electrodeposited zinc-nickel alloy coatings. *Materwiss. Werksttech.* **2018**, *49*, 472–482. [CrossRef]
74. Azizi, F.; Kahoul, A. Electrodeposition and corrosion behaviour of Zn–Co coating produced from a sulphate bath. *Trans. IMF* **2016**, *94*, 43–48. [CrossRef]
75. Garcia, J.R.; do Lago, D.C.B.; Cesar, D.V.; Senna, L.F. Pulsed cobalt-rich Zn–Co alloy coatings produced from citrate baths. *Surf. Coat. Technol.* **2016**, *306*, 462–472. [CrossRef]
76. Gharahcheshmeh, M.H.; Sohi, M.H. Electrochemical studies of zinc–cobalt alloy coatings deposited from alkaline baths containing glycine as complexing agent. *J. Appl. Electrochem.* **2010**, *40*, 1563–1570. [CrossRef]
77. Yogesha, S.; Bhat, K.U.; Hegde, A.C. Effect of Current Density on Deposit Characters of Zn-Co Alloy and their Corrosion Behaviors. *Synth. React. Inorg. Met. Nano-Metal Chem.* **2011**, *41*, 405–411. [CrossRef]
78. Selvaraju, V.; Thangaraj, V. Corrosion properties of mild steel surface modified by bright Zn-Co alloy electrodeposit from acetate electrolytic bath. *Mater. Res. Express* **2018**, *6*, 026501. [CrossRef]
79. Muresan, L.M. Electrodeposited Zn-Nanoparticles Composite Coatings for Corrosion Protection of Steel. In *Handbook of Nanoelectrochemistry*; Springer International Publishing: Cham, Switzerland, 2016; pp. 333–353.
80. Popoola, P.A.I.; Malatji, N.; Fayomi, O.S. Fabrication and Properties of Zinc Composite Coatings for Mitigation of Corrosion in Coastal and Marine Zone. In *Applied Studies of Coastal and Marine Environments*; InTech: London, UK, 2016; Volume 32, pp. 137–144.
81. Anwar, S.; Khan, F.; Zhang, Y.; Caines, S. Zn composite corrosion resistance coatings: What works and what does not work? *J. Loss Prev. Process Ind.* **2021**, *69*. [CrossRef]
82. Anwar, S.; Khan, F.; Zhang, Y. Corrosion behaviour of Zn-Ni alloy and Zn-Ni-nano-TiO$_2$ composite coatings electrodeposited from ammonium citrate baths. *Process Saf. Environ. Prot.* **2020**, *141*, 366–379. [CrossRef]
83. Ataie, S.A.; Zakeri, A. RSM optimization of pulse electrodeposition of Zn-Ni-Al2O3 nanocomposites under ultrasound irradiation. *Surf. Coat. Technol.* **2019**, *359*, 206–215. [CrossRef]
84. Li, X.; Liang, M.; Zhou, H.; Huang, Q.; Lv, D.; Li, W. Composite of indium and polysorbate 20 as inhibitor for zinc corrosion in alkaline solution. *Bull. Korean Chem. Soc.* **2012**, *33*, 1566–1570. [CrossRef]
85. Utu, I.D.; Muntean, R.; Mitelea, I. Corrosion and Wear Properties of Zn-Based Composite Coatings. *J. Mater. Eng. Perform.* **2020**, *29*, 5360–5365. [CrossRef]
86. Alipour, K.; Nasirpouri, F. Effect of Morphology and Surface Modification of Silica Nanoparticles on the Electrodeposition and Corrosion Behavior of Zinc-Based Nanocomposite Coatings. *J. Electrochem. Soc.* **2019**, *166*, D1–D9. [CrossRef]
87. Boshkova, N.; Kamburova, K.; Koprinarov, N.; Konstantinova, M.; Boshkov, N.; Radeva, T. Obtaining and Corrosion Performance of Composite Zinc Coatings with Incorporated Carbon Spheres. *Coatings* **2020**, *10*, 665. [CrossRef]
88. Tuaweri, T.J. Influence of Process Parameters on the Cathode Current Efficiency of Zn/SiO$_2$ Electrodeposition. *Int. J. Mech. Eng. Appl.* **2013**, *1*, 93. [CrossRef]
89. Tuaweri, T.J.; Wilcox, G.D. Behaviour of Zn-SiO$_2$ electrodeposition in the presence of N, N-dimethyldodecylamine. *Surf. Coat. Technol.* **2006**, *200*, 5921–5930. [CrossRef]
90. Ohgai, T.; Ogushi, K.; Takao, K. Morphology Control of Zn-SiO$_2$ Composite Films Electrodeposited from Aqueous Solution Containing Quaternary Ammonium Cations. *J. Phys. Conf. Ser.* **2013**, *417*, 012006. [CrossRef]
91. Vathsala, K.; Venkatesha, T.V. Zn–ZrO$_2$ nanocomposite coatings: Elecrodeposition and evaluation of corrosion resistance. *Appl. Surf. Sci.* **2011**, *257*, 8929–8936. [CrossRef]
92. Sajjadnejad, M.; Mozafari, A.; Omidvar, H.; Javanbakht, M. Preparation and corrosion resistance of pulse electrodeposited Zn and Zn–SiC nanocomposite coatings. *Appl. Surf. Sci.* **2014**, *300*, 1–7. [CrossRef]
93. Kazimierczak, H.; Szymkiewicz, K.; Rogal, Ł.; Gileadi, E.; Eliaz, N. Direct Current Electrodeposition of Zn-SiC Nanocomposite Coatings from Citrate Bath. *J. Electrochem. Soc.* **2018**, *165*, D526–D535. [CrossRef]
94. Punith Kumar, M.K.; Srivastava, C. Morphological and electrochemical characterization of electrodeposited Zn–Ag nanoparticle composite coatings. *Mater. Charact.* **2013**, *85*, 82–91. [CrossRef]

95. Praveen, B.M.; Venkatesha, T.V. Generation and corrosion behavior of Zn-nano sized carbon black composite coating. *Int. J. Electrochem. Sci.* **2009**, *4*, 258–266.
96. Punith Kumar, M.K.; Singh, M.P.; Srivastava, C. Electrochemical behavior of Zn–graphene composite coatings. *RSC Adv.* **2015**, *5*, 25603–25608. [CrossRef]
97. Rekha, M.Y.; Srivastava, C. Microstructure and corrosion properties of zinc-graphene oxide composite coatings. *Corros. Sci.* **2019**, *152*, 234–248. [CrossRef]
98. Radhamani, A.V.; Lau, H.C.; Ramakrishna, S. Nanocomposite coatings on steel for enhancing the corrosion resistance: A review. *J. Compos. Mater.* **2020**, *54*, 681–701. [CrossRef]
99. Setiawan, A.R.; Noorprajuda, M.; Ramelan, A.; Suratman, R. Preparation of Zn-ZrO$_2$ Nanocomposite Coating by DC and Pulsed Current Electrodeposition Technique with Low Current Density. *Mater. Sci. Forum* **2015**, *827*, 332–337. [CrossRef]
100. Malatji, N.; Popoola, A.P.I.; Fayomi, O.S.I.; Loto, C.A. Multifaceted incorporation of Zn-Al$_2$O$_3$/Cr$_2$O$_3$/SiO$_2$ nanocomposite coatings: Anti-corrosion, tribological, and thermal stability. *Int. J. Adv. Manuf. Technol.* **2016**, *82*, 1335–1341. [CrossRef]
101. Boshkov, N. Influence of Organic Additives and of Stabilized Polymeric Micelles on the Metalographic Structure of Nanocomposite Zn and Zn-Co Coatings. *Port. Electrochim. Acta* **2017**, *35*, 53–63. [CrossRef]
102. Arora, S.; Sharma, B.; Srivastava, C. ZnCo-carbon nanotube composite coating with enhanced corrosion resistance behavior. *Surf. Coat. Technol.* **2020**, *398*, 126083. [CrossRef]
103. Punith Kumar, M.K.; Rekha, M.Y.; Agrawal, J.; Agarwal, T.M.; Srivastava, C. Microstructure, morphology and electrochemical properties of ZnFe-Graphene composite coatings. *J. Alloys Compd.* **2019**, *783*, 820–827. [CrossRef]
104. Roventi, G. Electrodeposition of Zn-Ni-ZrO$_2$, Zn-Ni-Al$_2$O$_3$ and Zn-Ni-SiC Nanocomposite Coatings from an Alkaline Bath. *Int. J. Electrochem. Sci.* **2017**, 663–678. [CrossRef]
105. Hammami, O.; Dhouibi, L.; Berçot, P.; Rezrazi, E.M.; Triki, E. Study of Zn-Ni Alloy Coatings Modified by Nano-SiO$_2$ Particles Incorporation. *Int. J. Corros.* **2012**, *2012*, 301392. [CrossRef]
106. Ghaziof, S.; Gao, W. The effect of pulse electroplating on Zn–Ni alloy and Zn–Ni–Al$_2$O$_3$ composite coatings. *J. Alloys Compd.* **2015**, *622*, 918–924. [CrossRef]
107. Blejan, D.; Muresan, L.M. Corrosion behavior of Zn-Ni-Al$_2$O$_3$ nanocomposite coatings obtained by electrodeposition from alkaline electrolytes. *Mater. Corros.* **2013**, *64*, 433–438. [CrossRef]
108. Zheng, H.-Y.; An, M.-Z. Electrodeposition of Zn–Ni–Al$_2$O$_3$ nanocomposite coatings under ultrasound conditions. *J. Alloys Compd.* **2008**, *459*, 548–552. [CrossRef]
109. Conrad, H.; Golden, T.D. Electrodeposited Zinc-Nickel Nanocomposite Coatings. In *Nanocomposites—Recent Evolutions*; IntechOpen: London, UK, 2019; Volume 32, pp. 137–144.
110. Nemes, P.I.; Lekka, M.; Fedrizzi, L.; Muresan, L.M. Influence of the electrodeposition current regime on the corrosion resistance of Zn-CeO2 nanocomposite coatings. *Surf. Coat. Technol.* **2014**, *252*, 102–107. [CrossRef]
111. Sajjadnejad, M.; Ghorbani, M.; Afshar, A. Microstructure-corrosion resistance relationship of direct and pulse current electrodeposited Zn-TiO$_2$ nanocomposite coatings. *Ceram. Int.* **2015**, *41*, 217–224. [CrossRef]
112. Al-Dhire, T.M.; Zuhailawati, H.; Anasyida, A.S. Effect of current density on corrosion and mechanical properties of Zn-SiC composite coating. *Mater. Today Proc.* **2019**, *17*, 664–671. [CrossRef]
113. Karimi Azar, M.M.; Shooshtari Gugtapeh, H.; Rezaei, M. Evaluation of corrosion protection performance of electroplated zinc and zinc-graphene oxide nanocomposite coatings in air saturated 3.5wt.% NaCl solution. *Colloids Surf. A Physicochem. Eng. Asp.* **2020**, *601*, 125051. [CrossRef]
114. PraveenKumar, C.M.; Venkatesha, T.V.; Vathsala, K.; Nayana, K.O. Electrodeposition and corrosion behavior of Zn–Ni and Zn–Ni–Fe$_2$O$_3$ coatings. *J. Coat. Technol. Res.* **2012**, *9*, 71–77. [CrossRef]
115. Exbrayat, L.; Rébéré, C.; Ndong Eyame, R.; Steyer, P.; Creus, J. Corrosion behaviour in saline solution of pulsed-electrodeposited zinc-nickel-ceria nanocomposite coatings. *Mater. Corros.* **2017**, *68*, 1129–1142. [CrossRef]
116. Roventi, G.; Fratesi, R.; Della Guardia, R.A.; Barucca, G. Normal and anomalous codeposition of Zn-Ni alloys from chloride bath. *J. Appl. Electrochem.* **2000**, *30*, 173–179. [CrossRef]
117. Tome, L.I.N.; Baião, V.; da Silva, W.; Brett, C.M.A. Deep eutectic solvents for the production and application of new materials. *Appl. Mater. Today* **2018**, *10*, 30–50. [CrossRef]
118. Smith, E.L.; Abbott, A.P.; Ryder, K.S. Deep Eutectic Solvents (DESs) and Their Applications. *Chem. Rev.* **2014**, *114*, 11060–11082. [CrossRef]
119. Liu, F.; Deng, Y.; Han, X.; Hu, W.; Zhong, C. Electrodeposition of metals and alloys from ionic liquids. *J. Alloys Compd.* **2016**, *654*, 163–170. [CrossRef]
120. Abbott, A.P.; Dalrymple, I.; Endres, F.; Macfarlane, D.R. Why use Ionic Liquids for Electrodeposition? In *Electrodeposition from Ionic Liquids*; Wiley-VCH Verlag GmbH & Co. KGaA: Weinheim, Germany, 2008; pp. 1–13.
121. Marcus, Y. Application of Deep Eutectic Solvents. In *Deep Eutectic Solvents*; Springer Nature: Cham, Switzerland, 2019; pp. 111–151. [CrossRef]
122. Bernasconi, R.; Panzeri, G.; Accogli, A.; Liberale, F.; Nobili, L.; Magagnin, L. Electrodeposition from Deep Eutectic Solvents. In *Progress and Developments in Ionic Liquids*; InTech: London, UK, 2017.
123. An, M.-Z.; Yang, P.-X.; Su, C.-N.; Nishikata, A.; Tsuru, T. Electrodeposition of Cobalt in an Ionic Liquid Electrolyte at Ambient Temperature. *Chin. J. Chem.* **2008**, *26*, 1219–1224. [CrossRef]

124. Fashu, S.; Gu, C.D.; Zhang, J.L.; Huang, M.L.; Wang, X.L.; Tu, J.P. Effect of EDTA and NH$_4$Cl additives on electrodeposition of Zn-Ni films from choline chloride-based ionic liquid. *Trans. Nonferrous Met. Soc. China* **2015**, *25*, 2054–2064. [CrossRef]
125. Bernasconi, R.; Panzeri, G.; Firtin, G.; Kahyaoglu, B.; Nobili, L.; Magagnin, L. Electrodeposition of ZnNi Alloys from Choline Chloride/Ethylene Glycol Deep Eutectic Solvent and Pure Ethylene Glycol for Corrosion Protection. *J. Phys. Chem. B* **2020**. [CrossRef] [PubMed]
126. Lei, C.; Alesary, H.F.; Khan, F.; Abbott, A.P.; Ryder, K.S. Gamma-phase Zn-Ni alloy deposition by pulse-electroplating from a modified deep eutectic solution. *Surf. Coat. Technol.* **2020**, *403*, 126434. [CrossRef]
127. Bucko, M.; Tomic, M.V.; Maksimovic, M.; Bajat, J.B. The importance of using hydrogen evolution inhibitor during the Zn and Zn–Mn electrodeposition from ethaline. *J. Serbian Chem. Soc.* **2019**, *84*, 1221–1234. [CrossRef]
128. Bucko, M.; Stevanovic, S.; Bajat, J. Tailoring the corrosion resistance of Zn-Mn coating by electrodeposition from deep eutectic solvents. *Zast. Mater.* **2018**, *59*, 173–181. [CrossRef]
129. Li, R.; Dong, Q.; Xia, J.; Luo, C.; Sheng, L.; Cheng, F.; Liang, J. Electrodeposition of composition controllable Zn–Ni coating from water modified deep eutectic solvent. *Surf. Coat. Technol.* **2019**, *366*, 138–145. [CrossRef]
130. Pereira Oliveira, N.M. Metal Electrodeposition from Deep Eutectic Solvents. Ph.D. Thesis, Universidade Do Porto, Porto, Portugal, 2018. Available online: https://repositorio-aberto.up.pt/bitstream/10216/110784/2/252422.pdf (accessed on 20 January 2021).
131. Marín-Sánchez, M.; Ocón, P.; Conde, A.; García, I. Electrodeposition of Zn-Mn coatings on steel from 1-ethyl-3-methylimidazolium bis (trifluoromethanesulfonyl) imide ionic liquid. *Surf. Coat. Technol.* **2014**, *258*, 871–877. [CrossRef]
132. Bakkar, A.; Neubert, V. Electrodeposition onto magnesium in air and water stable ionic liquids: From corrosion to successful plating. *Electrochem. Commun.* **2007**, *9*, 2428–2435. [CrossRef]
133. Jiang, F.; Huang, K.; Shi, W.; Yang, X.; Zhang, Y. Corrosion behavior of zinc coating electroplated at high current from a deep eutectic solvent. *Mater. Res. Express* **2018**, *6*, 016402. [CrossRef]
134. Panzeri, G.; Muller, D.; Accogli, A.; Gibertini, E.; Mauri, E.; Rossi, F.; Nobili, L.; Magagnin, L. Zinc electrodeposition from a chloride-free non-aqueous solution based on ethylene glycol and acetate salts. *Electrochim. Acta* **2019**, *296*, 465–472. [CrossRef]
135. Zhao, L.; Liu, Q.; Gao, R.; Wang, J.; Yang, W.; Liu, L. One-step method for the fabrication of superhydrophobic surface on magnesium alloy and its corrosion protection, antifouling performance. *Corros. Sci.* **2014**, *80*, 177–183. [CrossRef]
136. Zhang, J.; Kang, Z. Effect of different liquid–solid contact models on the corrosion resistance of superhydrophobic magnesium surfaces. *Corros. Sci.* **2014**, *87*, 452–459. [CrossRef]
137. Qian, H.; Xu, D.; Du, C.; Zhang, D.; Li, X.; Huang, L.; Deng, L.; Tu, Y.; Mol, J.M.C.; Terryn, H.A. Dual-action smart coatings with a self-healing superhydrophobic surface and anti-corrosion properties. *J. Mater. Chem. A* **2017**, *5*, 2355–2364. [CrossRef]
138. Liu, J.; Fang, X.; Zhu, C.; Xing, X.; Cui, G.; Li, Z. Fabrication of superhydrophobic coatings for corrosion protection by electrodeposition: A comprehensive review. *Colloids Surf. A Physicochem. Eng. Asp.* **2020**, *607*, 125498. [CrossRef]
139. Zhang, D.; Wang, L.; Qian, H.; Li, X. Superhydrophobic surfaces for corrosion protection: A review of recent progresses and future directions. *J. Coat. Technol. Res.* **2016**, *13*, 11–29. [CrossRef]
140. Ferrari, M.; Benedetti, A.; Cirisano, F. Superhydrophobic coatings from recyclable materials for protection in a real sea environment. *Coatings* **2019**, *9*, 303. [CrossRef]
141. Wang, P.; Zhang, D.; Qiu, R.; Hou, B. Super-hydrophobic film prepared on zinc as corrosion barrier. *Corros. Sci.* **2011**, *53*, 2080–2086. [CrossRef]
142. Brassard, J.D.; Sarkar, D.K.; Perron, J.; Audibert-Hayet, A.; Melot, D. Nano-micro structured superhydrophobic zinc coating on steel for prevention of corrosion and ice adhesion. *J. Colloid Interface Sci.* **2015**, *447*, 240–247. [CrossRef]
143. Feng, Y.; Chen, S.; Frank Cheng, Y. Stearic acid modified zinc nano-coatings with superhydrophobicity and enhanced antifouling performance. *Surf. Coat. Technol.* **2018**, *340*, 55–65. [CrossRef]
144. Li, H.; Yu, S.; Han, X.; Zhang, S.; Zhao, Y. A simple method for fabrication of bionic superhydrophobic zinc coating with crater-like structures on steel substrate. *J. Bionic Eng.* **2016**, *13*, 622–630. [CrossRef]
145. Polyakov, N.A.; Botryakova, I.G.; Glukhov, V.G.; Red'kina, G.V.; Kuznetsov, Y.I. Formation and anticorrosion properties of superhydrophobic zinc coatings on steel. *Chem. Eng. J.* **2020**, 127775. [CrossRef]
146. Hu, C.; Xie, X.; Zheng, H.; Qing, Y.; Ren, K. Facile fabrication of superhydrophobic zinc coatings with corrosion resistanceviaan electrodeposition process. *New J. Chem.* **2020**, *44*, 8890–8901. [CrossRef]
147. Li, R.; Gao, Q.; Dong, Q.; Luo, C.; Sheng, L.; Liang, J. Template-free electrodeposition of ultra-high adhesive superhydrophobic Zn/Zn stearate coating with ordered hierarchical structure from deep eutectic solvent. *Surf. Coat. Technol.* **2020**, *403*, 126267. [CrossRef]
148. Wang, P.; Zhang, D.; Qiu, R.; Wu, J.; Wan, Y. Super-hydrophobic film prepared on zinc and its effect on corrosion in simulated marine atmosphere. *Corros. Sci.* **2013**, *69*, 23–30. [CrossRef]
149. Barati Darband, G.; Aliofkhazraei, M.; Khorsand, S.; Sokhanvar, S.; Kaboli, A. Science and Engineering of Superhydrophobic Surfaces: Review of Corrosion Resistance, Chemical and Mechanical Stability. *Arab. J. Chem.* **2020**, *13*, 1763–1802. [CrossRef]
150. Ferrari, M.; Benedetti, A.; Santini, E.; Ravera, F.; Liggieri, L.; Guzman, E.; Cirisano, F. Biofouling control by superhydrophobic surfaces in shallow euphotic seawater. *Colloids Surf. A Physicochem. Eng. Asp.* **2015**, *480*, 369–375. [CrossRef]
151. Popoola, A.P.I.; Aigbodion, V.S.; Fayomi, O.S.I. Surface characterization, mechanical properties and corrosion behaviour of ternary based Zn–ZnO–SiO$_2$ composite coating of mild steel. *J. Alloys Compd.* **2016**, *654*, 561–566. [CrossRef]

52. Chu, Q.; Liang, J.; Hao, J. Facile fabrication of a robust super-hydrophobic surface on magnesium alloy. *Colloids Surf. A Physicochem. Eng. Asp.* **2014**, *443*, 118–122. [CrossRef]
53. Wang, Z.; Shen, L.; Jiang, W.; Fan, M.; Liu, D.; Zhao, J. Superhydrophobic nickel coatings fabricated by scanning electrodeposition on stainless steel formed by selective laser melting. *Surf. Coat. Technol.* **2019**, *377*, 124886. [CrossRef]
54. Rivas-Esquivel, F.M.; Brisard, G.M.; Ortega-Borges, R.; Trejo, G.; Meas, Y. Zinc electrochemical deposition from ionic liquids and aqueous solutions onto indium tin oxide. *Int. J. Electrochem. Sci.* **2017**, *12*, 2026–2041. [CrossRef]
55. Prado, R.; Weber, C.C. Applications of Ionic Liquids. In *Application, Purification, and Recovery of Ionic Liquids*; Kuzmina, O., Hallett, J.P., Eds.; Elsevier BV: Amsterdam, The Netherlands, 2016; pp. 1–58. [CrossRef]
56. Odnevall, I.; Leygraf, C. The formation of $Zn_4Cl_2(OH)_4SO_4 \cdot 5H_2O$ in an urban and an industrial atmosphere. *Corros. Sci.* **1994**, *36*, 1551–1559. [CrossRef]
57. Ivan, C.C. Recent progress and required developments in atmospheric corrosion of galvanised steel and zinc. *Materials* **2017**, *10*, 1288. [CrossRef]
58. Walsh, F.C.; Wang, S.; Zhou, N. The electrodeposition of composite coatings: Diversity, applications and challenges. *Curr. Opin. Electrochem.* **2020**, *20*, 8–19. [CrossRef]

Article

Corrosion Fatigue Behavior of Twin Wire Arc Sprayed and Machine Hammer Peened ZnAl4 Coatings on S355 J2C + C Substrate

Michael P. Milz [1,*], Andreas Wirtz [2], Mohamed Abdulgader [3], Anke Kalenborn [1], Dirk Biermann [2], Wolfgang Tillmann [3] and Frank Walther [1]

1. Chair of Materials Test Engineering (WPT), TU Dortmund University, Baroper Str. 303, 44227 Dortmund, Germany; anke.kalenborn@tu-dortmund.de (A.K.); frank.walther@tu-dortmund.de (F.W.)
2. Institute of Machining Technology (ISF), TU Dortmund University, Baroper Str. 303, 44227 Dortmund, Germany; andreas.wirtz@tu-dortmund.de (A.W.); dirk.biermann@tu-dortmund.de (D.B.)
3. Chair of Materials Engineering (LWT), TU Dortmund University, Leonhard-Euler-Str. 2, 44227 Dortmund, Germany; mohamed.abdulgader@tu-dortmund.de (M.A.); wolfgang.tillmann@tu-dortmund.de (W.T.)
* Correspondence: michael.milz@tu-dortmund.de; Tel.: +49-231-755-8031

Abstract: Offshore installations, e.g., offshore wind turbines and pipelines, are exposed to various mechanical loads due to wind or waves and corrosive loads such as seawater or mist. ZnAl-based thermal sprayed coatings, often in conjunction with organic coatings, provide sufficient corrosion protection and are well established for applications in marine environments. In this study, machine hammer peening (MHP) is applied after twin wire arc spraying to improve corrosion fatigue behavior through increased hardness, reduced porosity, and roughness compared to as-sprayed coatings. Mn-alloyed structural steel S355 J2 + C with and without ZnAl4 coating as well as with MHP post-treated ZnAl4 coating were cyclically loaded in 3.5% NaCl solution. MHP leads to a uniform coating thickness with lower porosity and roughness. ZnAl4 coating and MHP post-treatment improved corrosion fatigue behavior in the high cycle fatigue regime with an increase of the stress amplitude, applied to reach a number of cycles 1.2×10^6, up to 115% compared to sandblasted specimens. Corrosive attack of the substrate steel was successfully avoided by using the coating systems. Stress- and microstructure-dependent corrosion fatigue damage mechanisms were evaluated by mechanical and electrochemical measurement techniques.

Keywords: corrosion fatigue; corrosion; twin wire arc spraying; machine hammer peening; ZnAl4 coating; marine application; corrosion protection

1. Introduction

Offshore wind turbines, pipelines, or bridges are examples of applications in the marine environment. In addition to requirements such as good weldability and cost-effectiveness, they must withstand various stresses resulting from superimposed mechanical and corrosive loads. Due to this, a combination of structural steel and corrosive metallic protective coatings has proved to be the best solution. Since repair costs offshore are up to 100 times higher than on land, reliable service life of at least 20 years should be guaranteed [1].

Corrosive protection of metallic coatings is based on various principles. Barrier coatings separate the substrate steel from the corrosive environment. Therefore, coatings must have sufficient adhesion to the substrate, adequate strength to resist first damage, and be suitably ductile to resist cracking [2]. Cathodic protection coatings additionally work as a sacrificial anode. Substrate material and coating form a galvanic couple with the metallic coating having a more negative electrode potential. If the coating is damaged and the substrate steel comes into contact with the corrosive medium, the coating will preferentially degrade.

In marine applications, it is common to use organic coatings such as paintings as the second layer above the metallic coatings to improve the corrosion protection because it acts as a barrier coating [3]. If impressed cathodic protection is also applied, the coating requires a high electrical resistance [2]. In addition to corrosive stresses, various mechanical stresses act on structural elements in a marine environment, e.g., due to wind, waves, and tide. In case of offshore wind turbines, these mechanical stresses result in about 10^9 load cycles with an estimated service life of 20 years [4].

ZnAl-based coatings, which provide sufficient cathodic protection, are well established for applications in marine environments with additional organic coatings [5]. Besides Aluminum, other alloying elements such as Manganese or Silicon also improve the corrosion resistance of pure Zinc coatings [6–8]. Furthermore, the corrosion products of ZnAl coatings work as a barrier between the corrosive medium and coating or substrate [9].

ZnAl-based coatings can be applied by galvanization or thermal spraying processes. Due to the electroplating bath used during the galvanization process, the dimensions of the components to be coated are limited. Brittle Fe–Zn intermetallic phases formed during galvanization reduce the mechanical performance [9]. For the thermal spraying process, the dimensions of the parts to be coated are not limited and the heat supplied is lower than provided during galvanization [10]. The thermal spraying process produces coatings with higher porosity than galvanic coatings, a lamellar structure, and residual stresses [11]. Residual stresses can be introduced thermally or kinetically, whereas high temperatures lead to tensile residual stresses that reduce the fatigue strength of components and higher kinetic energy results in compressive stresses, which are known as beneficial for fatigue strength [12,13]. Thermal-sprayed ZnAl coatings produced by the twin wire arc spraying (TWAS) process also exhibit compressive residual stresses due to the effect of kinetic energy, as the impact of larger and not fully molten particles causes the thermal effect to surpass the kinetic effect [14].

Mechanical post-treatments of coatings can lead to improvement of the corrosion resistance by eliminating porosity, densification, and homogenization of the coating structure [15]. Machine hammer peening (MHP) generally reduces porosity, increases hardness, and introduces compressive residual stresses in near-surface areas [16,17]. This possibility of adjusting near-surface properties by MHP was also proven for ZnAl coatings [14,18].

In previous papers [19,20] of this study, the properties of the coating systems depending on the manufacturing process were investigated in detail. The use of argon and compressed air as atomizing gas in the TWAS process result in a lower oxide content in the case of compressed air [19]. Nevertheless, the oxides in form of small layers within the coating lead to cracks after MHP. Basically, Zn and Al rich regions can be distinguished and in addition oxide layers are formed. Thermally assisted MHP process have reduced and prevented this cracking by simultaneously increasing the hardness, introducing higher residual stresses, and reducing the porosity of ZnAl coatings compared to MHP processes conducted at room temperature [20].

Punchi-Carbera et al. [21] investigated the fatigue and corrosion fatigue resistance of Ni-based coatings deposited on a high strength steel by high-velocity oxygen fuel (HVOF). The fatigue life of coated specimens tested in air was lower compared to uncoated specimens due to the presence of alumina particles on the surface, which act as crack initiators. Comparative tests in NaCl solution showed an increase in the corrosion fatigue life of the coated specimens, as the coating served as a corrosion protection layer.

In previous fatigue tests in air of structural steel S355 JRC + C without and with ZnAl coating and with MHP post-treated ZnAl coating, the ZnAl-coated and MHP post-treated specimens showed better fatigue strength in the high fatigue regime.

The aim of this study was to improve the corrosion fatigue properties by adjusting the coating properties. Therefore, mechanical compacting by MHP was performed, which should induce a reduction of porosity and roughness, as well as an increase of compressive residual stresses and hardness. Instrumented corrosion fatigue tests were carried out in

3.5%NaCl solution at various stress amplitudes. Mechanical and electrochemical measuring techniques were applied to determine the corrosion fatigue damage mechanisms.

2. Materials and Methods

2.1. Materials and Manufacturing

Three specimen conditions I–III of the structural Mn-alloyed steel S355 J2C + C were investigated. The base material was sandblasted (I) before applying a ZnAl4-coating (II) and subsequently MHP post-treated (III). The manufacturing process of the conditions I–III is described in detail below. All tests were performed using specimens with a geometry according to Figure 1.

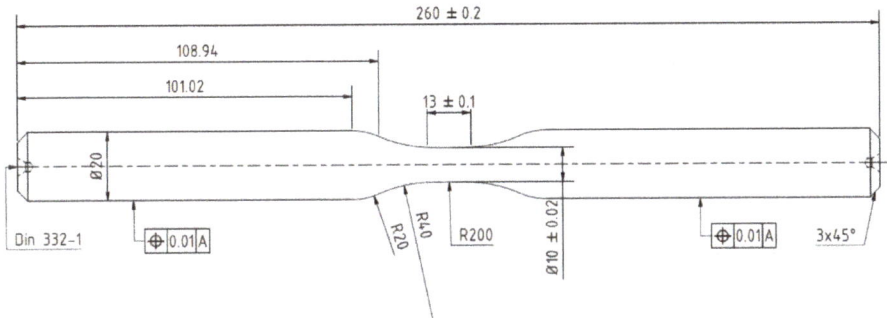

Figure 1. Specimen geometry, all dimensions in mm.

2.1.1. Substrate Material

The structural steel S355 J2C + C (1.0579) was sandblasted and used as substrate material for the coating systems. Tables 1 and 2 show the chemical composition and material properties of the steel according to the datasheet of the manufacturer.

Table 1. Cast analysis of S355 J2C + C according to the datasheet of the manufacturer.

Element	Fe	C	Si	Mn	P	S	Cu	Al	V
wt.-%	Bal.	0.1510	0.2235	1.2646	0.0134	0.0060	0.3131	0.0200	0.0043

Table 2. Physical properties of S355 J2C + C according to the datasheet of the manufacturer.

Tensile Strength R_m in MPa	Yield Strength R_e in MPa	Elongation at Fracture A_5 in %
672	615	12.7

Sandblasting was carried out to ensure sufficient adhesion for the following ZnAl4 coating. Therefore, Alumina oxide powder with EKF 24 (600–850 µm) size fraction, 4 bar blasting pressure, 100 mm stand-off distance, and 45° blasting angle has been used for manufacturing. An ultrasonic ethanol bath enables the removal of dust residues and the final cleaning of the sandblasted surfaces.

2.1.2. Twin Wire Arc Spray Process

Twin-wire arc spraying (TWAS) process was carried out using a spray unit Durum Duraspray 450 (Durum, Germany) to apply ZnAl4 coatings on sandblasted specimens. Contaminations of the surface were removed using ethanol. The process was exclusively performed within the rejuvenated specimen area. The chemical composition of the feedstock wires made of ZnAl4 is given in Table 3.

Table 3. Chemical composition of ZnAl4 feedstock wires according to the datasheet of the manufacturer.

Element	Zn	Al	Si	Fe	Pb	Cu	Sn
wt.-%	Bal.	3.5–4.5	≤0.030	≤0.005	≤0.003	≤0.002	≤0.001

The TWAS process was performed with 3.2 m/min wire feed rate, 22 V arc voltage, and 5 bar atomization gas pressure using atomization gas dry and compressed air. An industrial robot, ABB IRB 4600, and a rotating unit with the handling parameters 120 mm stand-off distance between the spray gun and substrate surface, v_s = 18,000 mm/min axial gun velocity, and s = 4 mm meander spacing were used. Coatings were made in 4 gun overruns in total. The specimens were masked, except of the rejuvenated specimen area during the spraying process to avoid the coating of unwanted areas.

2.1.3. Machine Hammer Peening Process

Coated specimens were machine hammer-peened using a high-performance turn-mill center Index G250 (Index-Werke, Esslingen, Germany) with a Pokolm FORGEFix Air MHP tool using a spherical solid carbide tappet with diameter d_p = 16 mm (Figure 2).

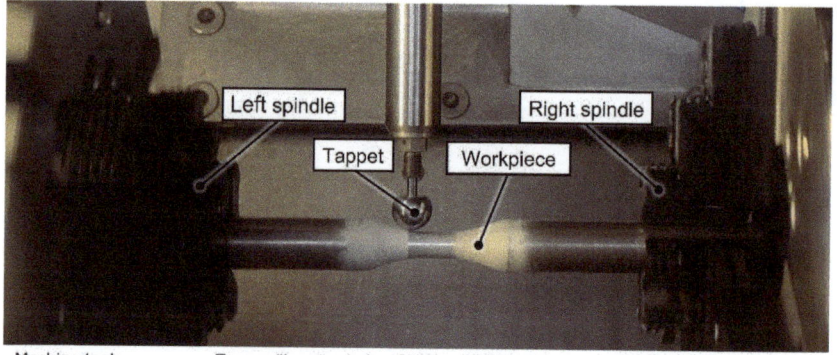

Machine tool:	Turn-mill center Index G250	MHP tool:	Pokolm FORGEFix Air
Hammering frequency:	$f \approx$ 210 Hz	Stroke:	h = 2.0 mm
Max. indentation depth:	$a_{i,max}$ = 0.25 mm	Feed velocity:	v = 2.0 m/min
Track distance:	l_p = 0.67 mm	Impact density:	$p_i \approx$ 9.4 mm^{-2}

Figure 2. Setup of the machine hammer peening (MHP) process using turn-mill center index G250.

The process parameter settings were chosen based on former studies [14,22]. A compressed air pressure p = 6 bar resulted in a hammering frequency $f \approx$ 210 Hz with a stroke set to h = 2.0 mm. A feed speed v_f = 2.0 m/min, a maximal indentation depth $a_{i,max} \approx$ 0.25 mm, and a line pitch l_p = 0.67 mm were used. The process parameter resulted in an impact density of about p_i = 9.4 indents/mm^2. Figure 3 shows the gauge length of sandblasted (I), ZnAl-coated (II), and MHP post-treated (III) specimens.

2.2. Material Characterization and Testing Methods

2.2.1. Macro- and Microstructure Investigations

Surface qualities of specimen conditions I–III were evaluated by roughness measurements and metallographic characterization. Mean roughness depth R_z, arithmetical mean roughness R_a, mean smoothing depth R_p, thickness, and hardness of the coatings were characterized before and after the MHP post-treatment. Roughness parameters were determined using a confocal white light microscope Nanofocus μsurf (Oberhausen, Germany) with a magnification 50× long objective and a robust Gaussian filter with a cutoff wavelength λ_c = 0.8 mm. Metallographically prepared cross-sections were used to determine substrate grain structure, coating morphology, and coating thickness by means of an Olympus BX51 optical microscope (Hamburg, Germany). For corrosion fatigue tests, the diameter of the gauge length was measured using the optical micrometer Keyence TM-040 (Neu-Isenburg, Germany). After each processing step, three measurements were carried out and the average values were calculated.

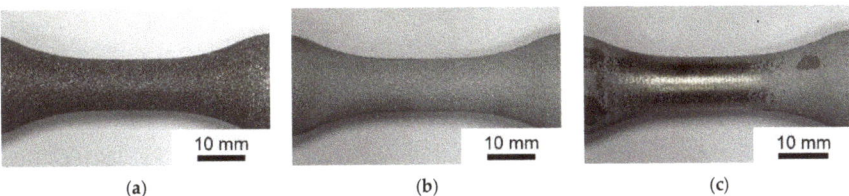

Figure 3. Gauge length of specimen conditions I–III: (**a**) sandblasted (I), (**b**) ZnAl-coated (II), (**c**) MHP post-treated (III).

2.2.2. Corrosion Fatigue Testing

For corrosion fatigue characterization, constant amplitude tests (CAT) were performed stress-controlled with a sinusoidal load-time function at a stress ratio R = −1 (fully-reversed loading). According to previous tests, stress amplitudes in the range of 280 to 360 MPa were selected. The fatigue stresses were calculated based on the initial diameter of sandblasted specimens, for all specimen conditions I–III. CAT ended with failure or reaching the limited number of cycles $N_{limit} = 2 \times 10^6$. The specimens were coated with polyurethane, except a 9 mm wide area in the middle of the gauge length, in order to create a defined test area for electrochemical measurements during fatigue loading.

A self-designed corrosion cell was used to perform corrosion fatigue tests in 3.5% NaCl solution, as shown in Figure 4. Continuous exchange and tempering of the corrosion medium was controlled by a pump and thermostat. Polymer seals were used to enable the sealing of the cell. As the installation of the extensometer inside the corrosion cell was not possible, an extensometer with 89 mm gauge length was applied at the specimens shafts outside the cell. Corrosion behavior and damage mechanisms during corrosion fatigue tests were evaluated using a three-electrode setup and a potentiostat interface 1000 (Gamry instruments, Warminster, PA, USA). Thereby, the specimen dealt as the working electrode, a silver chloride electrode was used as the reference electrode, and a graphite electrode served as the counter electrode. Before fatigue loading, the specimens were exposed for 30 to 60 min (30 min: sandblasted (I); 60 min: (ZnAl-coated (II), MHP post-treated (III)) to the medium in order to reach their specific open circuit potential (OCP).

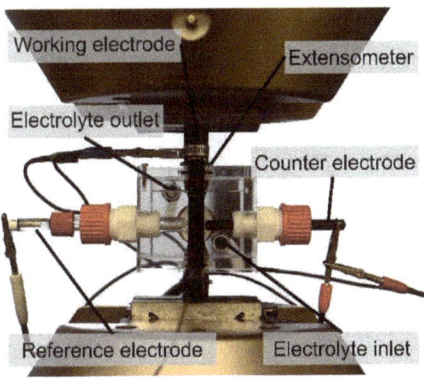

Figure 4. Experimental corrosion fatigue test setup.

3. Results and Discussion

3.1. Substrate Material and Coating Systems Characterization

The Mn-alloyed structural steel S355 J2C + C has a ferritic-pearlitic microstructure, as shown in light micrographs, Figure 5a,b. The sample was etched with 5% Nital solution. White areas are ferritic, dark areas pearlitice. The rough surface and plastic deformations at the specimen surface are a result of the sandblasting process, Figure 5c,d.

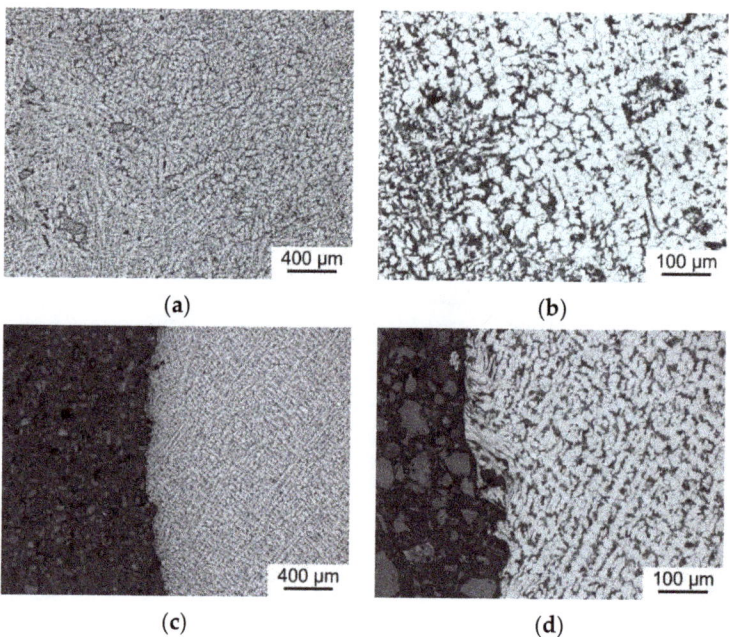

Figure 5. Polished cross-section of sandblasted (I) specimen, (**a,b**) microstructure of the center area, (**c,d**) microstructure of the edge area.

Light micrographs of ZnAl-coated (II) and MHP post-treated (III) specimens show an intact and continuous coating over the whole circumference, Figure 6a,c. A higher number of pores and a rougher surface can be determined for the ZnAl-coating (II) (Figure 6b) compared to the MHP post-treated (III) coating (Figure 6d). Furthermore, the adhesion of the coating to the substrate appears to be improved as a result of MHP process, as the width of gaps between the substrate and the coating (white arrows) are reduced, Figure 6b,d.

Three values of the coating thickness were determined at three locations for each specimen condition using light micrographs in order to calculate the average value and standard deviations; Table 4. The average coating thickness of all measurements differs by 9.9 µm. The significant value is the standard deviation, which is 13.1 µm higher for ZnAl-coated (II) specimens. Due to this, mechanical compacting by MHP leads to a more homogenous and constant coating thickness.

A smoothening of the surface by MHP, which was determined in optical micrographs (Figure 6), can be confirmed by axial roughness line scans for 250 µm width and 6.8 mm length (Figure 7). Two specimens were evaluated for each specimen condition I–III. Highest roughness values were determined for the sandblasted specimens; Table 4. The mean values of the arithmetical mean roughness R_a and mean roughness depth R_z have been reduced by more than 50% for MHP post-treated (III) specimen compared to the ZnAl-coated (II) specimen.

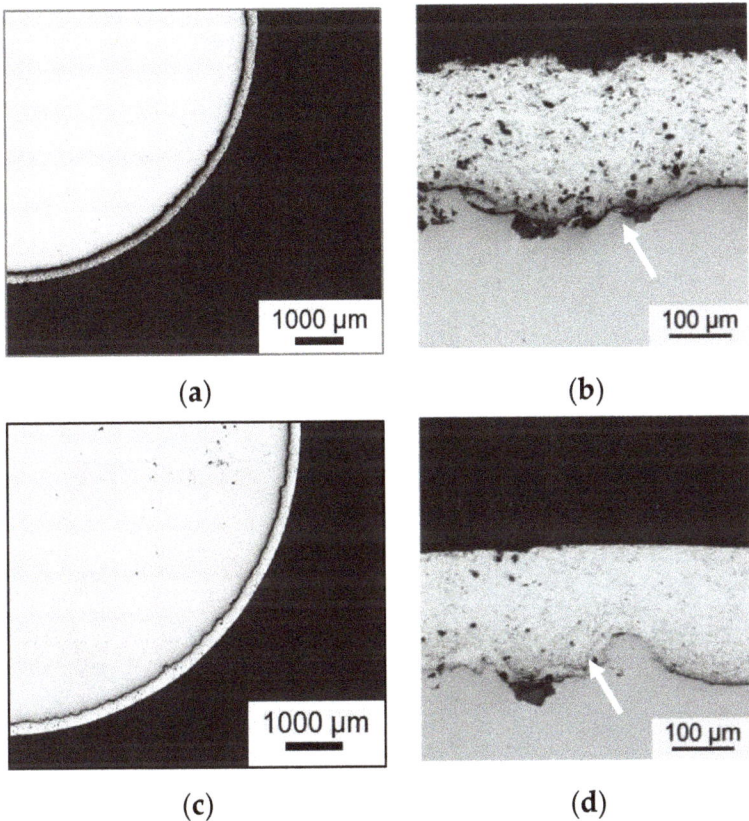

Figure 6. Polished cross-section of ZnAl-coated (II) specimen as (**a**) overview and (**b**) detail and of the MHP post-treated (III) specimen as (**c**) overview and (**d**) detail.

Table 4. Average coating thickness of all measurements and roughness values in gauge length: sandblasted (I), ZnAl-coated (II), and machine hammer-peened (III) specimens.

Specimen Surface	Average Coating Thickness in μm	Mean Roughness Depth Rz in μm	Arithm. Mean Roughness Ra in μm
Sandblasted (I)	-	93.8 ± 9.8 | 106.0 ± 12.2	14.9 ± 1.68 | 14.2 ± 1.1
ZnAl-coated (II)	224.5 ± 27.1	76.0 ± 6.4 | 79.1 ± 3.8	9.1 ± 0.7 | 9.6 ± 0.8
Machine hammer-peened (III)	214.6 ± 14.0	24.3 ± 2.0 | 26.1 ± 1.8	4.4 ± 0.4 | 4.7 ± 0.2

3.2. Corrosion Fatigue Testing

Corrosion fatigue tests were performed at five and six stress amplitudes, respectively, with a minimum of nine specimens tested per condition. Figure 8 shows the cyclic deformation curves for MHP post-treated (III) specimens at three stress amplitudes. As expected, higher stress amplitudes lead to increasing strain amplitudes and decreasing lifetimes.

OCP decreases at the test beginning because new deformation-induced surface areas are in contact with the electrolyte. The decrease is more pronounced for higher stress amplitudes. Minimum OCP values are reached at 10^4 to 3×10^4 cycles. The following increase is expected to be a result of corrosion fatigue damage mechanisms within the substrate steel, which shows a significantly higher OCP than the coatings; Figure 8. After fatigue crack initiation in the substrate, the OCP increase is more pronounced for ongoing fatigue loading, as more deformation- and damage-induced surface areas of the substrate steel get into contact with the electrolyte. In comparison, OCP of sandblasted (I) specimens start at a higher level and constantly decreases until failure, as corrosion fatigue damage mechanisms of the substrate are exclusively relevant; Figure 9.

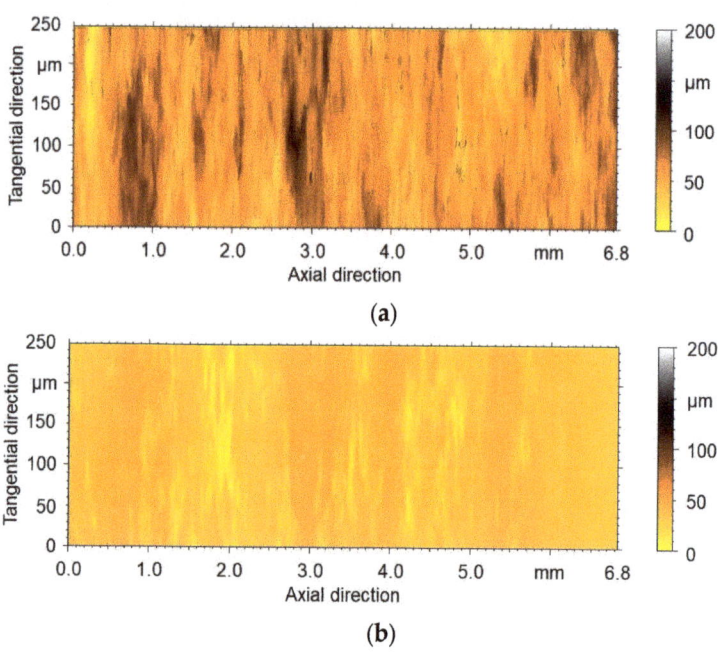

Figure 7. Axial roughness line scans for (**a**) ZnAl-coated (II) and (**b**) MHP post-treated (III) specimens.

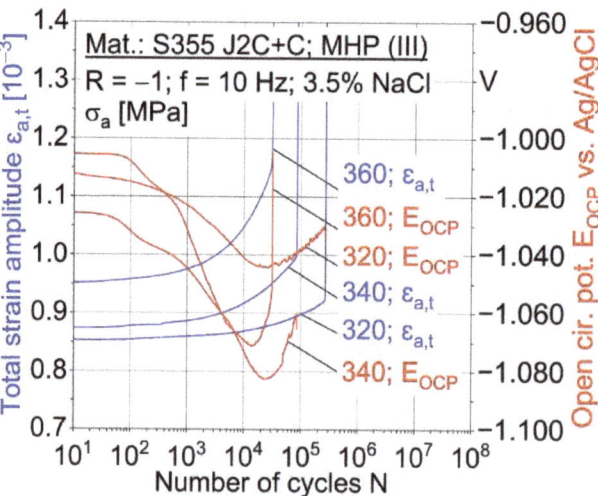

Figure 8. Cyclic deformation curves for corrosion fatigue tests of machine hammer-peened (III) specimens in 3.5% NaCl solution.

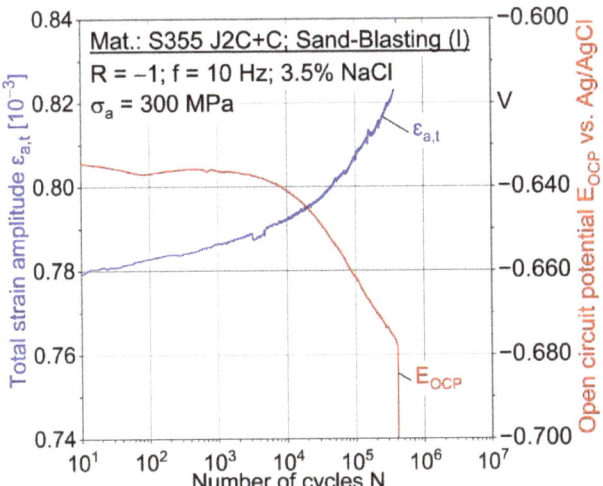

Figure 9. Corrosion fatigue tests of sandblasted (I) specimens in 3.5% NaCl solution.

A higher deformation rate due to higher stress amplitude also leads to more pronounced OCP changes. Consequently, the correlation of mechanical and electrochemical measurement techniques is well suitable to describe corrosion fatigue damage mechanisms of the material S355 J2C + C without and with ZnAl-coating (II) and MHP post-treatment (III).

Based on the change in OCP, the stress-dependent corrosion fatigue damage behavior can be evaluated in detail; Figure 10. The change in open circuit potential ΔE_{ocp} is calculated based on the specific potential after the exposure time without load and current potential. Higher stress amplitudes promote damage mechanisms, which can be related to the amount and width of cracks, leading to failure at lower lifetimes. The more pronounced decrease for higher stress amplitudes is expected to be a result of plastic deformation, crack initiation, and propagation within the coating. The following crack propagation within the substrate is also accelerated by higher stresses, resulting in a rapid OCP increase. The OCP minimum is reached earlier for higher stresses and is well suitable to indicate the beginning of corrosion fatigue damage within the substrate.

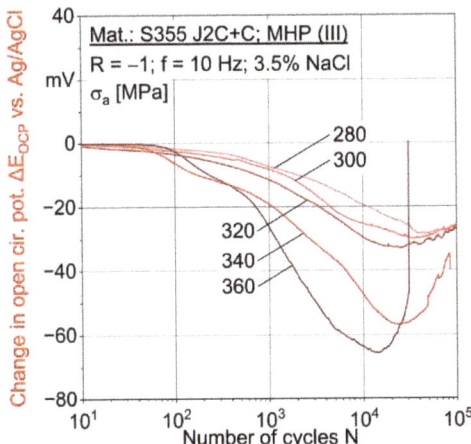

Figure 10. Change in open circuit potential within corrosion fatigue tests of machine hammer-peened (III) specimens in 3.5% NaCl solution.

Figure 11 shows the S-N diagram for specimen conditions sandblasted (I), ZnAl-coated (II) and machine hammer-peened (III). In the low cycle fatigue (LCF) range, the lifetime is nearly independent of specimen conditions. Measured numbers of cycles to failure differ in a small range of 24,000 to 38,000 (360 MPa), and 60,000 to 85,000 (340 MPa).

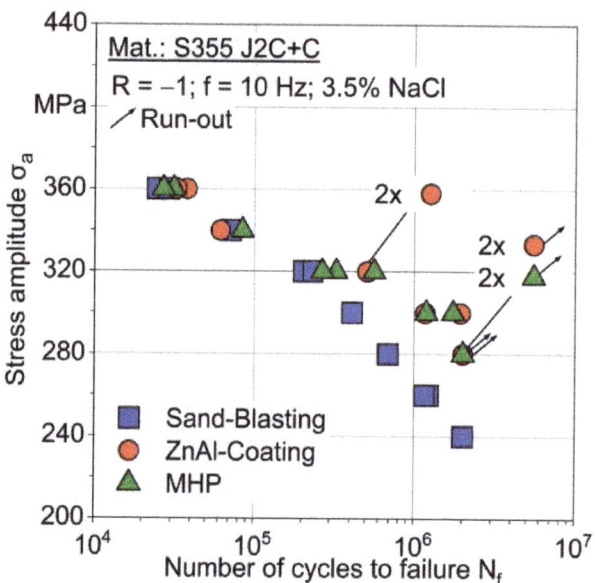

Figure 11. S-N diagram for specimen conditions: Sand-Blasting (I), ZnAl-Coating (II), and machine hammer-peened (III).

For decreasing stress amplitude and increasing testing time in a corrosive environment, the improvement of the corrosion fatigue performance due to ZnAl-coating (II) and MHP post-treatment (III) is more pronounced. The stress amplitude, applied to reach a number of cycles to failure $N_f = 1.2 \times 10^6$ can be increased from 260 to 300 MPa, i.e., +15%, due to additional ZnAl-coating (II) and MHP post-treatment (III). While specimens II and III run-out at 280 MPa, specimen I failed at 260 MPa. On comparing the fatigue performance of ZnAl-coated and MHP post-treated specimens within the investigated loading range, no significant differences were determined. A further improvement of the performance for longer test durations and more aggressive media by the MHP process is expected, since the surface is smooth and the number and size of pores within the coating are decreased.

For evaluation of the test media influence, additional fatigue tests were carried out in 5% NaCl solution with MHP post-treated (III) specimens; Figure 12. This increase in NaCl content does not lead to significant difference in corrosion fatigue behavior.

After corrosion fatigue testing, sandblasted (I) specimens loaded at 320 MPa show a significant corrosive attack within the gauge length and on the fracture surface (Figure 13a,b). In comparison, the corrosive attack of MHP post-treated (III) specimens, also tested at 320 MPa, is limited to the coating, as white corrosive products of ZnAl had formed (Figure 13c,d). Due to this, the corrosive attacks on the substrate seem to be successfully avoided by using the coating as corrosion protection. After 56 h corrosion fatigue testing at 280 MPa without fracture, cracks in the coating and corrosion products of the ZnAl-coating (II) can be detected for MHP post-treated (III) specimens (Figure 13e). Consequently, the ZnAl-coating (II) is well suited to act as a sacrificial anode and protect the underlying substrate from corrosion.

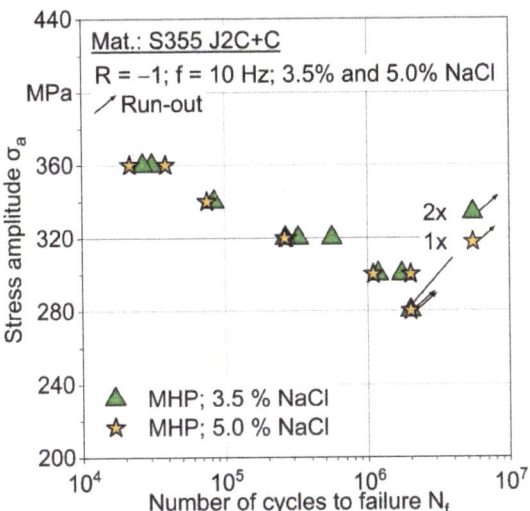

Figure 12. S-N diagram for 3.5% and 5.0% NaCl solution for specimen condition: Machine hammer-peened (III).

On the fracture surface of MHP post-treated (III) specimens, which were tested at 300 MPa in 3.5% NaCl solution, the crack initiation can be located at the specimen surface (Figures 14b,c and 15b). The fatigue fracture area, where the fatigue crack propagates under lower deformation features and with formation of fatigue striations (Figure 14d), can be distinguished from the region of sudden failure. The white arrows mark fatigue striations and a side crack along a fatigue striation. This behavior is characteristic for all specimen conditions and stress amplitudes (Figure 15a,b). The area of sudden failure is characterized by a ductile dimple fracture, which is common for structural steels (Figure 15c,d). A distinct corrosion attack cannot be observed for coated (II) and MHP post-treated (III) specimens.

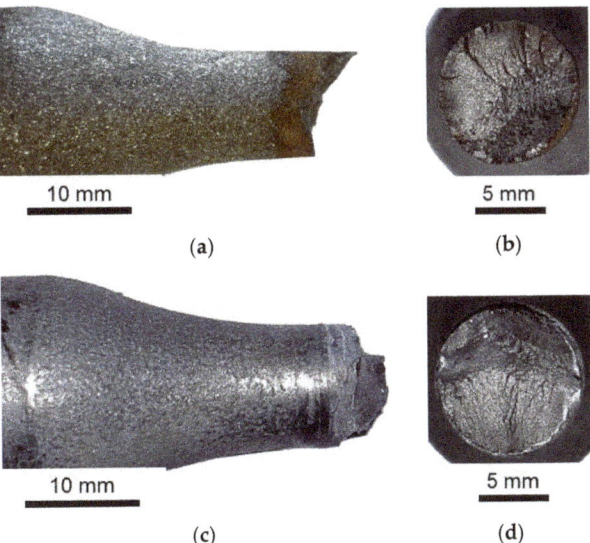

Figure 13. *Cont.*

(e)

Figure 13. Macrographs of specimens after corrosion fatigue testing, (**a**,**b**) sandblasted (I) loaded at 320 MPa, (**c**,**d**) MHP post-treated (III) loaded at 320 MPa, (**e**) machine hammer-peened (III) loaded at 280 MPa.

Figure 14. Fractography of machine hammer-peened (III) specimen loaded at 300 MPa and tested in 3.5% NaCl solution: (**a**) overview, (**b**) area of crack initiation with fatigue striations, (**c**) area of crack initiation with fatigue striations at higher magnification and (**d**) area of crack initiation with fatigue striations at higher magnification.

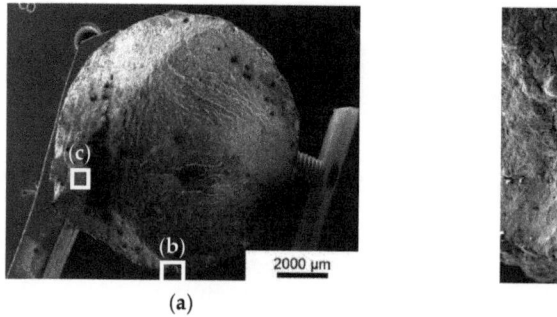

Figure 15. *Cont.*

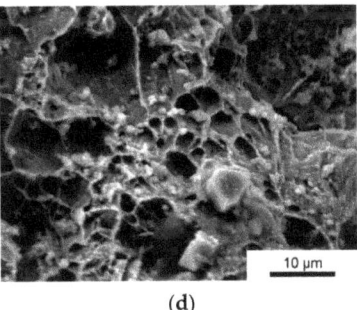

(c)   (d)

Figure 15. Fractography of ZnAl-coated (II) specimen tested in 3.5% NaCl solution at 300 MPa and: (**a**) overview, (**b**) area of crack initiation, (**c**) area of sudden failure and ductile dimple fracture and (**d**) area of sudden failure and ductile dimple fracture at higher magnification.

4. Conclusions and Outlook

The combination of twin wire arc spraying (TWAS) and machine hammer peening (MHP) resulted in coating thicknesses above 200 µm, covering the entire substrate material S355 J2C + C without cracks vertical to the substrate. MHP improved the roughness of the ZnAl coatings with a reduction of the arithmetical mean roughness and mean roughness depth by over 50%. The MHP process also created a more consistent coating thickness throughout the specimen and less porosity.

Direct correlations of material reactions by means of mechanical and electrochemical measurement techniques and corrosion fatigue damage mechanisms were determined for the three conditions: sandblasted (I), ZnAl-coated (II), and MHP post-treated (III). Higher total strain amplitudes for higher stress amplitudes lead to a more significant change in the open circuit potential (OCP). As a result of plastic deformation, newly generated surface gets in contact with the electrolyte and reduces the OCP in case of the coating systems II and III. Therefore, the initial decrease in OCP of MHP-treated (III) samples is expected to be related to the change in coating properties due to deformation, microcrack formation, and crack propagation within the coating. A significant increase of OCP for ZnAl-coated (II) and MHP post-treated (III) specimens indicated crack propagation, which reached the substrate steel. Further increase in OCP is a result of more substrate material in contact with the electrolyte due to crack initiation and propagation. The assumptions will be confirmed by interrupted corrosion fatigue tests to verify the relationship between cyclic deformation, fatigue crack mechanisms, and electrochemical measurements.

ZnAl-coated (II) and MHP post-treated (III) coating systems significantly improved corrosion fatigue behavior at lower stress amplitudes (260 to 320 MPa) compared to sandblasted (I) specimens. ZnAl coating (II) and MHP post-treatment (III) can increase the stress amplitude applied to achieve a defined lifetime compared to sandblasted (I) specimens. In the low cycle fatigue (LCF) regime, the corrosion fatigue performance of all conditions was similar, as the corrosive attack was negligible. There is no difference in corrosion resistance between ZnAl-coated (II) and MHP post-treated (III) specimens, as condition II already provides sufficient corrosion protection. Consequently, a further improvement of the corrosion fatigue performance in the high cycle fatigue (HCF) regime is expected for more severe corrosive loading due to more aggressive media or longer test durations. This study presents the high potential of the ZnAl-coating and MHP post-treatment, as most applications in the maritime sector are subjected to HCF loading.

In future studies, fatigue tests in air will be evaluated in combination with potentiodynamic polarization measurements of the three conditions in order to separate mechanical and corrosion damage effects. Additional corrosion fatigue tests in more aggressive medium and under anodic polarization will be used to present further improvement of the corrosions fatigue performance achieved by MHP processing.

Author Contributions: Conceptualization, M.P.M., A.W., M.A., D.B., W.T. and F.W.; methodology M.P.M.; software, M.P.M.; validation, M.P.M.; formal analysis, M.P.M.; investigation, M.P.M., A.W and M.A.; resources, D.B., W.T. and F.W.; data curation, M.P.M.; writing—original draft preparation, M.P.M.; writing—review and editing, M.P.M., A.W., M.A., A.K., D.B., W.T. and F.W.; visualization, M.P.M. and A.K.; supervision, D.B., W.T. and F.W.; project administration, D.B., W.T. and F.W.; funding acquisition, D.B., W.T. and F.W. All authors have read and agreed to the published version of the manuscript.

Funding: The authors gratefully acknowledge the funding by the German Research Foundation (Deutsche Forschungsgemeinschaft, DFG) of the project "Process parameters correlated characterization of the corrosion fatigue behavior of post-treated ZnAl-coated arc-sprayed systems" (project number 426365081).

Institutional Review Board Statement: Not applicable.

Informed Consent Statement: Not applicable.

Data Availability Statement: Not applicable.

Conflicts of Interest: The authors declare no conflict of interest.

References

1. Mühlberg, K. Corrosion protection of offshore wind turbines—A challenge for the steel builder and paint applicator. *J. Prot. Coat. Linings* **2010**, *27*, 20–33.
2. Masi, G.; Matteucci, F.; Tacq, J.; Balbo, A. *State of the Art Study on Materials and Solutions Against Corrosion in Offshore Structures*, 3rd ed.; North Sea Solutions for Innovation in Corrosion for Energy, 2019; pp. 37–40. Available online: http://nessieproject.com/library/reports-and-researches/NeSSIE%20Report%20Study%20on%20Materials%20and%20Solutions%20in%20Corrosion (accessed on 22 December 2021).
3. Momber, A.W.; Marquardt, T. Protective coatings for offshore wind energy devices (OWEAs): A review. *J. Coat. Technol. Res.* **2018**, *15*, 13–40. [CrossRef]
4. Schaumann, P.; Lochte-Holtgreven, S.; Steppeler, S. Special fatigue aspects in support structures of offshore wind turbines. *Mater. Werkst.* **2011**, *42*, 1075–1081. [CrossRef]
5. Panossian, Z.; Mariaca, L.; Morcillo, M.; Flores, S.; Rocha, J.; Peña, J.J.; Herrera, F.; Corvo, F.; Sanchez, M.; Rincon, O.D.; et al. Steel cathodic protection afforded by zinc, aluminium and zinc/aluminium alloy coatings in the atmosphere. *Surf. Coat. Technol.* **2005**, *190*, 244–248. [CrossRef]
6. Choe, H.B.; Lee, H.S.; Shin, J.H. Experimental study on the electrochemical anti-corrosion properties of steel structures applying the arc thermal metal spraying method. *Materials* **2014**, *7*, 7722–7736. [CrossRef] [PubMed]
7. Bobzin, K.; Oete, M.; Linke, T.F.; Schulz, C. Corrosion of wire arc sprayed ZnMgAl. *Mater. Corros.* **2015**, *66*, 520–526. [CrossRef]
8. Sugimura, S.; Liao, J. Long-term corrosion protection of arc spray Zn-Al-Si coating system in dilute chloride solutions and sulfate solutions. *Surf. Coat. Technol.* **2016**, *302*, 398–409. [CrossRef]
9. Shih, H.C.; Hsu, J.W.; Sun, C.N.; Chung, S.C. The lifetime assessment of hot-dip 5% Al-Zn coatings in chloride environments. *Surf. Coat. Technol.* **2002**, *150*, 70–75. [CrossRef]
10. Syrek-Gerstenkorn, B.; Paul, S.; Davenport, A.J. Sacrificial thermally sprayed aluminium coatings for marine environments: A review. *Coatings* **2020**, *10*, 267. [CrossRef]
11. Wielage, B.; Lampke, T.; Grund, T. Thermal spraying of wear and corrosion resistant surfaces. *Key Eng. Mater.* **2008**, *384*, 75–98. [CrossRef]
12. Radaj, D.; Vormwald, M. Einflüsse auf die Schwingfestigkeit. In *Ermüdungsfestigkeit Grundlagen für Ingenieure*, 3rd ed.; Springer: Berlin/Heidelberg, Germany, 2007; pp. 110–121.
13. Chen, Y.; Liang, X.; Liu, Y.; Xu, B. Prediction of residual stresses in thermally sprayed steel coatings considering the phase transformation effect. *Mater. Des.* **2010**, *31*, 3852–3858. [CrossRef]
14. Tillmann, W.; Abdulgader, M.; Hagen, L.; Biermann, D.; Timmermann, A.; Wirtz, A.; Walther, F.; Milz, M. Mechanical and microstructural properties of post-treated Zn4Al sprayed coatings using twin wire arc spraying. In Proceedings of the International Thermal Spray Conference, Virtual, 24–28 May 2021; Azarmi, F., Chen, X., Cizek, J., Cojocaru, C., Jodoin, B., Koivuluoto, H., Lau, Y., Fernandez, R., Ozdemir, O., Salami Jazi, H., et al., Eds.; ASTM International: Novelty, OH, USA, 2021; pp. 750–757. [CrossRef]
15. Wielage, B.; Grund, T.; Pokhmurska, H.; Rupprecht, C.; Lampke, T. Tailored surfaces by means of thermal spraying and post-treatment. *Key Eng. Mater.* **2008**, *384*, 99–116. [CrossRef]
16. Hacini, L.; Van Lê, N.; Bocher, P. Effect of impact energy on residual stresses induced by hammer peening of 304L plates. *J. Mater. Proc. Technol.* **2008**, *208*, 542–548. [CrossRef]
17. Adjassoho, B.; Kozeschnik, E.; Lechner, C.; Bleicher, F.; Goessinger, S.; Bauer, C. Induction of residual stresses and increase of surface hardness by machine hammer peening technology. In *Annals of DAAAM for 2012 & Proceedings of the 23rd International DAAAM Symposium*, Zadar, Croatia, 24–27 October 2012; DAAAM International: Vienna, Austria, 2012; Volume 23, pp. 697–702.

8. Timmermann, A.; Abdulgader, M.; Hagen, L.; Milz, M.; Wirtz, A.; Biermann, D.; Tillmann, W.; Walther, F. Charakterisierung lichtbogengespritzter, mittels Machine Hammer Peening nachbehandelter Korrosionsschutzschichten. *Therm. Spray Bull.* **2021**, *14*, 46–52.
9. Tillmann, W.; Abdulgader, M.; Wirtz, A.; Milz, M.; Biermann, D.; Walther, F. The effect of Argon as atomization gas on the microstructure, machine hammer peening post-treatment, and corrosion behavior of twin wire arc sprayed (TWAS) ZnAl4 coatings. *Coatings* **2022**, *12*, 32. [CrossRef]
10. Wirtz, A.; Abdulgader, M.; Milz, M.; Tillmann, W.; Walther, F.; Biermann, D. Thermally assisted machine hammer peening of arc-sprayed ZnAl-based corrosion protective coatings. *J. Manuf. Mater. Proc.* **2021**, *5*, 109. [CrossRef]
11. Puchi-Cabrera, E.S.; Staia, M.H.; Lesage, J.; Chicot, D.; La Barbera-Sosa, J.G.; Ochoa Pérez, E.A. Fatigue performance of a SAE 1045 steel coated with a Colmonoy 88 alloy deposited by HVOF thermal spraying. *Surf. Coat. Technol.* **2006**, *201*, 2038–2045. [CrossRef]
12. Timmermann, A.; Abdulgader, M.; Hagen, L.; Koch, A.; Wittke, P.; Biermann, D.; Tillmann, W.; Walther, F. Effect of machine hammer peening on the surface integrity of a ZnAl-based corrosion protective coating. In Proceedings of the MATEC Web of Conferences, Thessaloniki, Greece, 2–3 July 2020; p. 1008. [CrossRef]

Aggressiveness of Different Ageing Conditions for Three Thick Marine Epoxy Systems

Alexis Renaud [1], Victor Pommier [1], Jérémy Garnier [1], Simon Frappart [2], Laure Florimond [3], Marion Koch [4], Anne-Marie Grolleau [2], Céline Puente-Lelièvre [5] and Touzain Sebastien [1,*]

[1] LaSIE UMR CNRS 7356, La Rochelle Université, Av. Michel Crépeau, 17042 La Rochelle, France; alexis.renaud@univ-lr.fr (A.R.); victorpierre.pommier@gmail.com (V.P.); j.garnier874@laposte.net (J.G.)
[2] Naval Group, Technocampus Océan, 5 rue de l'Halbrane, 44340 Bouguenais, France; simon.frappart@dcnsgroup.com (S.F.); anne-marie.grolleau@dcnsgroup.com (A.-M.G.)
[3] General Electric, 11 rue Arthur III-CS 86325, CEDEX 2, 44263 Nantes, France; laure.florimond@ge.com
[4] Chantiers de l'Atlantique, BD de la Légion d'Honneur, 44600 Saint-Nazaire, France; marion.koch@stx.fr
[5] IRT Jules Verne, CHEM Chaffault, Le Chaffault, 44340 Bouguenais, France; celine.puente-lelievre@irt-jules-verne.fr
* Correspondence: stouzain@univ-lr.fr

Abstract: Three different coated steel systems were aged in natural or artificial seawater, in neutral salt spray (NSS), and using alternate immersion tests in order to evaluate the aggressiveness of the different ageing conditions. Commercial epoxy coatings were applied onto steel (S355NL), hot-galvanized steel (HDG), and Zn-Al15 thermal spraying coated steel. The defect-free systems were immersed in artificial seawater at 35 °C for 1085 days and in natural seawater for 1200 days and were characterized by electrochemical impedance spectroscopy (EIS). Panels with artificial defects were immersed for 180 days in artificial seawater and, regarding adhesion, were evaluated according to ISO 16276-2. In parallel, the three coated systems were submitted to cyclic neutral salt spray (NSS) for 1440 h: defect-free panels were regularly evaluated by EIS, while the degree or corrosion was measured onto panels with artificial defect. After NSS, defect-free panels were immersed in artificial seawater at 35 °C for further EIS investigations. Finally, alternate immersion tests were performed for 860 days for the three defect-free coated systems and for 84 days for panels with a defect. The results showed that, for defect-free panels, immersions in natural or artificial seawater and NSS did not allowed us to distinguish the three different systems that show excellent anticorrosion properties. However, during the alternate immersion test, the organic coating system applied onto HDG presented blisters, showing a greater sensitivity to this test than the two other systems. For panels with a defect, NSS allowed to age the coatings more rapidly than monotone conditions, and the coating system applied onto steel presented the highest degree of corrosion. Meanwhile, the coating systems applied onto HDG and the thermal spray metallic coating showed similar behavior. During the alternate immersion test, the three coated systems with a defect showed clearly different behaviors, therefore it was possible to rank the three systems. Finally, it appeared that the alternate immersion test was the most aggressive condition. It was then proposed that a realistic thermal cycling and an artificial defect are needed when performing ageing tests of thick marine organic coating systems in order to properly rank/evaluate the different systems.

Keywords: epoxy marine coatings; accelerated ageing tests; thermal cycling; EIS; corrosion degree

1. Introduction

The sea is a potentially huge source of energy, but the development of Marine Renewable Energy (MRE) technologies able to harness that energy is at a relatively early stage [1]. Part of the problem is that seawater is an aggressive medium, and that metallic structures are submitted to corrosion. An effective solution to protect against corrosion is organic coatings [2,3] which can be applied with high thicknesses and/or can be associated with cathodic protection [4]. Among them, epoxy-based marine coatings have been widely applied for the anticorrosion of metallic materials due to their low cost and their efficiency in aggressive mediums such as seawater [5,6]. However, all organic coating formulations have to be tested before being applied in order to offer a guarantee in terms of lifetime and effectiveness [7] because they suffer from degradation from the environment. In the past, much research has been devoted to evaluating the performance of organic coatings using natural or artificial tests [2,8–16].

Because anticorrosive coatings for marine environment are usually very robust, accelerated ageing tests are needed in order to promote the coating degradation. The physicochemical properties of the coatings are monitored during these tests, and many fields in material chemistry and physics are involved [17]. The coatings are often tested using standardized methods, such as the salt fog, the climatic chamber, the Q-UV chamber, thermal cycling, etc., and, in some cases, a combination of these methods is proposed. Many accelerated ageing cycles are used in practice to try to find a balance between various tests and obtain the most natural possible exposition [18]; however, very often, there is not a good correlation between accelerated ageing tests and degradation in a natural environment [19]. Research is still in progress in order to find reliable protocols and/or techniques to better understand organic coating degradation [20–29]. However, it is necessary to keep in mind that the degradation mechanisms have to be the same as the real life conditions [30], and that only the intensity of the degradation factors (temperature, humidity, UV radiation, mechanical stresses, etc.) can be increased, respecting the limit properties of the organic material. It is known that the continuous salt fog test is unrealistic [30–32], and a recent work showed that it did not correlate with other tests [33]. On the other hand, a simple UV-accelerated test is not sufficient to simulate the natural exposure and it is important to study the synergic effect of different accelerated factors (for example, UV radiation plus thermal cycling) on many different types of organic coatings [10,32]. In fact, fast temperature changes with high humidity and low humidity periods appear as the most harmful accelerated ageing conditions [9,12,34]. Recently, Qu et al. [35] used different wetting and drying cycles for the ageing of a coating system constituted by a thick epoxy primer and an acrylic polyurethane topcoat. They found that, the longer the wetting time in one cycle, the more serious the degradation of the coating samples. For offshore applications, organic coatings can be tested according to ISO 20340 standard, which alternates UV/condensation (ISO 11507), neutral salt spray (ISO9227), and a freezing phase ($-20\,^\circ$C), with a total duration of 4200 h. However, Le Bozec et al. [33] showed that there was no impact from the freezing phase and that the ISO 20340 cycle showed a satisfying correlation with test variants where UV/condensation was replaced by a wet and dry transition at a similar temperature. Davalos-Monteiro et al. [36] also compared ISO 20340 conditions to natural exposure testing, but attempts to correlate natural exposure results with cyclic test results proved unsuccessful. The authors highlighted the requirement of performing further research to provide an indication of whether a correlation with natural exposure might or might not be possible.

As one can see, the perfect accelerated ageing test that fully represents service conditions may not exist because of the very numerous factors that have to be considered, including the chemical nature of organic coatings. However, attempts to find realistic ageing conditions that one allow to accelerate the coating degradation and to rank different systems must be undertaken [37]. This work aimed to study the ageing of an epoxy coating system applied onto three different metallic substrates, using different ageing tests in natural or artificial seawater, neutral salt spray, and an alternate immersion test, in order to evaluate the aggressiveness of the different ageing conditions. Coated systems with and without initial artificial defect were considered and monotonous ageing conditions were compared to cyclic ageing conditions in order to define the most aggressive ageing condition for thick marine epoxy coatings. The evaluation of the coating degradation was performed by visual inspection (corrosion degree ISO 4628-8) for panels with defect and by electrochemical characterization (barrier properties, ISO 16773) for defect-free panels.

2. Experimental Part

2.1. Materials

Organic coatings were applied onto steel panels (S355NL) by airless spraying process under controlled climatic conditions, according to ISO 8502-4 standard by Cryo-West company (Trignac, France). The panel size was 150 mm × 100 mm × 5 mm. Three different surface preparations were used: steel panels (S355NL) submitted to blast-cleaning (Sa 2.5 according to ISO 8501-1 standard; Cryo-West company, Trignac, France), steel panels (S355NL) coated by a hot-galvanized layer (200 µm thick; France Galva company, Carquefou, France) and steel panels (S355NL) coated by a Zn-Al15 thermal spaying coat (150 µm thick; Cryo-West company, Trignac, France). Commercial paints (Hempel company, Trignac, France) were used and applied as presented in Table 1. Hempadur 45,703 is an epoxy-polyamide primer, Hempadur 45,753 is an epoxy-polyamide/amine primer and Hempadur 15,570 is an epoxy/polyamide adduct paint.

Table 1. Name of the systems, composition of the systems, and total dry film thicknesses of the studied organic coating systems.

System Name	Paint System	Total Dry Film Thickness
Organic system ROX	Hempadur 45703: 250 µm + Hempadur 45753: 250 µm	500 µm
Galvanized system RG	Hempadur 15570: 50 µm + Hempadur 45703: 200 µm + Hempadur 45753: 250 µm	500 µm
Thermal spray system RM	Hempadur 45703: 30 µm + Hempadur 45703: 250 µm + Hempadur 45753: 250 µm	530 µm

All coatings were dried at ambient air for at least 14 days, according to the technical specifications. For panels submitted to electrochemical measurements (electrochemical impedance spectroscopy, or EIS), a hole (diameter 2 mm) was drilled in the top edge of the panel and a metallic wire was inserted. The electrical connection was then embedded in a polyurethane resin to avoid moisture (Figure 1).

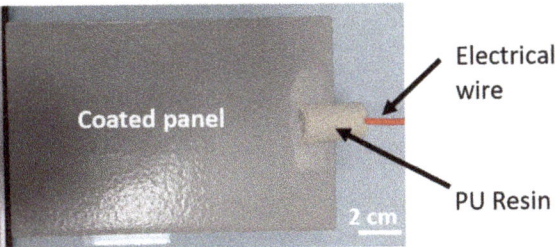

Figure 1. Coated panel with electrical connection.

2.2. Ageing Tests

Defect-free samples were immersed in natural seawater (Port des Minimes, La Rochelle, France) for 1200 days (29,000 h). Samples were regularly brought back to the laboratory for EIS measurements in order to evaluate the barrier properties of the organic coating systems, according to ISO 16773 standard. Six identical panels (ROX1nat to ROX6nat; RG1nat to RG6nat; RM1nat to RM6nat) were tested for reproducibility.

Defect-free samples were also immersed in the laboratory at a controlled temperature, using NaCl 35 g·L^{-1} solution for 1085 days (26,000 h). The ageing temperature was 35 °C in order to accelerate degradation mechanisms but without damaging the polymer. Actually, the glass transition temperature T_g was measured by DSC measurements (DSC Q100 TA Instruments, New Castle, DC, USA) for each organic coating and was higher than 65 °C. Even with moisture ingress and a decrease in T_g, which is usually about 20 °C due to plasticization [38,39], the ageing temperature is well below the humid T_g. Moreover artificial seawater was not chosen in order to avoid the presence of Ca^{2+} and Mg^{2+} ions which leads to the formation of calcareous deposits [40,41] onto uncoated metallic surfaces. As RM and RG systems present a zinc layer under the organic coatings, the steel substrate is cathodically polarized in case of a defect in the organic coating and the calcareous deposit can delay the degradation of the substrate/coating interface for RM and RG systems, which is not the case for the ROX system, therefore a comparison between the different systems is not possible. Finally, six identical defect-free panels (ROX1 to ROX6; RG1 to RG6; RM1 to RM6) were tested for reproducibility.

For each system, a defect (X-cut panels according to ISO 16272-2 standard) was created through the organic coating system till the metallic surface. The defect was a V shape, realized with a scribe (ISO 2409 standard) with Elcometer 1538 (40 mm length, two scribes with a 30° angle, Figure 2). The panels were then immersed in NaCl 35 g·L^{-1} solution for 181 days (4344 h) at 35 °C. Regularly, the samples were taken out and the degree of corrosion was evaluated according to ISO 4628-8 standard. Six identical defect-free panels for each system were tested for reproducibility.

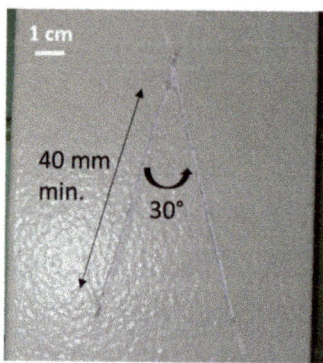

Figure 2. X-cut panel.

Defect-free and X-cut panels were submitted to cyclic neutral salt spray (ASCOTT CC450IP, LABOMAT, France) for 60 days (1440 h). The cycle A from ISO 11997-1 standard was applied and is constituted as: 2 h of salt spray at 35 °C + 4 h of a dry period (20–30% Relative Humidity RH) at 60 °C + 2 h of a humid period (RH > 95%) at 50 °C. Four identical defect-free panels were evaluated by EIS measurements every week and, after 60 days, they were immersed in NaCl 35 g·L^{-1} solution for further EIS investigations. Four identical X-cut panels were also tested for 60 days with the cyclic neutral salt spray and the degree of corrosion was regularly evaluated.

Alternate immersion tests were also performed to better simulate marine environment. The coated samples were immersed in NaCl 35 g·L^{-1} solution in a glass vessel that was inserted in a climatic chamber (model WK 180/40, WEISS TECHNIK, Martillac, France) for thermal cycling for 5 days: 24 h at 5 °C then 24 h at 50 °C then 24 h at 5 °C then 24 h at 50 °C and finally 24 h at 5 °C. After 5 days, the samples were removed from the solution and placed in the climatic chamber for 2 days at 50 °C and 98%RH. The total duration for the alternate immersion test was at least 20,000 h. Every week, EIS measurements were performed with 6 identical defect-free panels and the degree of corrosion was evaluated for 4 identical X-cut panels.

2.3. EIS Measurements

Electrochemical impedance spectroscopy (EIS) measurements were performed in NaCl 35 g·L^{-1} solution using a REF600 potentiostat and a Paint Test Cell PTC1 (16 cm^2) from GAMRY (GAMRY, Warminster PA, USA), with a calomel reference electrode and a graphite counter electrode. Defect-free coated panels with the PTC1 cell were placed in an oven (acting as a Faraday cage) at 35 °C for at least 30 min before EIS measurement so that all EIS data were obtained at 35 °C and the different ageing conditions can be compared. The frequency domain was swept from 200 kHz down to 50 mHz, with a 30 mV RMS perturbation at the free corrosion potential and 11 points per decade. From EIS spectra, the low frequency (0.1 Hz) impedance modulus $|Z|_{0.1\,Hz}$ was measured and divided by the coating thickness e to obtain the reduced impedance modulus $|Z|_{0.1\,Hz}/e$ that allowed to us compare samples with different thicknesses. From the high-frequency domain, the capacitance of the organic coating system C_f was calculated as:

$$C_f = \frac{1}{2\pi f \cdot Z_i} \quad (1)$$

where f is 10 kHz and Z_i is the imaginary part of the impedance. Then, the water uptake was calculated with the well-known Brasher and Kingsbury [42] equation, where one considers the volume fraction of pigment [43]:

$$\chi_V(\%) = \frac{(1-f_P^v)\cdot \log\left(\frac{C_f(t)}{C_f(t=0)}\right)}{\log \varepsilon_w} \quad (2)$$

where ε_w is the water permittivity, f_P^v is the volume fraction of the pigment ($f_P^v = 0.46$), $C_f(t)$ is the measured capacitance at time t, and $C_f(t=0)$ is the measured capacitance at initial time.

All the results are presented as function of the reduced time $\tau = \frac{\sqrt{time}}{e}$ (s$^{1/2}$·cm^{-1}) in order to remove the thickness effect [44,45] and to compare results obtained from samples with different thicknesses.

2.4. Degree of Corrosion

X-cut samples were regularly removed from the ageing test and observed with a microscope USB CONRAD DP-M14 camera (CONRAD, Haubourdin, France). Then, the pictures were analyzed using the software ImageJ to measure the mean scribe width (at least 6 points).

3. Results and Discussion

3.1. Thickness Measurements

Before applying the PTC1 cell onto the defect-free panels for EIS measurements, thickness (at least 10 points) was measured using an Elcometer gauge (Elcometer, La Chapelle Saint Mesmin, France) in order to find a zone with a homogeneous thickness, since EIS measurements can be greatly affected by a thickness gradient [46,47]. The mean thicknesses are given in Table 2.

These experimental values are quite different from the expected thickness about 500–550 μm (cf. Table 1) since thicknesses higher than 700 μm are frequently found. This relates the difficulty to obtain homogeneous thicknesses onto small surfaces with industrial process. It is also a strong proof that the thickness influence has to be removed from the considered parameters in order to compare all the systems.

Table 2. Name and measured thickness of the defect-free coated panels tested in the different ageing conditions.

Ageing Conditions	Panel Name	Thickness (µm)	Panel Name	Thickness (µm)	Panel Name	Thickness (µm)
Immersion at 35 °C in NaCl solution	ROX1	544	RG1	555	RM1	630
	ROX2	658	RG2	514	RM2	634
	ROX3	618	RG3	528	RM3	577
	ROX4	535	RG4	479	RM4	615
	ROX5	578	RG5	539	RM5	612
	ROX6	683	RG6	505	RM6	650
Immersion in seawater	ROX1_nat	490	RG1_nat	490	RM1_nat	594
	ROX2_nat	542	RG2_nat	542	RM2_nat	569
	ROX3_nat	524	RG3_nat	524	RM3_nat	599
	ROX4_nat	584	RG4_nat	584	RM4_nat	653
	ROX5_nat	514	RG5_nat	514	RM5_nat	640
	ROX6_nat	554	RG6_nat	554	RM6_nat	541
Cyclic NSS	ROX1_NSS	651	RG1_NSS	725	RM1_NSS	667
	ROX2_NSS	521	RG2_NSS	536	RM2_NSS	679
	ROX3_NSS	483	RG3_NSS	719	RM3_NSS	696
	ROX4_NSS	517	RG4_NSS	711	RM4_NSS	669
Alternate cycling	ROX1_AC	594	RG1_AC	720	RM1_AC	756
	ROX2_AC	571	RG2_AC	705	RM2_AC	741
	ROX3_AC	600	RG3_AC	648	RM3_AC	690
	ROX4_AC	530	RG4_AC	671	RM4_AC	708
	ROX5_AC	594	RG5_AC	662	RM5_AC	796
	ROX6_AC	640	RG6_AC	681	RM6_AC	708

3.2. Defect-Free Samples Immersed at 35 °C

EIS raw data obtained every 1 h for 50 h for panel ROX5 are shown in Figure 3. A one-time constant R/C behavior is clearly obtained with a capacitive domain in the high frequency region (phase values near −90°) and a resistive domain (phase values tending to 0°) for lower frequencies. The same typical behavior was obtained for other ROX panels and also for RM and RG systems.

This shows a rapid decrease in the barrier effect of the organic coating system due to water uptake and a decrease in the low frequency impedance modulus $|Z|_{0.1\,Hz}$ [8,48]. In the high frequency domain, the capacitive behavior indicates that the impedance value is dominated by its imaginary part, therefore the film capacitance C_f can be readily calculated using Equation (1). Figure 4 presents the reduced low-frequency impedance modulus evolution with ageing time for ROX, RG, and RM systems.

For the three systems, the six identical panels show a very good reproducibility. For all systems, there is a rapid decrease in $|Z|_{0.1\,Hz}/e$, about two decades after one week, which means a decrease in the barrier effect due to water ingress. The decrease in $|Z|_{0.1\,Hz}/e$ is the consequence of the progressive penetration of the electrolyte in the polymeric network and through resistive paths (pores, microscopic defects), which increases the conductivity of the organic layers [49,50]. The electrochemical behavior of the paint, which is highly capacitive at the beginning of the immersion period, becomes significantly resistive in the low frequency domain. Such behavior is typical for defect-free organic coatings exposed to aqueous electrolytes and has been widely described in the literature [51–53].

Then, for the ROX system, the reduced modulus values slightly increase before tending to a plateau, and final values are close to 10 GΩ·cm (5 × 10^8 Ω·cm^2) for 26,000 h of immersion (1085 days). These three different periods can be better seen when the reduced modulus is plotted vs. the reduced time. For RG panels, after the initial decrease in $|Z|_{0.1\,Hz}/e$ from 1 TΩ·cm to 10 GΩ·cm, there is a long increase up to 50 GΩ·cm and a slight decrease after 20,000 h of immersion, with final values of about 30 GΩ·cm (1.5 × 10^9 Ω·cm^2) for 26,000 h of immersion (1085 days). For RM systems, a slight increase in the reduced modulus is observed after the initial decrease, similar to ROX system, except the values remain quite constant afterwards, with final values close to 20 GΩ·cm (10^9 Ω·cm^2) for 26,000 h of immersion (1085 days). This can be explained by the presence of the additional coat onto RM systems (Table 1). For all systems, the slight increase in the reduced modulus value can be explained by a post-reticulation process [54,55] of the polymer matrix that enhances the barrier effect. Moreover, the porosity or the microscopic defects of the organic layers may be partially blocked by the swelling of the matrix or by the generation of few corrosion products [56,57]. The main observation is that, for the considered immersion period (1085 days), the coating systems present similar high impedance values that do not allow us to differentiate them, and all systems can be ranked as excellent, according to [12].

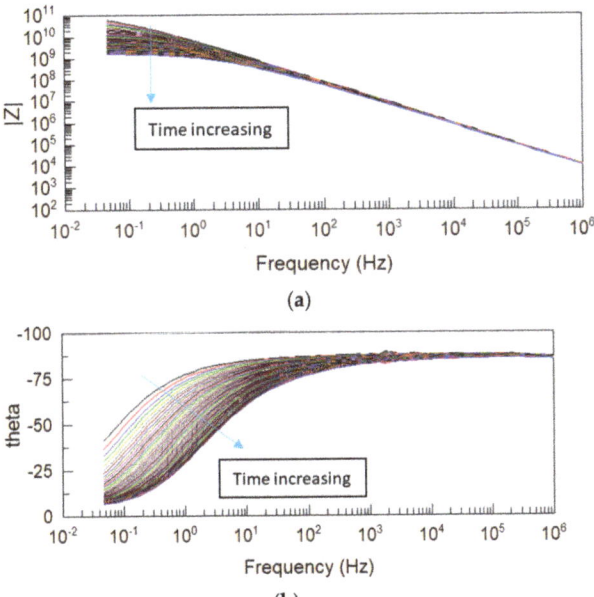

Figure 3. Typical EIS raw data obtained for panel ROX5 (a) impedance modulus $|Z|$ in ohm (Ω) and (b) the phase Theta in degree (°).

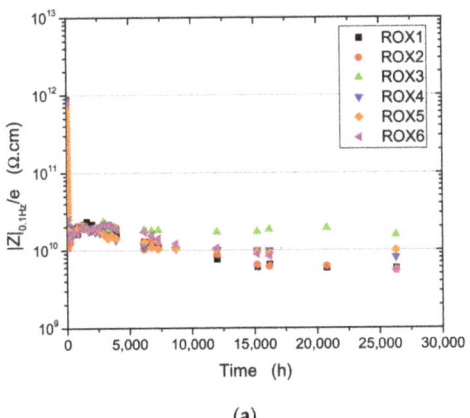

(a)

Figure 4. *Cont.*

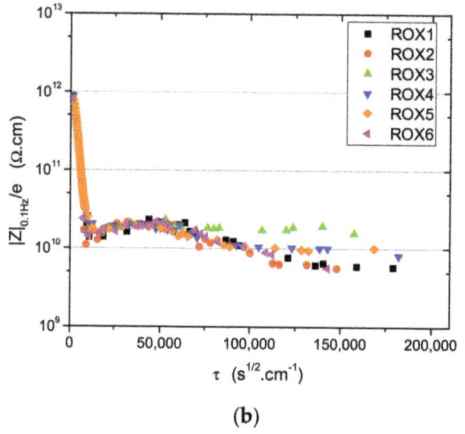

(b)

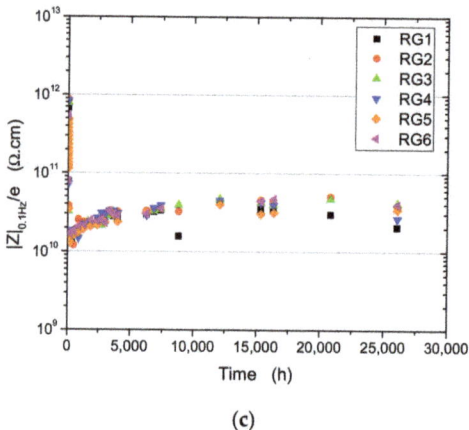

(c)

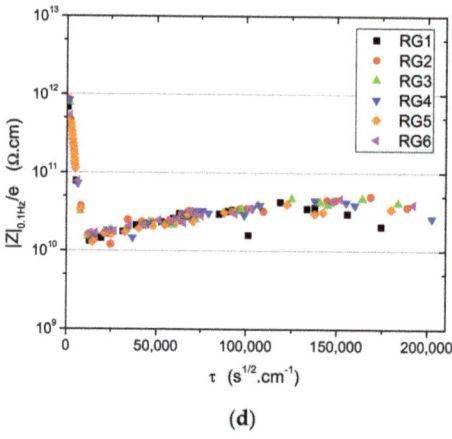

(d)

Figure 4. *Cont.*

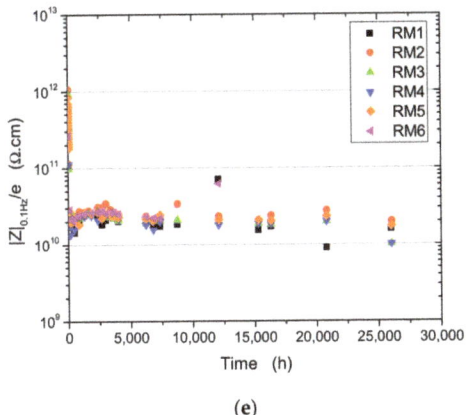

(e)

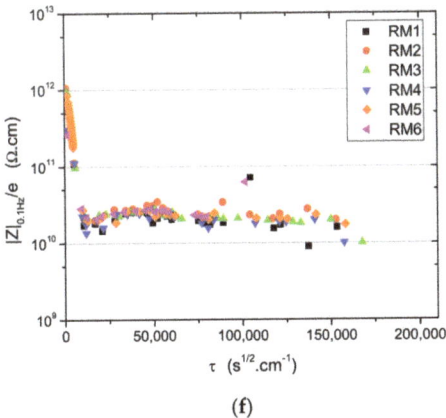

(f)

Figure 4. Evolution of the reduced impedance modulus $|Z|_{0.1\,Hz}/e$ vs. ageing time for ROX system (**a**), RG system (**c**), and RM system (**e**) and evolution of the reduced impedance modulus $|Z|_{0.1\,Hz}/e$ vs. reduced time for ROX system (**b**), RG system (**d**), and RM system (**f**) during the immersion at 35 °C in laboratory.

Water uptake values were calculated according to Equation (2) and plotted vs. the reduced time (Figure 5). For each system, the volume water uptake values are reproducible and are quite similar for all the systems for the ageing period. For all systems, the volume water uptake increases until 12,000 $s^{1/2} \cdot cm^{-1}$ (about 7 days), which perfectly corresponds to the decrease in the reduced impedance modulus previously observed (Figure 4). Then, the values remain quite constant until 25,000 $s^{1/2} \cdot cm^{-1}$ (about 30 days), with values between 3.5 and 4%, showing a pseudo-Fickian behavior. After the plateau, one can observed a slight increase that could be related to the presence of water at the pigment/polymer interface, paint delamination, swelling, or water accumulation in the coating in a heterogeneous way [58–60]. Indeed, the long-term water uptake of the samples may be associated with the swelling of the organic layers, increasing the thickness of the coating systems, and therefore the overall impedance, as previously discussed. Such behavior is in good accordance with many results described in the literature obtained on model epoxy coatings [61], as well as on commercial organic paints [62–64]. The main result is that all systems absorb similar water content, which is not surprising, since the same coating formulations are applied onto the metallic panels.

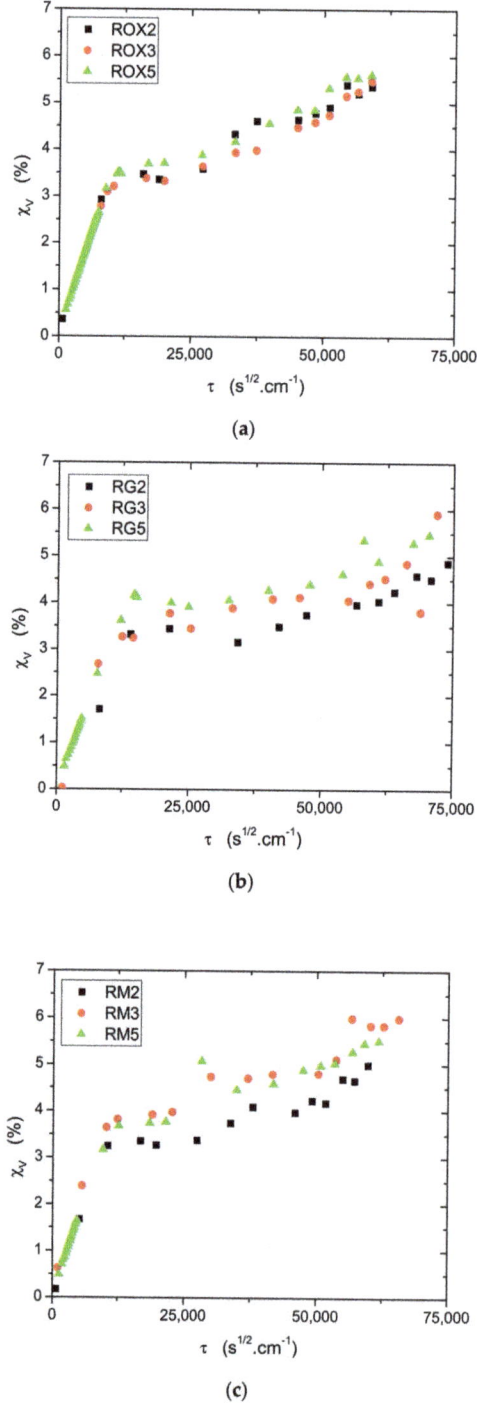

Figure 5. Evolution of the volume water uptake values with reduced time for (**a**) ROX system, (**b**) RG system, and (**c**) RM system during the immersion at 35 °C in laboratory.

From these EIS measurements, it appears that the three systems behave quite similarly and that the immersion at 35 °C for 1085 days does not allow one to observe significant degradation and/or difference between the three systems. The same observation was made by other authors for epoxy coatings tested in aerated ASTM solution for 2 years of testing [65].

3.3. Defect-Free Samples Immersed in Natural Seawater

The evolution of the reduced impedance modulus for panels immersed in natural seawater is presented in Figure 6. For each system, a typical curve obtained for the same system from the ageing at 35 °C in laboratory is added to the plot for comparison with natural conditions.

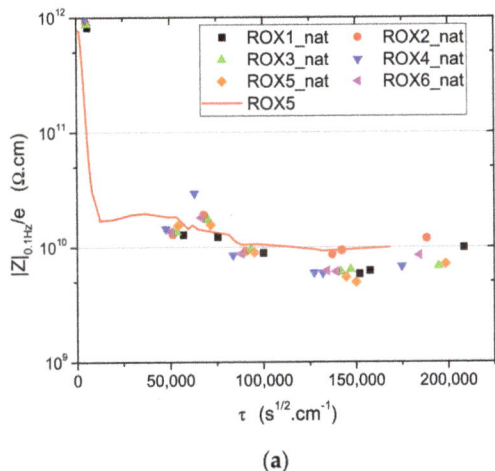

(a)

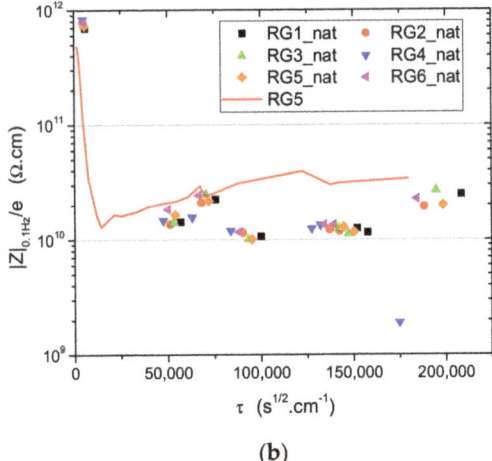

(b)

Figure 6. Cont.

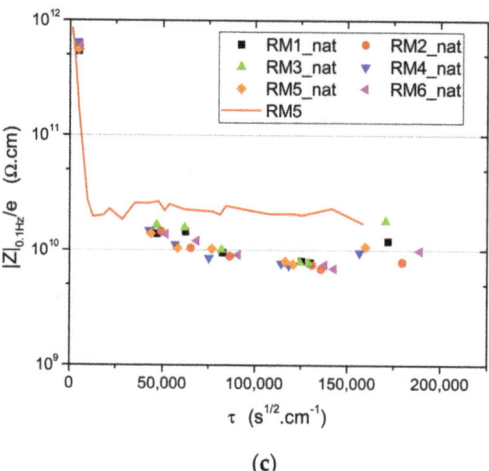

(c)

Figure 6. Evolution of $|Z|_{0.1\,Hz}/e$ vs. reduced time for the ageing in natural seawater for (**a**) ROX system, (**b**) RG system, and (**c**) RM system.

Initial impedance modulus values are high, but they rapidly decrease with ageing time, as for the ageing at 35 °C in laboratory. For the three systems, the reduced impedance modulus slightly decreases with ageing time, showing final values that are globally lower than those observed during ageing at 35 °C. This can be explained by the low natural thermal cycles [2] that induce internal stresses [22,34,66,67] within the coating, leading to a faster decrease in the barrier efficiency. The main result is that the three systems behave quite similarly and that the natural immersion for 1200 days does not allow one to observe the significant degradation and/or difference between the three systems. The same observation was made by other authors for epoxy coatings tested for 7 months in natural seawater [8].

3.4. Defect-Free Samples in Cyclic NSS then Immersed at 35 °C

The evolution of the reduced impedance modulus for panels submitted to cyclic NSS is presented in Figure 7. For each system, a typical curve obtained for the same system from the ageing at 35 °C in laboratory is added to the plot for comparison with cyclic NSS conditions.

EIS measurements performed show high constant impedance values (>1 TΩ·cm ie) ($|Z|_{0.1\,Hz} \approx 50$ GΩ·cm^2) for all systems during the cyclic NSS period (1440 h). This means that the barrier effect is not modified by the cyclic ageing conditions and can be easily explained by the dry period at 60 °C that deletes the water uptake, as has been already mentioned by other studies [68]. Moreover, no defect (blister and/or delamination) was visually observed. It can therefore be concluded that much more time is needed to observe an eventual degradation of the coating using the cyclic NSS. After this initial period in cyclic NSS, panels were placed at 35 °C in NaCl solution. With this constant immersion, the reduced impedance modulus values decrease, showing a decrease in the barrier efficiency induced by water uptake. For the RG system, the final values of $|Z|_{0.1\,Hz}/e$ are similar to those observed during constant immersion at 35 °C, while, for the RM and ROX systems, the final values are higher.

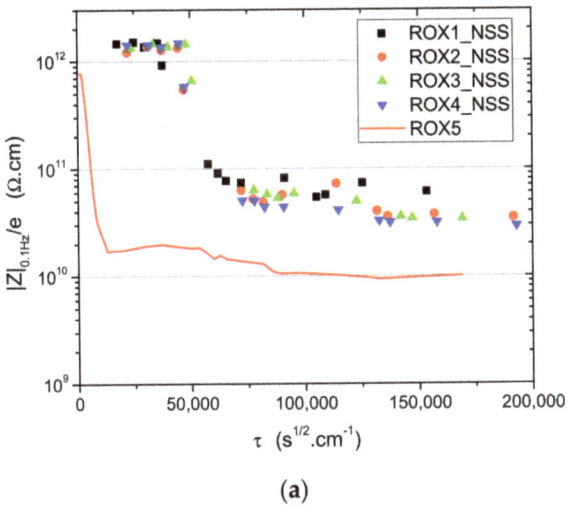

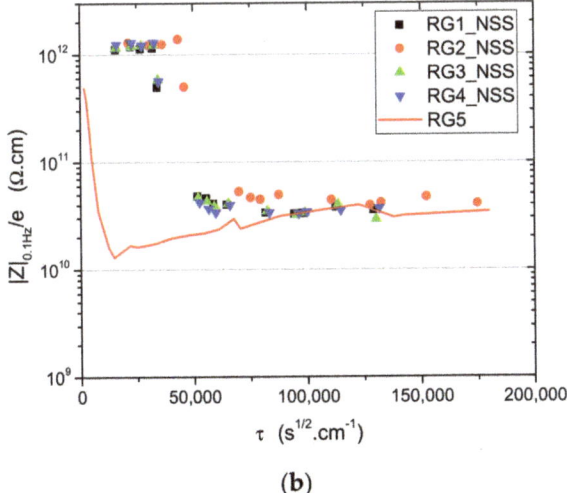

Figure 7. *Cont.*

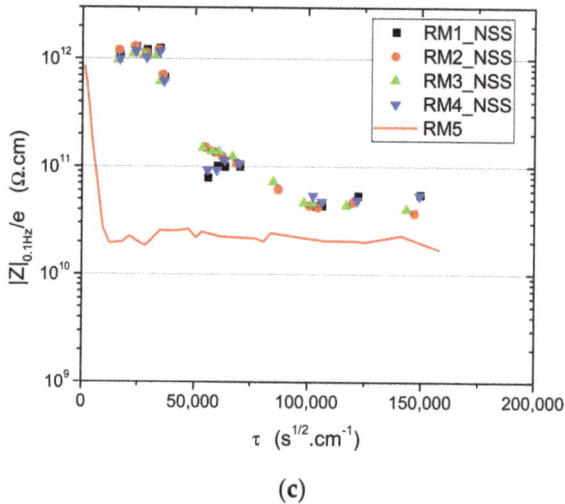

Figure 7. Evolution of $|Z|_{0.1\,Hz}/e$ vs. reduced time for the ageing in cyclic NSS for (**a**) ROX system, (**b**) RG system, and (**c**) RM system.

The initial period in cyclic NSS contains a dry period at 60 °C, which is close to the T_g value (between 65 °C and 70 °C) of the systems. In such conditions, post-reticulation and/or physical ageing [69–72] processes can take place and modify the polymer network, leading to a lower water uptake and then to a better barrier effect. That was observed for the ROX and RM systems, which are constituted by the same coating formulations, but not for the RG system. In fact, the RG system is constituted by another primer that could be more sensitive to thermal effects, leading to a lower barrier effect. Unfortunately, it was not possible to emphasize the thermal sensitivity of this primer in this work, since only coated panels were available from the industrial partners and not the paint formulations. Still, it is worth noting that, despite their relatively low thickness compared to the overall system, primers play a key role in the durability of protective paints. Indeed, the performance and the stability of the primer govern the quality of the adhesion between the metallic substrate and the coating system, which is essential to prevent and limits the propagation of corrosion. Many scientific papers have reported the importance of the adhesion regarding the durability of protective systems, and it can be improved thanks to several surface pretreatments [73,74] or through the use of the appropriate primer [75,76]. In the present case, the step at 60 °C from the NSS may deteriorate the primer and/or the interface between the primer and the substrate, leading to a loss of adhesion at the expense of the barrier properties.

3.5. Defect-Free Samples in Alternate Cycling

The evolution of the reduced impedance modulus for panels submitted to alternate cycling is presented in Figure 8. For each system, a typical curve obtained for the same system from the ageing at 35 °C in a laboratory is added to the plot for comparison with alternate cycling conditions.

For almost all coated panels, the reduced impedance modulus values drastically decrease after one week of the alternate cycling conditions. This is a strong difference with cyclic NSS, where a dry period was maintained for 4 h at 60 °C. In the alternate cycling conditions, the panels are either immersed or in contact with air at a relative humidity higher than 98%. In such conditions, no removal of the water uptake is expected, therefore the barrier effect decreases. After this initial decrease, the reduced impedance modulus slightly increases and tends to plateau with values higher than those observed with the immersion at 35 °C in NaCl solution. Again, thermal cycling seems to enhance the barrier effect, which can be explained, as done previously, by post-reticulation and/or physical ageing processes. If the ROX and RM systems present similar reduced impedance modulus values during the whole ageing period, the RG system rapidly presented some important blisters showing the significant degradation of the organic coating system. With such observations, the test was stopped for the RG system.

As previously seen with cyclic NSS, the RG system appears to be particularly sensitive to thermal cycling in humid conditions. The appearance of blisters confirms the loss of adhesion that was previously suspected for the RG samples exposed to NSS. Moreover, it confirms the fact that higher ageing temperatures may favor the loss of adhesion. Alternate cycling was able to request the Achilles heel of the system and to show a clear difference between RG samples and the other systems. This difference, which was already suspected from NSS results, is much more significant with alternate cycling. It may be due to the longer periods, where samples are exposed to high humidity combined with high temperature, and to the higher temperature amplitude during this test, inducing stresses at the interface between the paint system and the substrate. The same conclusions were obtained by others studies dealing with thermal cycling [30,32,77]. Unfortunately, it seems that more time is needed with these ageing conditions to differentiate the ROX and RM systems.

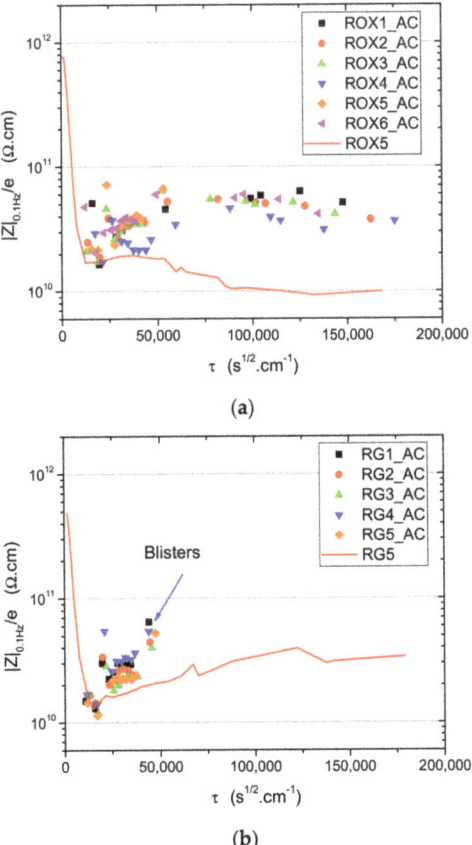

Figure 8. *Cont.*

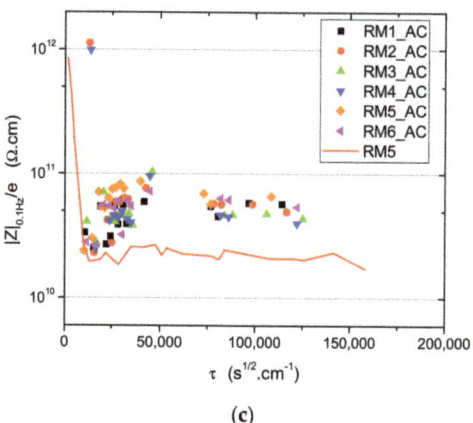

(c)

Figure 8. Evolution of $|Z|_{0.1\,Hz}/e$ vs. reduced time for the ageing in alternate cycling for (**a**) ROX system, (**b**) RG system, and (**c**) RM system.

3.6. X-Cut Samples Immersed at 35 °C

The evolution of the corrosion degree for the three systems immersed at 35 °C in NaCl solution is presented in Figure 9. For each system, the presented curve is the mean curve obtained from six identical panels.

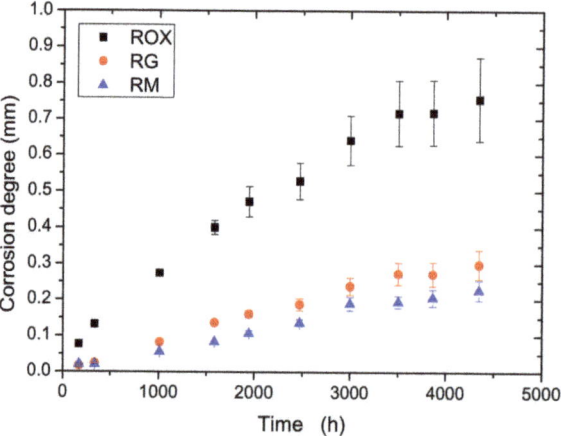

Figure 9. Evolution of the corrosion degree for X-cut panels immersed in NaCl solution at 35 °C.

For the RM and RG systems, the corrosion degree values slightly increase and are quite similar for both systems, with values lower than 0.3 after 180 days of immersion. For the ROX system, the corrosion degree increases more rapidly, with values close to 0.8 after 180 days of immersion. These results show that the ROX system is less performant than the RM and RG systems, where zinc layers are deposited onto the steel substrate. It can be reminded that no specific surface treatment was applied onto the steel substrate before applying the organic coatings, which can explain such results. In these ageing conditions, and for X-cut samples, the contribution of the sacrificial layers (galvanized or thermal sprayed) and their ability to limit the propagation of corrosion are emphasized, as has been already shown by other studies [78,79], while the quality of the paint system (adhesion, barrier properties) is less crucial.

3.7. X-Cut Samples in Cyclic NSS

The evolution of the corrosion degree for the three systems submitted to cyclic NSS is presented in Figure 10. For each system, the presented curve is the mean curve obtained from four identical panels.

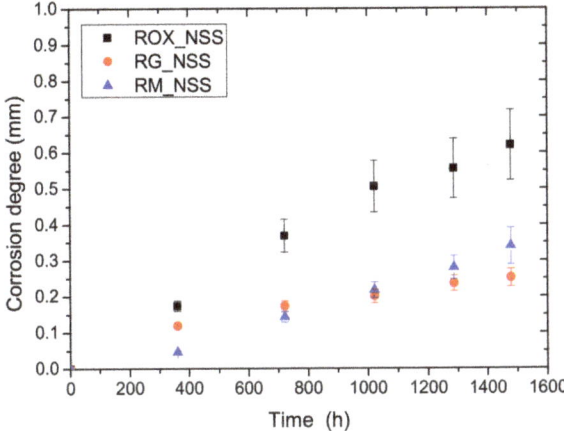

Figure 10. Evolution of the corrosion degree for X-cut panels submitted to cyclic NSS.

For RM and RG systems, the corrosion degree values slightly increase and are quite similar for both systems, with values lower than 0.3 after 1440 h (60 days) of ageing. For the ROX system, the corrosion degree increases more rapidly, with values close to 0.6 after 60 days of cyclic NSS. These results show again that the ROX system is less performant than RM and RG systems, for which no clear difference can be made after 1440 h of cycling NSS. In the same manner as immersion in NaCl, the NSS test emphasizes the contribution of the Zn-based sacrificial layers. These results are in agreement with previous studies [80] that showed that metallic Zn-based coatings increased the time for the appearance of reddish corrosion products in the edges and incision during the salt spray chamber test.

3.8. X-Cut Samples in Alternate Cycling

The evolution of the corrosion degree for the three systems submitted to alternate cycling is presented in Figure 11. For each system, the presented curve is the mean curve obtained from four identical panels.

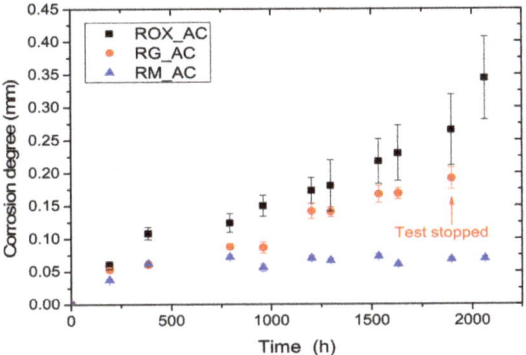

Figure 11. Evolution of the corrosion degree for X-cut panels submitted to alternate cycling.

With the alternate cycling, the three systems clearly present different behaviors. Again the ROX system presents a more rapid increase in the corrosion degree, with values at 0.35 after 2064 h (86 days). For the RM system, the corrosion degree slightly increases and remains close to 0.07, while the RG system presents an intermediate behavior until 79 days and drastically degrades afterwards, with a large delamination of the coating, therefore the test had to be stopped. As seen previously with defect-free panels, the RG system appears to be particularly sensitive to humid thermal cycles, especially regarding adhesion.

The alternate cycling applied to X-cut panels was then able to emphasize the contribution of the sacrificial layers in the same way as immersion in NaCl or NSS testing, but also to highlight the fragility of the adhesion, as it was observed on defect-free samples. Among the ageing conditions presented in this work, the alternate cycling should then be the preferred test to fully discriminate different paint systems, as it involves corrosion and delamination phenomena.

4. Conclusions

From the results of this work, the following conclusions can be proposed about the efficiency of the different ageing tests to differentiate the behaviors of thick marine coatings applied onto steel (ROX system), hot-galvanized steel (RG), and Zn-Al15 thermal spraying coated steel (RM):

When applied to defect-free samples, immersion tests in NaCl 35 $g \cdot L^{-1}$ solution at a constant temperature (35 °C) were only able to show the decrease in the paint resistivity for all systems. No significant difference was observed between the three systems after 1085 days of immersion. Much more time or lower thicknesses may be needed in such tests to expect degradation and eventual differentiation between different systems. When X-cut samples were immersed in NaCl, only the contribution of the sacrificial Zn layers has been shown.

The immersion in natural seawater for 1200 days showed no significant difference for the three systems. This test with true natural conditions is therefore not adapted for the rapid evaluation of thick coating performances.

No significant difference was observed between the defect-free systems during the cyclic NSS test for 1440 h. Only adhesion issues could have been suspected from these ageing conditions, but the phenomena were not clear enough. For panels with defects, the NSS cyclic test was clearly able to show the effect of the sacrificial Zn layers in the same way as the immersion tests.

When applied to defect-free panels, the alternate cycling test was able to clearly emphasize adhesion issues between the paint system and the substrate, as it has been seen with the blistering of RG samples. Moreover, when applied to X-cut panels, corrosion propagation was also involved, and this test allowed us to highlight the contribution of sacrificial Zn layers, as observed in the RM systems.

Thermal cycling is the only test which has been able to clearly differentiate defect-free systems by stressing the adhesion of the coatings. However, care must be taken when defining the thermal cycles so that the maximum temperature remains lower than the wet glass transition temperature. Finally, in the framework of this study, for thick, highly pigmented coatings, X-cut panels and alternate cycling have been the only ageing conditions to fully discriminate different paint systems by involving both corrosion and adhesion loss mechanisms. It can be noted that a long testing time is still needed to rank different thick organic coating systems, therefore more studies have to be performed in order to find rapid realistic ageing conditions. For example, steps of exposition to UV-light could also be introduced in the ageing cycles in order to induce the chemical degradation of the organic compounds and bring the testing to a more realistic level. Mechanical stresses should also be considered, depending on the targeted applications.

Further investigations should also be carried out to study paints applied on other metals, and to understand the degradation of more complex systems such as paints containing corrosion inhibitors or self-healing coatings.

Author Contributions: Investigation, A.R., V.P., J.G. and T.S.; resources, C.P.-L.; writing—original draft preparation, T.S.; writing—review and editing, A.R. and T.S.; supervision, S.F., M.K., A.-M.G.; project administration, L.F. All authors have read and agreed to the published version of the manuscript.

Funding: This research was funded by IRT Jules Verne (French Institute in Research and Technology in Advanced Manufacturing Technologies for Composite, Metallic and Hybrid Structures).

Institutional Review Board Statement: Not applicable.

Informed Consent Statement: Not applicable.

Data Availability Statement: Not applicable.

Acknowledgments: This study was part of the ADUSCOR project managed by IRT Jules Verne (French Institute in Research and Technology in Advanced Manufacturing Technologies for Composite, Metallic and Hybrid Structures). The authors wish to associate the partners of this project, respectively NAVAL GROUP, CHANTIERS DE L'ATLANTIQUE, GENERAL ELECTRIC, CETIM and IRT JULES VERNE.

Conflicts of Interest: The authors declare no conflict of interest.

References

1. Borthwick, A.G.L. Marine Renewable Energy Seascape. *Engineering* **2016**, *2*, 69–78. [CrossRef]
2. Fredj, N.; Cohendoz, S.; Feaugas, X.; Touzain, S. Ageing of marine coating in natural and artificial seawater under mechanical stresses. *Prog. Org. Coat.* **2012**, *74*, 391–399. [CrossRef]
3. Song, G.-L.; Feng, Z. Modification, Degradation and Evaluation of a Few Organic Coatings for Some Marine Applications. *Corros. Mater. Degrad.* **2020**, *1*, 408–442. [CrossRef]
4. Le Thu, Q.; Takenouti, H.; Touzain, S. EIS characterization of thick flawed organic coatings aged under cathodic protection in seawater. *Electrochim. Acta* **2006**, *51*, 2491–2502. [CrossRef]
5. Funke, W. The role of adhesion in corrosion protection by organic coatings. *J. Oil Colour Chem. Assoc.* **1979**, *68*, 229–232.
6. Wicks, Z.W.; Jones, F.N.; Pappas, S.P.; Wicks, A.D. *Organic Coatings: Science and Technology*, 2nd ed.; John Wiley & Sons, Inc.: Hoboken, NJ, USA, 1999.
7. Sørensen, P.; Kiil, S.; Dam-Johansen, K.; Weinell, C. Anticorrosive coatings: A review. *J. Coat. Technol. Res.* **2009**, *6*, 135–176. [CrossRef]
8. Mansfeld, F.; Xiao, H.; Han, L.T.; Lee, C.C. Electrochemical impedance and noise data for polymer coated steel exposed at remote marine test sites. *Prog. Org. Coat.* **1997**, *30*, 89–100. [CrossRef]
9. Fedrizzi, L.; Bergo, A.; Fanicchia, M. Evaluation of accelerated aging procedures of painted galvanised steels by EIS. *Electrochim. Acta.* **2006**, *51*, 1864–1872. [CrossRef]
10. Deflorian, F.; Rossi, S.; Fedrizzi, L.; Zanella, C. Comparison of organic coating accelerated tests and natural weathering considering meteorological data. *Prog. Org. Coat.* **2007**, *59*, 244–250. [CrossRef]
11. Croll, S.G.; Shi, X.; Fernando, B.M.D. The interplay of physical aging and degradation during weathering for two crosslinked coatings. *Prog. Org. Coat.* **2008**, *61*, 136–144. [CrossRef]
12. Bierwagen, G.P.; He, L.; Li, J.; Ellingson, L.; Tallman, D.E. Studies of a new accelerated evaluation method for coating corrosion resistance—Thermal cycling testing. *Prog. Org. Coat.* **2000**, *39*, 67–78. [CrossRef]
13. Yang, X.F.; Li, J.; Croll, S.G.; Tallman, D.E.; Bierwagen, G.P. Degradation of low gloss polyurethane aircraft coatings under UV and prohesion alternating exposures. *Polym. Degrad. Stab.* **2003**, *80*, 51–58. [CrossRef]
14. Corfias, C.; Pebere, N.; Lacabanne, C. Characterization of a thin protective coating on galvanized steel by electrochemical impedance spectroscopy and a thermostimulated current method. *Corros. Sci.* **1999**, *41*, 1539–1555. [CrossRef]
15. Kendig, M.; Mansfeld, F.; Tsai, S. Determination of the long term corrosion behavior of coated steel with A.C. impedance measurements. *Corros. Sci.* **1983**, *23*, 317–329. [CrossRef]
16. Zhang, W.; Chen, X.-Z.; Yin, P.-F.; Xu, Z.-K.; Han, B.; Wang, J. EIS study on the deterioration process of organic coatings under immersion and different cyclic wet-dry ratios. *Appl. Mech. Mater.* **2012**, *161*, 58–66. [CrossRef]
17. Yang, C.; Xing, X.; Li, Z.; Zhang, S. A Comprehensive Review on Water Diffusion in Polymers Focusing on the Polymer–Metal Interface Combination. *Polymers* **2020**, *12*, 138. [CrossRef]
18. Schulz, U.; Trubiroha, P.; Schernau, U.; Baumgart, H. Effects of acid rain on the appearance of automotive paint systems studied outdoors and in a new artificial weathering test. *Prog. Org. Coat.* **2000**, *40*, 151–165. [CrossRef]
19. Wernståhl, K.M. Service life prediction of automotive coatings, correlating infrared measurements and gloss retention. *Polym. Degrad. Stab.* **1996**, *54*, 57–65. [CrossRef]
20. Margarit-Mattos, I.C.P. EIS and organic coatings performance: Revisiting some key points. *Electrochim. Acta* **2020**, *354*, 136725. [CrossRef]

21. Da Silva, T.C.; Mallarino, S.; Touzain, S.; Margarit-Mattos, I.C.P. DMA, EIS and thermal fatigue of organic coatings. *Electrochim. Acta* **2019**, *318*, 989–999. [CrossRef]
22. Olivier, M.-G.; Romano, A.-P.; Vandermiers, C.; Mathieu, X.; Poelman, M. Influence of the stress generated during an ageing cycle on the barrier properties of cataphoretic coatings. *Prog. Org. Coat.* **2008**, *63*, 323–329. [CrossRef]
23. Oliveira, J.L.; Skilbred, A.W.B.; Loken, A.; Henriques, R.R.; Soares, B.G. Effect of accelerated ageing procedures and flash rust inhibitors on the anti-corrosive performance of epoxy coatings: EIS and dynamic-mechanical analysis. *Prog. Org. Coat.* **2021**, *159*, 106387. [CrossRef]
24. Allahar, K.N.; Bierwagen, G.P.; Gelling, V.J. Understanding ac-dc-ac accelerated test results. *Corros. Sci.* **2010**, *52*, 1106–1114. [CrossRef]
25. Bistac, S.; Vallat, M.F.; Schultz, J. Durability of steel/polymer adhesion in an aqueous environment. *Int. J. Adhes. Adhes.* **1998**, *18*, 365–369. [CrossRef]
26. Hollaender, J. Rapid assessment of food/package interactions by electrochemical impedance spectroscopy (EIS). *Food Addit. Contam.* **1997**, *14*, 617–626. [CrossRef] [PubMed]
27. Suay, J.J.; Rodríguez, M.T.; Izquierdo, R.; Kudama, A.M.; Saura, J.J. Rapid Assessment of Automotive Epoxy Primers by Electrochemical Techniques. *J. Coat. Technol.* **2003**, *75*, 103–111. [CrossRef]
28. Miszczyk, A.; Darowicki, K. Accelerated ageing of organic coating systems by thermal treatment. *Corros. Sci.* **2001**, *43*, 1337–1343. [CrossRef]
29. Zargarnezhad, H.; Asselin, E.; Wong, D.; Lam, C.N.C. A critical review of the time-dependent performance of polymeric pipeline coatings: Focus on hydration of epoxy-based coatings. *Polymers* **2021**, *13*, 1517. [CrossRef] [PubMed]
30. Bierwagen, G.; Tallman, D.; Li, J.; He, L.; Jeffcoate, C. EIS studies of coated metals in accelerated exposure. *Prog. Org. Coat.* **2003**, *46*, 149–158. [CrossRef]
31. Bierwagen, G.P.; Jeffcoate, C.S.; Li, J.; Balbyshev, S.; Tallman, D.E.; Mills, D.J. The use of electrochemical noise methods (ENM) to study thick, high impedance coatings. *Prog. Org. Coat.* **1996**, *29*, 21–29. [CrossRef]
32. Valentinelli, L.; Vogelsang, J.; Ochs, H.; Fedrizzi, L. Evaluation of barrier coatings by cycling testing. *Prog. Org. Coat.* **2002**, *45*, 405–413. [CrossRef]
33. LeBozec, N.; Thierry, D.; le Calvé, P.; Favennec, C.; Pautasso, J.-P.; Hubert, C. Performance of marine and offshore paint systems: Correlation of accelerated corrosion tests and field exposure on operating ships. *Mater. Corros.* **2015**, *66*, 215–225. [CrossRef]
34. Ochs, H.; Vogelsang, J. Effect of temperature cycles on impedance spectra of barrier coatings under immersion conditions. *Electrochim. Acta* **2004**, *49*, 2973–2980. [CrossRef]
35. Qu, S.; Ju, P.; Zuo, Y.; Zhao, X.; Tang, Y. The effect of various cyclic wet-dry exposure cycles on the failure process of organic coatings. *Int. J. Electrochem. Sci.* **2019**, *14*, 10754–10755. [CrossRef]
36. Thirion, P. Proprietes viscoelastiques des reseaux a l'etat caoutchoutique. *Eur. Polym. J.* **1974**, *10*, 1093–1101. [CrossRef]
37. Mills, D.J.; Jamali, S.S. The best tests for anti-corrosive paints. And why: A personal viewpoint. *Prog. Org. Coat.* **2017**, *102*, 8–17. [CrossRef]
38. Apicella, A.; Tessieri, R.; De Cataldis, C. Sorption modes of water in glassy epoxies. *J. Memb. Sci.* **1984**, *18*, 211–225. [CrossRef]
39. Apicella, A.; Nicolais, L. Effect of water on the properties of epoxy matrix and composite. In *Epoxy Resins and Composites I*; Springer: Berlin/Heidelberg, Germany, 1985; pp. 69–77. [CrossRef]
40. Barchiche, C.; Deslouis, C.; Festy, D.; Gil, O.; Refait, P.; Touzain, S.; Tribollet, B. Characterization of calcareous deposits in artificial seawater by impedance techniques: 3—Deposit of $CaCO_3$ in the presence of Mg (II). *Electrochim. Acta* **2003**, *48*, 1645–1654. [CrossRef]
41. Deslouis, C.; Doncescu, A.; Festy, D.; Gil, O.; Maillot, V.; Touzain, S.; Tribollet, B. Kinetics and characterisation of calcareous deposits under cathodic protection in natural sea water. *Mater. Sci. Forum* **1998**, *289–292*, 1163–1180. [CrossRef]
42. Brasher, D.M.; Kingsbury, A.H. Electrical measurements in the study of immersed paint coatings on metal. I. Comparison between capacitance and gravimetric methods of estimating water-uptake. *J. Appl. Chem.* **1954**, *4*, 62–72. [CrossRef]
43. Fredj, N.; Cohendoz, S.; Mallarino, S.; Feaugas, X.; Touzain, S. Evidencing antagonist effects of water uptake and leaching processes in marine organic coatings by gravimetry and EIS. *Prog. Org. Coat.* **2010**, *67*, 287–295. [CrossRef]
44. Boisseau, A.; Davies, P.; Thiebaud, F. Sea water ageing of composites for ocean energy conversion systems: Influence of glass fibre type on static behaviour. *Appl. Compos. Mater.* **2012**, *19*, 459–473. [CrossRef]
45. Dang, D.N.; Cohendoz, S.; Mallarino, S.; Feaugas, X.; Touzain, S. Effects of curing program on mechanical behavior and water absorption of DGEBA/TETa epoxy network. *J. Appl. Polym. Sci.* **2013**, *129*, 2451–2463. [CrossRef]
46. Cattarin, S.; Comisso, N.; Musiani, M.; Tribollet, B. The Impedance of an Electrode Coated by a Resistive Film with a Graded Thickness. *Electrochem. Solid-State Lett.* **2008**, *11*, C27–C30. [CrossRef]
47. Touzain, S. Some comments on the use of the EIS phase angle to evaluate organic coating degradation. *Electrochim. Acta* **2010**, *55*, 6190–6194. [CrossRef]
48. Bouvet, G.; Nguyen, D.D.; Mallarino, S.; Touzain, S. Analysis of the non-ideal capacitive behavior for high impedance organic coatings. *Prog. Org. Coat.* **2014**, *77*, 2045–2053. [CrossRef]

29. Nguyen, A.S.; Musiani, M.; Orazem, M.E.; Pébère, N.; Tribollet, B.; Vivier, V. Impedance analysis of the distributed resistivity of coatings in dry and wet conditions. *Electrochim. Acta* **2015**, *179*, 452–459. [CrossRef]
30. Mišković-Stanković, V.B.; Dražić, D.M.; Teodorović, M.J. Electrolyte penetration through epoxy coatings electrodeposited on steel. *Corros. Sci.* **1995**, *37*, 241–252. [CrossRef]
31. Walter, G.W. The application of impedance methods to study the effects of water uptake and chloride ion concentration on the degradation of paint films—II. Free films and attached/free film comparisons. *Corros. Sci.* **1991**, *32*, 1085–1103. [CrossRef]
32. Duval, S.; Keddam, M.; Sfaira, M.; Srhiri, A.; Takenouti, H. Electrochemical impedance spectroscopy of epoxy-vinyl coating in aqueous medium analyzed by dipolar relaxation of polymer. *J. Electrochem. Soc.* **2002**, *149*, B520. [CrossRef]
33. Amirudin, A.; Thieny, D. Application of electrochemical impedance spectroscopy to study the degradation of polymer-coated metals. *Prog. Org. Coat.* **1995**, *26*, 1–28. [CrossRef]
34. Zhou, J.; Lucas, J.P. Hygrothermal effects of epoxy resin. Part I: The nature of water in epoxy. *Polymer* **1999**, *40*, 5505–5512. [CrossRef]
35. Irigoyen, M.; Aragon, E.; Perrin, F.X.; Vernet, J.L. Effect of UV aging on electrochemical behavior of an anticorrosion paint. *Prog. Org. Coat.* **2007**, *59*, 259–264. [CrossRef]
36. Pebere, N.; Picaud, T.; Duprat, M.; Dabosi, F. Evaluation of corrosion performance of coated steel by the impedance technique. *Corros. Sci.* **1989**, *29*, 1073–1086. [CrossRef]
37. Mansfeld, F. Use of electrochemical impedance spectroscopy for the study of corrosion protection by polymer coatings. *J. Appl. Electrochem.* **1995**, *25*, 187–202. [CrossRef]
38. Deflorian, F.; Fedrizzi, L.; Rossi, S.; Bonora, P.L. Organic coating capacitance measurement by EIS: Ideal and actual trends. *Electrochim. Acta* **1999**, *44*, 4243–4249. [CrossRef]
39. Van Westing, E.P.M.; Ferrari, G.M.; De Wit, J.H.W. The determination of coating performance with impedance measurements-II. Water uptake of coatings. *Corros. Sci.* **1994**, *36*, 957–977. [CrossRef]
40. Nguyen, V.N.; Perrin, F.X.; Vernet, J.L. Water permeability of organic/inorganic hybrid coatings prepared by sol-gel method: A comparison between gravimetric and capacitance measurements and evaluation of non-Fickian sorption models. *Corros. Sci.* **2005**, *47*, 397–412. [CrossRef]
41. Dang, D.N.; Peraudeau, B.; Cohendoz, S.; Mallarino, S.; Feaugas, X.; Touzain, S. Effect of mechanical stresses on epoxy coating ageing approached by Electrochemical Impedance Spectroscopy measurements. *Electrochim. Acta* **2014**, *124*, 80–89. [CrossRef]
42. Mišković-Stanković, V.B.; Dražić, D.M.; Kačarević-Popović, Z. The sorption characteristics of epoxy coatings electrodeposited on steel during exposure to different corrosive agents. *Corros. Sci.* **1996**, *38*, 1513–1523. [CrossRef]
43. Fedrizzi, L.; Deflorian, F.; Boni, G.; Bonora, P.L.; Pasini, E. EIS study of environmentally friendly coil coating performances. *Prog. Org. Coat.* **1996**, *29*, 89–96. [CrossRef]
44. Perez, C.; Collazo, A.; Izquierdo, M.; Merino, P.; Novoa, X.R. Characterisation of the barrier properties of different paint systems. Part I. Experimental set-up and ideal Fickian diffusion. *Prog. Org. Coat.* **1999**, *36*, 102–108. [CrossRef]
45. Murray, J.N.; Hack, H.P. Testing organic architectural coatings in ASTM synthetic seawater immersion conditions using EIS. *Corrosion* **1992**, *48*, 671–685. [CrossRef]
46. Perera, D.Y. Effect of thermal and hygroscopic history on physical ageing of organic coatings. *Prog. Org. Coat.* **2002**, *44*, 55–62. [CrossRef]
47. Perera, D.Y. Physical ageing of organic coatings. *Prog. Org. Coat.* **2003**, *47*, 61–76. Available online: http://www.sciencedirect.com/science/article/B6THD-48NC0P4-1/2/dd9b3f7922d3afe1c28c0df3162bd9a0 (accessed on 30 September 2021). [CrossRef]
48. Bierwagen, G.P.; Wang, X.; Tallman, D.E. In situ study of coatings using embedded electrodes for ENM measurements. *Prog. Org. Coat.* **2003**, *46*, 163–175. [CrossRef]
49. Kong, E. Physical aging in epoxy matrices and composites. In *Epoxy Resins and Composites IV*; Dušek, K., Ed.; Springer: Berlin/Heidelberg, Germany, 1986; pp. 125–171. [CrossRef]
50. Elkebir, Y.; Mallarino, S.; Trinh, D.; Touzain, S. Effect of physical ageing onto the water uptake in epoxy coatings. *Electrochim. Acta* **2020**, *337*, 135766. [CrossRef]
51. Hutchinson, J.M. Physical aging of polymers. *Prog. Polym. Sci.* **1995**, *20*, 703–760. [CrossRef]
52. Le Guen-Geffroy, A.; Le Gac, P.-Y.; Habert, B.; Davies, P. Physical ageing of epoxy in a wet environment: Coupling between plasticization and physical ageing. *Polym. Degrad. Stab.* **2019**, *168*, 108947. [CrossRef]
53. Thai, T.T.; Druart, M.-E.; Paint, Y.; Trinh, A.T.; Olivier, M.-G. Influence of the sol-gel mesoporosity on the corrosion protection given by an epoxy primer applied on aluminum alloy 2024 –T3. *Prog. Org. Coat.* **2018**, *121*, 53–63. [CrossRef]
54. Deflorian, F.; Fedrizzi, L. Adhesion characterization of protective organic coatings by electrochemical impedance spectroscopy. *J. Adhes. Sci. Technol.* **1999**, *13*, 629–645. [CrossRef]
55. Ecco, L.G.; Fedel, M.; Deflorian, F.; Becker, J.; Iversen, B.B.; Mamakhel, A. Waterborne acrylic paint system based on nanoceria for corrosion protection of steel. *Prog. Org. Coat.* **2016**, *96*, 19–25. [CrossRef]
56. Bonora, P.L.; Deflorian, F.; Fedrizzi, L. Electrochemical impedance spectroscopy as a tool for investigating underpaint corrosion. *Electrochim. Acta* **1996**, *41*, 1073–1082. [CrossRef]

77. Park, J.H.; Lee, G.D.; Ooshige, H.; Nishikata, A.; Tsuru, T. Monitoring of water uptake in organic coatings under cyclic wet–dry condition. *Corros. Sci.* **2003**, *45*, 1881–1894. [CrossRef]
78. Fedrizzi, L.; Rodriguez, F.J.; Rossi, S.; Deflorian, F. Corrosion study of industrial painting cycles for garden furniture. *Prog. Org. Coat.* **2003**, *46*, 62–73. [CrossRef]
79. Bajat, J.B.; Mišković-Stanković, V.B.; Bibić, N.; Dražić, D.M. The influence of zinc surface pretreatment on the adhesion of epoxy coating electrodeposited on hot-dip galvanized steel. *Prog. Org. Coat.* **2007**, *58*, 323–330. [CrossRef]
80. Rosales, B.M.; Di Sarli, A.R.; De Rincón, O.; Rincón, A.; Elsner, C.I.; Marchisio, B. An evaluation of coil coating formulations in marine environments. *Prog. Org. Coat.* **2004**, *50*, 105–114. [CrossRef]

Article

Cathodic Protection of Complex Carbon Steel Structures in Seawater

Philippe Refait [1,*], Anne-Marie Grolleau [2], Marc Jeannin [1] and René Sabot [1]

1. LaSIE, UMR 7356 CNRS, La Rochelle Université, 17000 La Rochelle, France
2. CESMAN/CM Naval Group Research, Marine Corrosion & Cathodic Protection Department, 50104 Cherbourg-en-Cotentin, France
* Correspondence: prefait@univ-lr.fr; Tel.: +33-5-46-45-82-27

Abstract: Cathodic protection efficiency of complex carbon steel structures in confined seawater environment was studied using a specific experimental device. Schematically, this device consisted of a Plexiglas matrix, crossed by a channel 50 cm long, 5 mm deep, 1.5 to 5 cm wide, which moreover included four bends at 90°. Seawater flowed continuously inside the channel over 12 steel coupons embedded in the Plexiglas matrix. Cathodic protection was applied at a constant potential of −1060 mV vs. Ag/AgCl-seawater with respect to a reference electrode located outside the channel, at the seawater flow entry. The potential of four selected coupons was monitored over time via a microelectrode set close to each coupon. It varied significantly with the distance separating the coupons from the channel entry. At the end of the 3.5-month experiment, a polarization curve was acquired. The residual corrosion rate under cathodic protection was estimated via the extrapolation of the anodic Tafel line. It varied from <1 µm yr^{-1} to 16 µm yr^{-1}, depending on the potential reached by the coupon (between −900 and −1040 mV vs. Ag/AgCl-seawater) at the end of the experiment and on the properties of the calcareous deposit formed on the steel surface.

Keywords: seawater; cathodic protection; carbon steel; calcareous deposit; marine renewable energy

1. Introduction

Marine Renewable Energy (MRE) devices such as offshore wind turbines or ocean current turbines involve carbon steel structures with complex geometries and/or inner channels of small dimension where seawater may flow. In most cases, cathodic protection (CP) is envisioned to ensure the durability of the immersed part of the MRE devices. Modelling, predicting and quantifying the efficiency of CP for such complex structures may be challenging. Consequently, specific experimental data may be necessary for a reliable modelling of the CP system. An experimental device was designed to simulate a complex steel structure where seawater would flow inside a channel of small dimension. The steel structure inside the channel was mimicked by a series of 12 steel coupons connected to each other. Four of them could be disconnected individually from the system to be studied separately. In the channel, due to the small cross-sectional area, an important ohmic drop should increase the potential of the metal so that deep inside the channel an insufficient cathodic potential could prevail.

In seawater, the increase of the interfacial pH due to CP induces the formation of a mineral layer on the steel surface. At the potentials usually used for CP of carbon steel, this layer is mainly composed of aragonite $CaCO_3$ [1–4], and thus called calcareous deposit. The increase of pH at the steel/seawater interface is due to the increase of the cathodic reaction rate, a direct consequence of the cathodic polarization. In most cases, the main reaction involved is the reduction of dissolved O_2:

$$O_2 + 2H_2O + 4e^- \rightarrow 4OH^- \qquad (1)$$

The production of OH⁻ ions increases the pH, which modifies the inorganic carbonic equilibrium at the steel/seawater interface:

$$OH^- + HCO_3^- \rightarrow H_2O + CO_3^{2-} \qquad (2)$$

This process finally leads to the precipitation of $CaCO_3$ on the steel surface:

$$Ca^{2+} + CO_3^{2-} \rightarrow CaCO_{3(s)} \qquad (3)$$

The growth of the calcareous deposit promotes a physical barrier against O_2 diffusion so that the current density required for CP decreases with time. This phenomenon could strongly influence the ohmic drop inside a channel of small dimension, such as the one considered in the present study, and thus modify drastically the potential of the metal deep inside the channel.

To our best knowledge, only one study devoted to a similar topic was reported [5]. It aimed to compare experimental results with a computational analysis based on boundary element modelling. Our purpose was to provide more detailed experimental results. In particular, the aim was to monitor over time the potential of the various steel coupons present inside the channel and to assess the efficiency of CP for each coupon. A 3.5-month experiment was then carried out. At the end of the experiment, a surface analysis by X-ray diffraction was achieved to characterize the calcareous deposit formed on the steel coupons. Finally, to estimate CP efficiency, polarization curves were acquired for various coupons along the channel. As performed in previous studies dealing with cathodic protection of steel in soil [6–9], the residual corrosion rate of the metal at the protection potential E_{CP} was estimated by extrapolation of the anodic Tafel straight line down to E_{CP}.

2. Materials and Method

2.1. Description of the Experimental Device Used for the Study

The experimental device was designed so that seawater flows in a channel 50 cm long, 5 mm deep and 5 cm wide, except in a 12 cm long central region, where its width was reduced to 1.5 cm. The channel was machined in one of the two Plexiglas plates (2 cm thick) that constituted the body of the device, as illustrated in Figure 1.

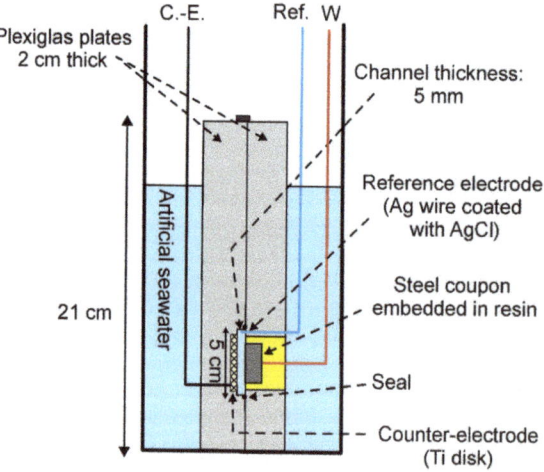

Figure 1. Schematic view (front) of the experimental device showing the respective positions of the steel coupon (grey rectangle) embedded in resin (in yellow), the Ag/AgCl microelectrode (in blue) and the Ti disk used as counter-electrode (hatched rectangle). C.-E., Ref. and W denote the three connections to the potentiostat.

In one of the Plexiglas plates, 12 large holes were drilled to insert 12 carbon steel coupons embedded in a resin matrix, as displayed in Figure 2. These coupons are denoted C1 to C12, with C1 being the coupon at the entry of the channel, where seawater comes in. A copper wire was welded on the rear side of each coupon so that the coupon could be connected to potentiostat #1 that ensured CP. Two kinds of coupons were used, with eight 3-cm diameter coupons in the 5-cm wide parts of the channel and four 1.5-cm diameter coupons in the 1.5-cm wide central part of the channel (Figure 2). The device was set in a tank (120 cm long × 25 cm deep × 15 cm wide) composed of three sections, separated by Plexiglas walls ensuring the sealing of each section. The first section, which can be called the "input section", was filled with artificial seawater. A counter-electrode (titanium grid), and an Ag/AgCl-seawater reference electrode (E_{ref} = +0.250 V vs. SHE) were set in this section and connected to potentiostat #1 used for CP. Five additional large (7 cm × 7 cm × 0.8 cm) steel plates were also immersed in the "input section" to constitute the outer part of the "simulated steel structure". They were connected to potentiostat #1 like the coupons inside the channel and thus protected similarly. The cathodic protection was applied at a constant potential of −1060 mV vs. Ag/AgCl-seawater (i.e., −1050 mV vs. SCE) with respect to the reference electrode set in the "input section". This potential value is the lowest value that could reach a steel structure connected to an Al-Zn-In galvanic anode.

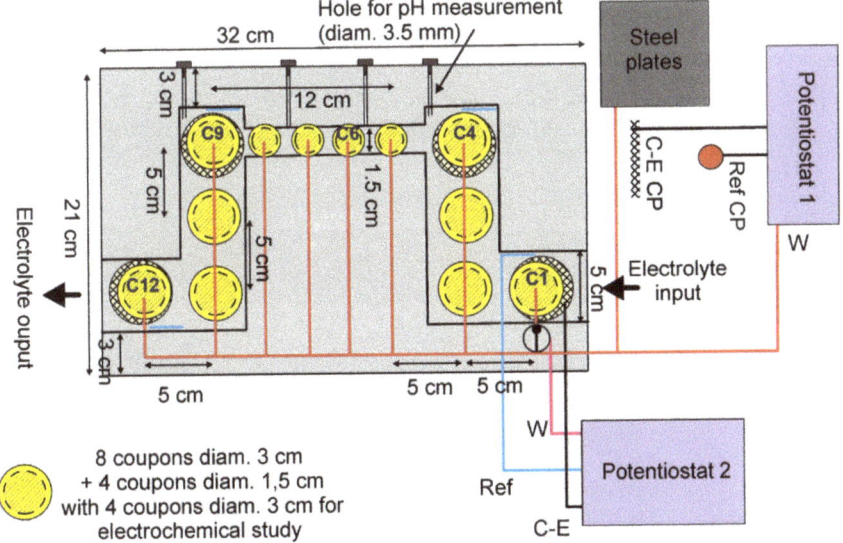

Figure 2. Schematic view (side) of the experimental device, showing the electrical connections and the positions of the four Ag/AgCl microelectrodes (blue lines) set in the channel close to coupons C1, C4, C9 and C12.

The central section of the tank, where the Plexiglas device was set, was not filled with seawater, to facilitate the visual observation of the coupons inside the channel. The third section, i.e., the "output section", was filled with artificial seawater like the "input section". An aquarium pump was placed at the bottom of this section and connected to the channel of the device via a 3-mm diameter plastic tube. The pump ensured a continuous seawater flow inside the channel at a rate, controlled weekly during the 3.5-month experiment, measured between 16 L h^{-1} and 28 L h^{-1}. The corresponding flow velocity was then between 1.8 cm s^{-1} and 3.1 cm s^{-1} in the main part of the channel and between 5.9 cm s^{-1} and 10.4 cm s^{-1} in the smaller central section.

The pumped seawater was transported up into a second tank, initially filled with a large volume of seawater. The seawater level of this thank remained constant as the excess seawater overflowed into the "input section" of the first tank. The overall volume of seawater was equal to 20 L. Half of this volume, i.e., 10 L, was renewed after 2 weeks, after 1 month and after 2 months of experiment. A picture of the entire experimental system is shown in Figure 3.

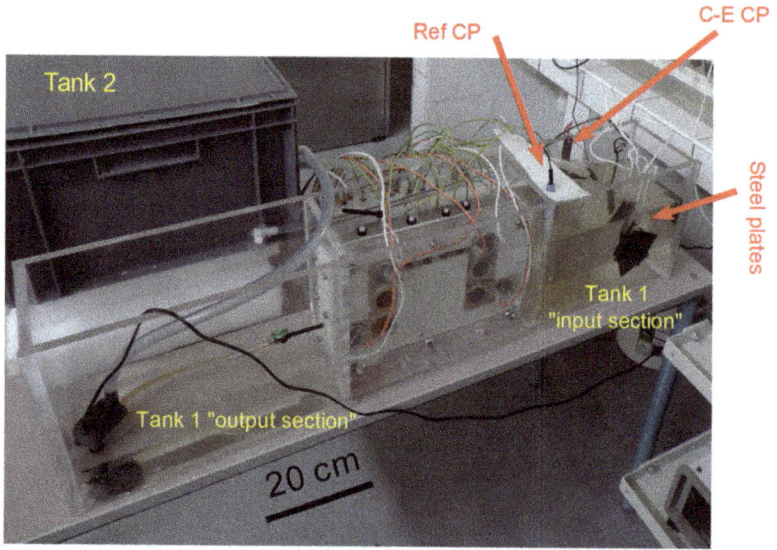

Figure 3. Image of the "real" experimental system while operating.

Four of the coupons, namely C1, C4, C9 and C12, could be disconnected individually and studied separately using a second potentiostat. Close to each coupon, an Ag wire covered with an AgCl layer was set inside the channel to be used as a local Ag/AgCl-seawater reference electrode specific to each coupon. Opposite to each of these four steel coupons, a Ti disk was inserted in the Plexiglas plate where the channel was machined to be used as a local counter-electrode (Figure 1).

Finally, four small holes (3.5-mm diameter) were drilled on top of the device so that a micro pH electrode could be inserted inside the channel for local pH measurements close to coupons C4, C6, C7 and C9 (Figure 2). The holes were closed with small rubber lids except when pH measurements were carried out.

The artificial seawater used here was based on the ATSM D1141 standard [10]. Its composition was NaCl (0.42 mol L^{-1}), MgCl$_2 \cdot$6H$_2$O (0.055 mol L^{-1}), Na$_2$SO$_4 \cdot$10H$_2$O (0.029 mol L^{-1}), CaCl$_2 \cdot$2H$_2$O (0.011 mol L^{-1}), KCl (0.009 mol L^{-1}) and NaHCO$_3$ (0.003 mol L^{-1}). Its pH was adjusted at 8.1 ± 0.1 by addition of small amounts of a 0.1 mol L^{-1} NaOH solution. Its conductivity was measured at 55.4 ± 0.2 mS/cm, i.e., a typical value for seawater.

S235-JR carbon steel rods (3-cm and 1.5-cm diameter) were used to prepare the coupons. The steel composition (wt %) was 98.2% Fe, 0.122% C, 0.206% Si, 0.641% Mn, 0.016% P, 0.031% S, 0.118% Cr, 0.02% Mo, 0.105% Ni and 0.451% Cu. The steel surface was abraded with silicon carbide (grade 180, particle size 80 µm), rinsed with deionized water, and carefully dried just before the coupons were set in the experimental device. The large steel plates immersed in the "input section" of tank #1 (Figures 2 and 3) were made of the same steel and their surfaces were prepared the same way.

The experiment was performed twice for 3.5 months, after an initial shorter experiment of 21 days, at room temperature (22 ± 2 °C) in each case. These experiments gave similar results but the article is focused on the last of the three experiments because, in particular, additional information about coupon C6 was only acquired in this case (see Section 2.2).

2.2. Electrochemical Measurements

When the ohmic drop is important, the "real" potential of the metal differs significantly from the applied potential E_{app}. This potential is the potential corrected from ohmic drop, $E_{IR\ free}$, expressed as:

$$E_{IR\ free} = E_{app} - RI \qquad (4)$$

In Equation (4), R is the resistance of the electrolyte that separates the reference electrode from the working electrode (steel coupon), and I the current required for CP. For a cathodic current, the value of I is considered negative, which implies that $E_{IR\ free} > E_{app}$.

In our experimental conditions, E_{app} is equal to -1060 mV vs. Ag/AgCl-seawater. It corresponds to the electric voltage between each coupon under CP and the reference electrode used for CP, immersed in the "input section" of tank #1.

The potential $E_{IR\ free}$ of coupons C1, C4, C9 and C12 could be measured using potentiostat #2 via the determination of the electric voltage between the coupon and the paired Ag/AgCl-seawater microelectrode set nearby inside the channel. These measurements were performed daily during the first 20 days, weekly until day 81 and one last time at the end of the experiment (day 109). In this last day, an additional measurement was performed for coupon C6 using a hole designed for pH measurement to set an Ag/AgCl-seawater microelectrode close to the coupon.

On the last day of experiment, polarization curves were acquired for coupons C1, C4, C9 and C12 using the associated counter-electrode and Ag/AgCl-seawater microelectrode. It was also achieved for C6, using an Ag/AgCl-seawater microelectrode as explained above and the closest counter-electrode, i.e., the one facing coupon C4. The curves were obtained from $E_{IR\ free}(1) = -1.10$ V vs. Ag/AgCl-seawater to $E_{IR\ free}(2) = -0.60$ V vs. Ag/AgCl-seawater, at a scan rate $dE/dt = 0.1$ mV/s. Potentiostats #1 and #2 were both VSP potentiostats (BioLogic, Seyssinet-Pariset, France). The polarization curves were measured with the water still flowing in the channel and CP was not interrupted before the polarization curve was acquired. The lowest potential (-1.1 V vs. Ag/AgCl-seawater) was chosen to be only slightly smaller than the applied potential E_{app}, so that water reduction remained negligible. The highest potential (-0.6 V vs. Ag/AgCl-seawater) was chosen so that in any case a sufficiently large anodic region could be investigated.

All pH measurements were performed with a Mettler-Toledo micro pH electrode (Mfr # 51343160) and a SevenExcellence pH meter S400 (Mettler-Toledo SAS, Viroflay, France). They were carried out at day 28 while the seawater flow inside the channel had, exceptionally, stopped, and every two weeks with seawater flowing normally inside the channel.

2.3. X-ray Diffraction Analysis

After the experiment, the surface of the coupons was analysed by X-ray diffraction (XRD). The coupons were directly set in the sample holder and the analysis was carried out with a classical powder diffractometer (Brucker AXS® D8-Advance), using Cu-Kα wavelength ($\lambda = 0.15406$ nm) in Bragg-Brentano geometry. The acquisition was performed at 40 kV and 40 mA, from $2\theta = 10°$ to $2\theta = 70°$, with an angular interval of 0.04° and a counting time of 3 s per angular position.

The various phases were identified using the ICDD-JCPDS-PDF-2 database (ICDD, Newtown Square, PA, USA) via the files 01-075-2230 (aragonite = $CaCO_3$), 00-005-0628 (halite = NaCl), 01-044-1415 (lepidocrocite = γ-FeOOH) and 03-065-4899 (α-Fe).

3. Results

3.1. pH Measurements

Table 1 lists the pH values measured inside the channel at the vicinity of coupons C4, C7 and C9 and outside the channel, in the "input section" of tank #1. The value measured outside the channel was in any case between 7.8 and 8.0. At days 33 and 53, the measurements were performed with seawater flowing normally inside the channel and the values obtained for coupons C4 and C7 were identical (or did not differ significantly) to those measured outside the channel. In contrast, the pH measured near coupon C9 was always slightly higher.

Table 1. Local pH measurements. The accuracy was estimated at ±0.1 pH via three successive measurements.

Measurement Time	pH of Seawater [1]	pH Close to C4	pH Close to C7	pH Close to C9
Day 28 (flow stopped)	7.9	8.7	9.2	9.5
Day 33	7.8	7.7	7.8	8.0
Day 53	8.0	8.0	8.0	8.5

[1] pH measured outside the channel, in the "input section" of tank #1 (see Figure 2).

On day 28, the seawater flow was accidentally stopped. It was actually observed that the 3-mm diameter tube connecting the channel to the aquarium pump (Figure 3) was clogged with rust particles coming from the insufficiently protected coupons. The measured pH values revealed that the pH was increasing all along the channel, from 8.7 close to C4 to 9.5 close to C9. This increase of pH is the consequence of CP, which increases the cathodic reaction rate and so the production of OH^- ions, as expressed by reaction (1) When the seawater flow was stopped, OH^- ions could be only transported by migration and diffusion. The observed increase of pH shows that the production of OH^- ions was sufficiently fast so that the local OH^- ions concentration could increase. Conversely, when seawater flowed inside the channel, the produced OH^- ions were also carried away by advection and could not accumulate at the vicinity of the coupons surface.

The results obtained near coupon C9 are, however, slightly different as the pH is also higher when seawater flows inside the channel. A detailed view of the experimental design is shown in Figure 4. It reveals that pH was actually measured in a corner of the channel far from the electrolyte pathway. The effects of advection in this confined region of the channel are thus insufficient to avoid the accumulation of OH^- ions.

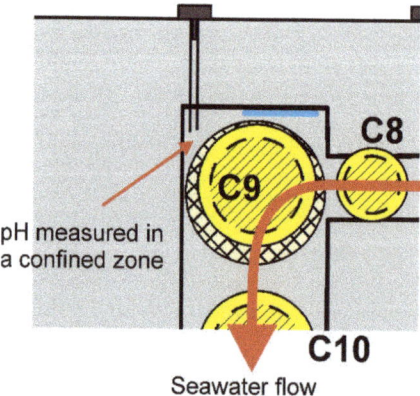

Figure 4. Schematic representation of the channel around coupon C9. The blue line corresponds to the Ag/AgCl microelectrode. The large bended red arrow displays (schematically) the water flow.

The pH was monitored all along the experiment (approximately every two weeks). For each coupon, it fluctuated around the values given in Table 1 for days 33 and 53, i.e., did not change significantly over time.

3.2. Potential Measurements

The potential corrected from ohmic drop, $E_{IR\ free}$, was measured as described in Section 2.2. for coupons C1, C4, C9 and C12. The evolution over time of these potentials is displayed in Figure 5.

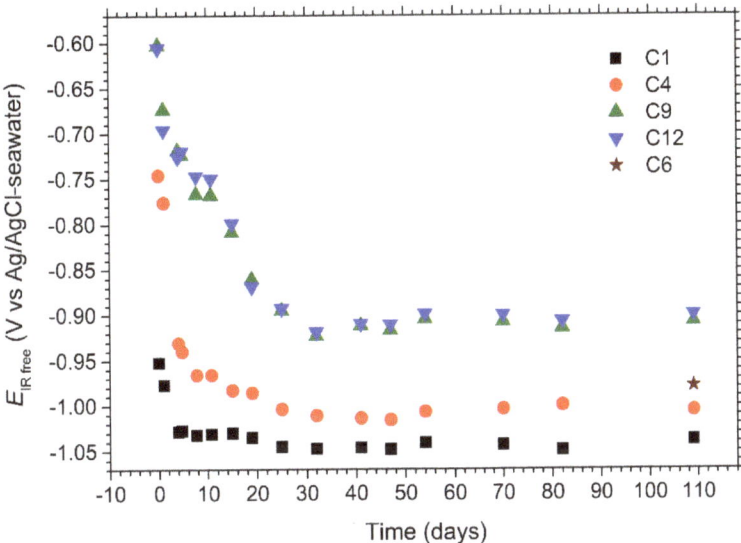

Figure 5. Evolution over time of the potential, measured with respect to the paired Ag/AgCl microelectrode, of coupons C1, C4, C9 and C12. A measurement was also performed at the end of the experiment for coupon C6 (see Section 2.2).

First, it can be seen that $E_{IR\ free}$ increases inside the channel from C1 to C12, whatever the considered time. The additional value measured for C6 at day 109 falls also between those obtained for C4 and C9, which validates the methodology used to obtain data for C6. This variation is due to the ohmic drop, which increases inside the channel as the distance from the reference electrode used to control the applied potential E_{app} increases.

Secondly, it is observed that:

$$\{E_{IR\ free}(C9) - E_{IR\ free}(C4)\} > \{E_{IR\ free}(C4) - E_{IR\ free}(C1)\} > \{E_{IR\ free}(C12) - E_{IR\ free}(C9)\}$$

This shows that the ohmic drop is the highest between C9 and C4. These two coupons are separated by the central region of the channel where the cross-sectional area is the smallest. Consequently, the electrical resistance of the electrolyte in this part of the channel is the highest. It must be recalled that the resistance R of a given volume of an electrolyte with resistivity ρ is given by:

$$R = \rho L/A \qquad (5)$$

In this equation, L is the length of the electrolyte volume and A its cross-sectional area, perpendicular to the current flow (assumed uniform). For similar L and ρ, R is then inversely proportional to A. In the main part of the channel, $A = 2.5\ cm^2$, while in the central part, $A = 0.75\ cm^2$. The ratio between the two cross-sectional areas is then 3.3. At day 41, the potentials are -1.046, -1.014 and -0.912 V vs. Ag/AgCl-seawater for coupons C1, C4, and C9, respectively. This leads to $\{E_{IR\ free}(C9) - E_{IR\ free}(C4)\} = 102$ mV and $\{E_{IR\ free}(C4) - E_{IR\ free}(C1)\} = 32$ mV, thus a ratio of 3.2 between both potential differences. Note that the distance between C4 and C9 is larger than that separating C1 from C4, and that the cross section area is larger in the bends, so that the theoretical ratio between both potential differences would be actually slightly higher than 3.7.

Conversely, the cross-sectional area (and the distance L) is the same between C12–C9 and C4–C1. The associated resistance is then the same. The ohmic drop is RI and depends on the current. The counter-electrode for CP is immersed in the "input section" of tank #1 so that the current, which flows between each coupon towards the counter-electrode, flows inside the channel. Consequently, the current that flows in the C12–C9 section corresponds to the sum of the currents required for the CP of coupons C12, C11 and C10. In contrast, the current that flows in the C4–C1 section is the sum of all currents (except that of C1) This explains why the ohmic drop is higher between C4 and C1 than between C12 and C9

Note that the current that flows in the C9-C4 section is then smaller than the current that flows in the C4-C1 section. This explains why the measured ratio between the corresponding potential differences (i.e., 3.2) is smaller than the theoretical ratio only based on the variation of the resistance R inside the channel (~3.7).

Thirdly, it is observed that the values of the potentials were initially high and decreased with time during the first 32 days. At the beginning, the values ranged from −0.95 V vs. Ag/AgCl-seawater for C1 to −0.6 V vs. Ag/AgCl-seawater for C9 and C12. This implies that C1 was correctly protected as soon as CP was applied while the protection was insufficient for C9 and C12. This could be visually appreciated as illustrated in Figure 6. The picture was taken five days after the beginning of the experiment. Coupons C11 and C2 clearly illustrate the effects of the increase of $E_{IR\ free}$ inside the channel. C11 is obviously entirely covered by a rust layer, i.e., is not correctly protected, while C2 is covered by a whitish layer, i.e., the calcareous deposit, which demonstrates that this coupon is indeed protected against corrosion.

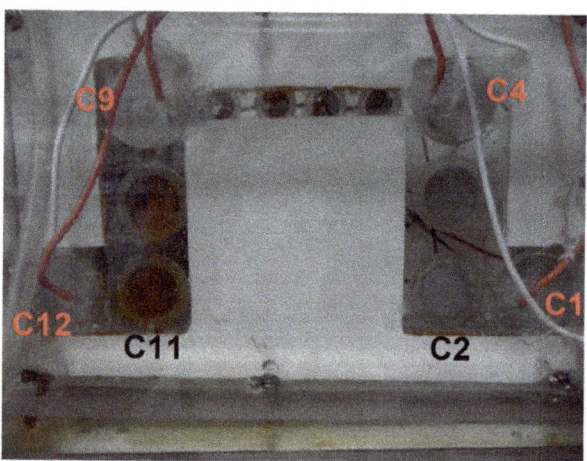

Figure 6. Image of the channel showing the surface of the steel coupons after 5 days of experiment. Coupons C1 (right of the image), C4, C9 and C12 are hidden by the corresponding counter-electrode set in the opposite side of the channel.

The potential of each coupon decreased with time during the first 32 days and stabilized afterwards. At the end of the experiment, all the $E_{IR\ free}$ values were below −0.85 V vs. Ag/AgCl-seawater, i.e., all the coupons could be considered as correctly protected according to CP standards, e.g., [11]. This decrease in potential is due to the decrease in the ohmic drop, which is necessarily due to a decrease in the current required for CP because the other parameter, i.e., the resistance of the electrolyte circuit, does not vary significantly (if at all). This decrease with time in the current required for CP is a well-known effect of calcareous deposition [1–4]. The mineral layer hinders the diffusion of O_2, decreases the active area of the metal, and thus slows down the cathodic reaction rate.

As shown in Figure 5, the potential of C1 decreased rapidly from −0.95 to −1.03 V vs. Ag/AgCl-seawater, and a calcareous deposit could form in a few days on its surface [3,4]. As shown in Figure 6, the whole surface of C2 was covered with such a mineral layer after 5 days. The decrease in the current flowing from C3, C2 and C1 consequently led to a decrease in the ohmic drop for C4 and its potential dropped from −0.75 to −0.97 V vs. Ag/AgCl-seawater after ten days. C4 was then itself progressively covered with a calcareous deposit. The current required for CP decreased in turn for C4, then C5, C6 and so on, so that the potential of all coupons finally decreased down to an acceptable value.

To illustrate the reproducibility of the results, the potentials measured at day 18 during the three experiments (21-day experiment and two 3.5-month experiments) are given in Table 2. It can be seen that at that time, the values measured for C12 vary significantly from one experiment to the other. Coupon C12 is the farthest from the entry of the channel and its behavior depends on that of each of the other coupons. However, the values finally measured for C12 at the end of the two 3.5-month experiments were similar (about −0.9 V vs. Ag/AgCl-seawater).

Table 2. Potential values (V vs. Ag/AgCl-seawater) measured at day 18 for each of the 3 experiments.

C1	C4	C9	C12
−1.033	−0.949	not measured	−0.760
−1.015	−0.976	−0.854	−0.836
−1.034	−0.985	−0.861	−0.869

3.3. Voltammetry

The polarization curves obtained for coupons C1, C4, C6 and C9 are displayed in Figure 7. The curve obtained for C12, very similar to that of C9, was omitted for clarity.

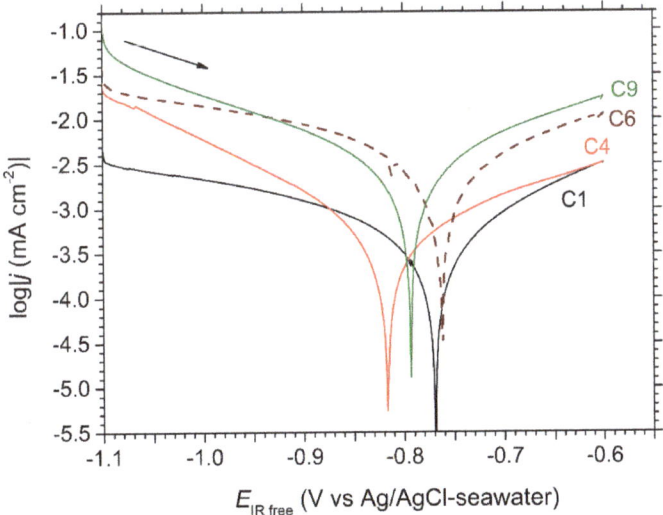

Figure 7. Polarization curves (log |j| vs. E) acquired at the end of the experiment for coupons C1 (black line), C4 (red line), C6 (brown dotted line) and C9 (green line).

First, it is observed that the corrosion potential varies between −0.82 V vs. Ag/AgCl-seawater and −0.76 V vs. Ag/AgCl-seawater, without any apparent link with the position of the coupons inside the channel. In contrast, both anodic and cathodic parts of the curves are shifted to higher current densities from C1 to C9. This evolution can be attributed to the calcareous deposit, which may be more protective for the coupons that were polarized at a lower potential during a longer time. Numerous studies devoted to calcareous deposition on steel in seawater demonstrated that a decrease in potential, in the range between −0.85 to −1.05 V vs. Ag/AgCl-seawater, led to denser and thicker layers [3,4,12]. A longer polarization time, at a given potential, also increases the thickness and decreases the porosity of the calcareous layer [3,4,13].

The anodic part of the log $|j|$ vs. $E_{IR\ free}$ curve proved linear in any case, with similar anodic Tafel slopes, for C4, C6 and C9, (i.e., 281, 287 and 272 mV/decade) and a significantly different slope for C1 (220 mV/decade). As for the cathodic part, two behaviors can be observed. In the case of C1 and C6, the polarization curve bends progressively, from E_{cor} to more cathodic potentials, so that the slope of the curve becomes very small below -0.9 V vs. Ag/AgCl-seawater. This shows that the kinetic of the cathodic reaction, i.e., mainly O_2 reduction, is strongly influenced by mass transport. Thus, the polarization curve tends towards a diffusion plateau at the lowest potentials. In the case of C4 and C9, the influence of mass transport is less important, and the slope of the polarization curve remains high at the lowest potentials. This difference is thoroughly discussed in Section 4.

The cathodic reaction being partially controlled by diffusion, the polarization curves did not obey Tafel law in the cathodic domain. Consequently, only the anodic part of the curve was considered for an interpretation of the voltammetry results based on the Tafel method. The anodic Tafel lines were then drawn and extrapolated down to E_{CP}, i.e., the final potential reached by the coupon at the end of the CP experiment. This extrapolation led to an estimation of the corrosion current density j_{cor}, i.e., the value $j_A(E_{cor})$ of the anodic current density at $E_{IR\ free} = E_{cor}$. Similarly, the value $j_A(E_{CP})$ of the anodic current density at the potential applied during CP gave an estimation of the residual anodic current density, i.e., the residual corrosion rate, expected low, reached under CP (see references [6–9] for more details). The obtained anodic Tafel lines, drawn for each coupon between E_{cor} and E_{CP}, are displayed in Figure 8. Using Faraday's law, the obtained j_{cor} and $j_A(E_{CP})$ values could finally be converted to corrosion rates. All the obtained values (potential, current density, and corrosion rate) are listed in Table 3.

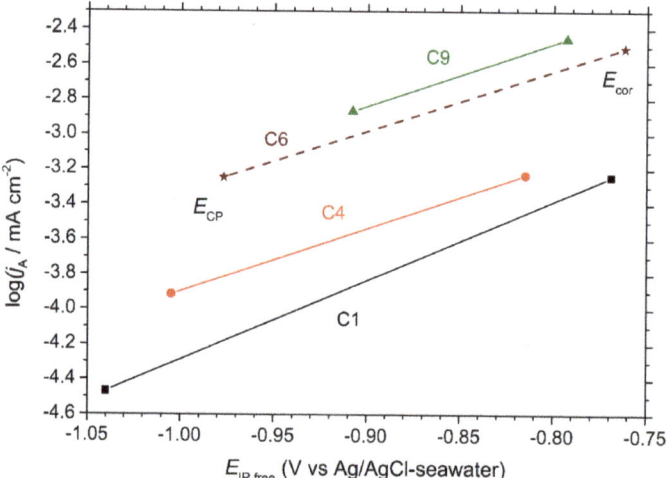

Figure 8. Anodic Tafel lines deduced graphically from the anodic branch of the polarization curves, extrapolated down to the last measured E_{CP} value, and drawn between E_{CP} and E_{cor} for coupons C1 (black line), C4 (red line), C6 (brown dotted line) and C9 (green line).

Table 3. Voltammetry measurements and data obtained via the extrapolation of the anodic Tafel line down to E_{CP}. E_{cor} is given in V vs. Ag/AgCl-seawater, j_{cor} and $j_a(E_{CP})$ in mA cm^{-2}, and the corrosion rate and residual corrosion rate in µm yr^{-1}.

Determined Parameter (Accuracy)	C1	C4	C6	C9
E_{cor} (±1 mV)	−0.769	−0.815	−0.762	−0.793
j_{cor} (±10%)	5.73 × 10^{-4}	5.85 × 10^{-4}	31.0 × 10^{-4}	35.0 × 10^{-4}
Corrosion rate (±10%)	7	7	36	41
E_{CP} (±5 mV)	−1.040	−1.005	−0.977	−0.908
$j_a(E_{CP})$ (±20%)	0.35 × 10^{-4}	1.21 × 10^{-4}	5.63 × 10^{-4}	13.5 × 10^{-4}
Residual corrosion rate at E_{CP} (±20%)	0.4	1.5	6.5	16

First, as can be seen in Figure 8 and read in Table 3, the corrosion rates of C1 and C4, almost identical, are significantly lower than those of C6 and C9, which are quite similar. This effect can be attributed to the calcareous deposit [4], which constitutes a more protective barrier against corrosion for coupons C1 and C4 polarized at lower potentials than C6 and C9 all through the experiment (see for instance Figure 5 to compare C1 and C4 with C9). The corrosion rate estimated for C1 and C4, i.e., 7 µm yr^{-1}, is actually very low, which indicates that the calcareous deposit provides an efficient protection against corrosion, at least a short time after the interruption of CP. Besides, it cannot be excluded that the increase of the interfacial pH promoted the formation of a nanometric passive or pseudo-passive layer on the steel surface.

The estimated residual corrosion rate increases from C1 to C9. As illustrated by Figure 8, this rate is not only linked to the E_{CP} values, but also to the respective positions of the anodic Tafel lines. Let us consider C6 and C9. In this case, the anodic Tafel lines are close but the E_{CP} value of C6 is significantly lower than that of C9. Consequently, CP is more efficient for C6 because the potential of this coupon is more cathodic. If we now compare C6 with C4, we can see that the E_{CP} values are not so different, but the anodic Tafel line of C4 is located at much lower current density values. In this case, CP is more efficient for C4 mainly because of the positioning of its anodic Tafel line. As explained earlier, the decrease of both anodic and cathodic current densities is due to the calcareous deposit. In other words, at the end of the experiment, CP was more efficient for C4, if compared to C6, because C4 had been previously polarized at a lower average cathodic potential, which had led to the formation of a more protective calcareous deposit (and maybe a more protective nanometric pseudo-passive layer). In the case of coupon C1, the anodic Tafel slope, different from that characteristic of coupons C4, C6 and C9 (Figure 8), is another factor that explains the very low residual corrosion rate (Table 3).

3.4. XRD Analysis

The XRD patterns obtained for coupons C7 and C9 are displayed in Figure 9.

In any case, the surface of the coupons, once extracted from the experimental device, was gently rinsed with deionized water. For the coupons covered with a fluffy layer of orange corrosion products (C11 and C12 for instance, see Figure 6), this rinsing removed most of the corrosion products. These products were analyzed separately and consisted mainly of lepidocrocite γ-FeOOH (data not shown).

Therefore, the XRD analysis reveals only the nature of the mineral layer formed on the steel surface below the orange corrosion products (if present). For all coupons, this mineral layer proved to be composed only of aragonite $CaCO_3$, as illustrated for C7, C9 and C10 in Figure 9. This is consistent with previous works dealing with calcareous deposition on steel immersed in seawater under cathodic protection [1–4].

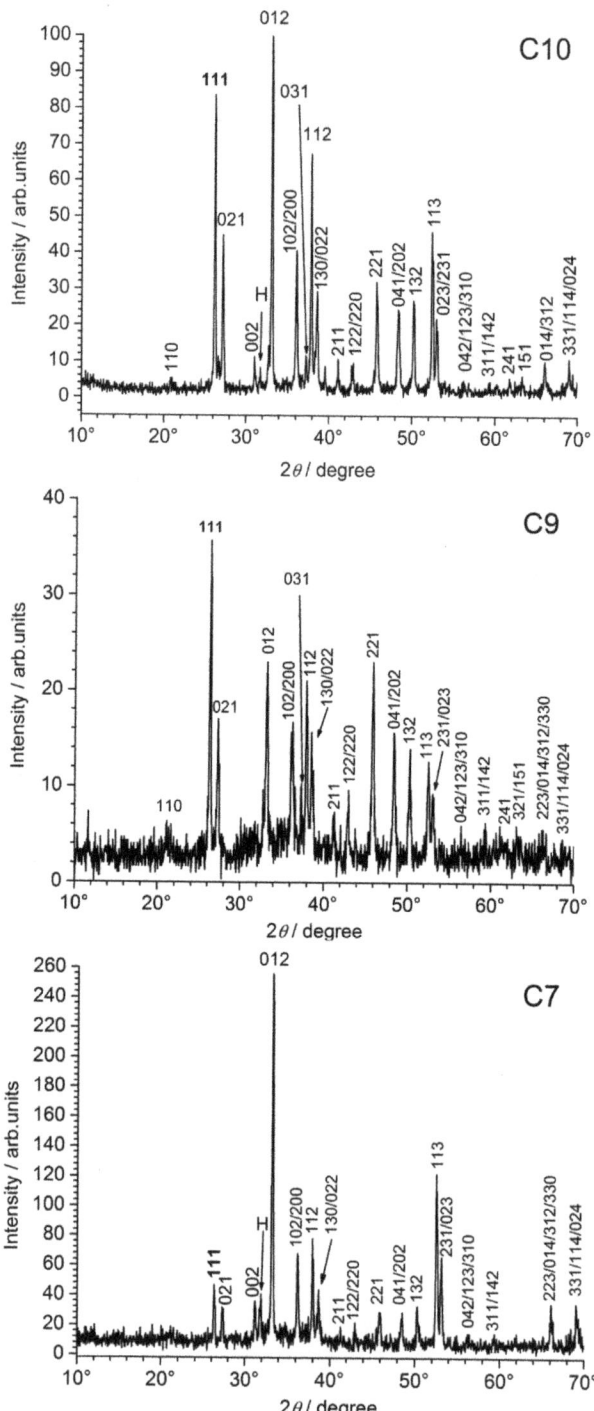

Figure 9. XRD analysis of coupons C10, C9 and C7. The diffraction lines of aragonite are denoted with the corresponding Miller index. H = main diffraction line of halite (NaCl).

When compared to the corresponding ICDD-JCPDS file, it appeared that the aragonite crystals exhibited a preferential orientation, revealed by an increased intensity of the 012 diffraction line. This phenomenon proved much more pronounced for coupons C5–C8, i.e., for the coupons set in the smaller part of the channel. The XRD pattern of coupon C7, compared to those of C9 and C10 in Figure 9, clearly illustrates this result.

4. Discussion

The experimental device was designed so that, inside a channel of small dimension filled with seawater, the ohmic drop could be important. As a result, the coupon C12 set the farthest inside the channel had an initial potential, corrected from ohmic drop, equal to $E_{IR\ free}$ = −0.61 V vs. Ag/AgCl-seawater. The applied potential was equal to −1.06 V vs. Ag/AgCl-seawater, so that the ohmic drop was 0.45 V for C12. At the beginning, coupon C12 was not protected. Its potential however decreased during the first month and remained constant at $E_{IR\ free}$ = −0.91 ± 0.01 V vs. Ag/AgCl-seawater from day 32 to day 109. The decrease observed in the first month is the consequence of the formation of a calcareous deposit on the surface of the coupons located close to the channel entry, which were correctly protected as soon as CP was applied. For instance, the initial value of the potential was $E_{IR\ free}$ = −0.95 V vs. Ag/AgCl-seawater for C1. Because of the progressive formation of the calcareous deposit, the ohmic drop decreased with time, allowing more and more coupons to be correctly protected and covered with a calcareous deposit.

At day 28, when the seawater flow inside the channel was interrupted, pH measurements demonstrated that the pH was increasing inside the channel, from C4 to C7 to C9 (Section 3.1). Because the interfacial pH is linked to the cathodic reaction rate, this result shows that the cathodic current density was increasing from C4 to C7 to C9. The already mature calcareous deposit formed on the surface of C4 implied a strongly decreased current demand. In contrast, the potential of C9 had just reached its minimal value approximately at day 28 (see Figure 5), which implies that the current demand was important. The deposit formed on C9 at higher average potential was moreover less "efficient" than the deposit formed on C4.

A steady state could, however, be reached for the whole system, approximately at day 32. This result shows that, once the deposit has formed on each coupon, and reached its "final" state (in terms of thickness and porosity), the current demand from each coupon became constant. The overall current flowing at a given position in the channel remained constant and the ohmic drop could not decrease any further. This means that the potential reached by the coupons located far inside the channel, i.e., C9–C12, could have been modelled through a theoretical computational approach only if the influence of the calcareous deposit forming on each coupon could have been modelled. Even though the modelling of calcareous deposition was already reported [13,14], this seems challenging in the present case because the effects of the calcareous deposit depend on the potential of the metal, which, in this particular problem, depends itself on the properties of the deposit formed previously on other coupons.

Moreover, coupon C6 set in the smaller part of the channel, though following the same trend observed for other studied coupons (C1, C4, C9 and C12), was characterized by two specific features. The first one relates to voltammetry: the cathodic part of the polarization curve for C6 differs from that of the polarization curves of C4 and C9. Secondly, as for the other coupons in the smaller part of the channel, XRD revealed a strong preferential orientation of the aragonite particles constituting the calcareous deposit. It can then be forwarded that the properties of the deposit formed on C6 differed from those of coupons C4 and C9. In the smaller part of the channel, the flow velocity was higher (up to 10 cm s^{-1} according to the measured flow rate), which is known to have an impact on calcareous deposition [4,12].

The efficiency of CP, assessed via voltammetry experiments, was observed to be associated not only to the cathodic level reached by a coupon at a given time, but also to the previous level of CP, that governs the protective efficiency of the calcareous deposit.

This last point is also illustrated by the low corrosion rates estimated for coupons C1 and C4 when CP was stopped (i.e., at $E = E_{cor}$). This protective ability was already observed to be important for aragonite layers [15]. It may also be partially due to the formation of a passive or pseudo-passive layer at the steel surface, favored by the increase of the interfacial pH. Such a layer would have more likely formed on the coupons polarized at the lowest potentials, and thus typically on coupon C1. This could explain the typical features of the polarization curve of coupon C1 (see Figures 7 and 8), i.e., a slightly different anodic Tafel slope and (ii) a stronger effect on O_2 diffusion in the cathodic domain (with respect to C4 and C9).

5. Conclusions

- This study demonstrated that reliable experimental data were important to understand, model and predict the behavior of steel structures involving complex geometries and seawater pathways with small dimensions. It emphasized the crucial role of calcareous deposition in the expansion of CP inside the meanders of the system.
- The formation of the deposit on the coupons located at the entry side of the channel induced a decrease of the current demand for these coupons, thus a decrease of the ohmic drop for the coupons located farther in the channel. After one month, all the coupons were correctly protected, with a potential lower or equal than -900 mV vs Ag/AgCl-seawater in any case.
- The CP efficiency proved linked to the physical and chemical properties of the deposit. The residual corrosion rate under cathodic protection varied in the present case from <1 µm yr^{-1} to 16 µm yr^{-1}, depending not only on the potential reached by the coupon at the end of the experiment (between -900 and -1040 mV vs. Ag/AgCl-seawater), but also on the properties of the calcareous deposit formed on the steel surface.
- For the coupons polarized at the lowest potential for the longest time (e.g., C1), the beneficial influence of a nanometric passive or pseudo-passive layer at the steel surface, favored by the increase of the interfacial pH, is forwarded.

Author Contributions: Conceptualization, A.-M.G., M.J. and P.R.; Methodology, R.S., M.J. and P.R.; Validation, P.R., R.S. and M.J.; Formal Analysis, P.R., R.S. and M.J.; Investigation, P.R., R.S. and M.J.; Data Curation, P.R., R.S. and M.J.; Writing—Original Draft Preparation, P.R.; Writing—Review and Editing, A.-M.G., M.J., R.S. and P.R.; Visualization, P.R.; Supervision, A.-M.G. and P.R.; Project Administration, A.-M.G. and P.R.; Funding Acquisition, A.-M.G. All authors have read and agreed to the published version of the manuscript.

Funding: This research received no external funding.

Institutional Review Board Statement: Not applicable.

Informed Consent Statement: Not applicable.

Data Availability Statement: The data presented in this study are available on request from the corresponding author.

Acknowledgments: This study was financially supported by the IRT Jules Verne (Nantes, France) as part of the ADUSCOR research program.

Conflicts of Interest: The authors declare no conflict of interest.

References

1. Humble, R.A. Cathodic protection of steel in sea water with magnesium anodes. *Corrosion* **1948**, *4*, 358–370. [CrossRef]
2. Mantel, K.E.; Hartt, W.H.; Chen, T.Y. Substrate, Surface Finish and Flow Rate Influences Upon Calcareous Deposit Structure and Properties. In Proceedings of the Corrosion NACE Conference 1990, NACE International, Houston, TX, USA, 9–14 March 1990; p. 374.
3. Barchiche, C.; Deslouis, C.; Festy, D.; Gil, O.; Refait, P.; Touzain, S.; Tribollet, B. Characterization of Calcareous Deposits in Artificial Sea Water by Impedance Techniques. 3- Deposit of $CaCO_3$ in the presence of Mg(II). *Electrochim. Acta* **2003**, *48*, 1645–1654. [CrossRef]

5. Hoseinieh, S.M.; Shahrabi, T.; Ramezanzadeh, B.; Farrokhu Rad, M. The role of porosity and surface morphology of calcium carbonate deposits on the corrosion behavior of unprotected API 5L X52 rotating disk electrodes in artificial seawater. *J. Electrochem. Soc.* **2016**, *163*, C515–C529. [CrossRef]
6. Krupa, M.; Hogan, E.; Lemieux, E.; Seelinger, A. Cathodic Protection of Highly Complex and Shielded Components. In Proceedings of the Corrosion NACE Conference 2006, NACE International, Houston, TX, USA, 12–16 March 2006; p. 06109.
7. Barbalat, M.; Lanarde, L.; Caron, D.; Meyer, M.; Vittonato, J.; Castillon, F.; Fontaine, S.; Refait, P. Electrochemical determination of residual corrosion rates of steel under cathodic protection in soils. *Corros. Sci.* **2012**, *55*, 246–253. [CrossRef]
8. Barbalat, M.; Caron, D.; Lanarde, L.; Meyer, M.; Fontaine, S.; Castillon, F.; Vittonato, J.; Refait, P. Estimation of residual corrosion rates of steel under cathodic protection in soils via voltammetry. *Corros. Sci.* **2013**, *73*, 222–229. [CrossRef]
9. Nguyen, D.D.; Lanarde, L.; Jeannin, M.; Sabot, R.; Refait, P. Influence of soil moisture on the residual corrosion rates of buried carbon steel structures under cathodic protection. *Electrochim. Acta* **2015**, *176*, 1410–1419. [CrossRef]
10. Olubambi, P.A.; Mjwana, P.; Jeannin, M.; Refait, P. Study of overprotective-polarization of steel subjected to cathodic protection in unsaturated soil. *Materials* **2021**, *14*, 4123. [CrossRef] [PubMed]
11. *ASTM D1141-98(2021)*; Standard Practice for Preparation of Substitute Ocean Water. ASTM International: West Conshohocken, PA, USA, 2021.
12. *BS EN 12473:2014*; General Principles of Cathodic Protection in Sea Water. The British Standards Institution: London, UK, 2016.
13. Wolfson, S.L.; Hartt, W.H. An initial investigation of calcareous deposits upon cathodic steel surfaces in sea water. *Corrosion* **1981**, *37*, 70–76. [CrossRef]
14. Yan, J.-F.; Nguyen, T.V.; White, R.E.; Griffin, R.B. Mathematical modeling of the formation of calcareous deposits on cathodiccally protected steel in seawater. *J. Electrochem. Soc.* **1993**, *140*, 733–742. [CrossRef]
15. Sun, W.; Liu, G.; Wang, L.; Li, Y. A mathematical model for modeling the formation of calcareous deposits cathodically protected steel in seawater. *Elec. Acta* **2012**, *78*, 597–608. [CrossRef]
16. Möller, H. The influence of Mg^{2+} on the formation of calcareous deposits on freely corroding low carbon steel in seawater. *Corros. Sci.* **2007**, *49*, 1992–2001. [CrossRef]

MDPI
St. Alban-Anlage 66
4052 Basel
Switzerland
Tel. +41 61 683 77 34
Fax +41 61 302 89 18
www.mdpi.com

Corrosion and Materials Degradation Editorial Office
E-mail: cmd@mdpi.com
www.mdpi.com/journal/cmd

www.ingramcontent.com/pod-product-compliance
Lightning Source LLC
LaVergne TN
LVHW070732100526
838202LV00013B/1220